BRIDGING THE HUDSON

The Poughkeepsie Railroad Bridge
and Its Connecting Rail Lines

A Many-Faceted History

ALSO BY CARLETON MABEE

The American Leonardo:
A Life of Samuel F. B. Morse
(Pulitzer Prize)

The Seaway Story
(A History of the Great Lakes-St. Lawrence Canals)

Black Freedom:
The Nonviolent Abolitionists from 1830 through the Civil War
(Anisfield-Wolf Award)

Black Education in New York State:
From Colonial to Modern Times
(John Ben Snow Prize)

WITH JAMES A. FLETCHER:
A Quaker Speaks from the Black Experience:
The Life and Selected Writings of Barrington Dunbar

WITH SUSAN MABEE NEWHOUSE:
Sojourner Truth: Slave, Prophet, Legend
(Outstanding Book Award of the
Gustavus Myers Center
for the Study of Human Rights)

EDITED BY JOHN K. JACOBS:
Listen to the Whistle:
An Anecdotal History of the Wallkill Valley Railroad
in Ulster and Orange Counties, New York

BRIDGING THE HUDSON

The Poughkeepsie Railroad Bridge
and Its Connecting Rail Lines

A Many-Faceted History

by
CARLETON MABEE

PURPLE MOUNTAIN PRESS
Fleischmanns, New York

Bridging the Hudson: The Poughkeepsie Railroad Bridge and Its Connecting Rail Lines
A Many-Faceted History

First Edition
2001

Published by Purple Mountain Press, Ltd.
P.O. Box 309, Fleischmanns, New York 12430-0309
845-254-4062, 845-254-4476 (fax)
purple@catskill.net
http://www.catskill.net/purple

Copyright © 2001 by Carleton Mabee

All rights reserved under International and Pan-American Copyright Conventions.
No part of this book may be reproduced or transmitted in any form
without permission in writing from the publisher.

Library of Congress Control Number:
2001 135764

International Standard Book Numbers:
1-930098-24-3 limited hardcover edition
1-930098-25-1 paperback edition

Cover photograph copyright © 2001 by Ted Spiegel

Cover photo credits
Bridge at sunset: Ted Spiegel
Bridge construction: Local History Collection, Adriance Memorial Library, Poughkeepsie
View across bridge: J. W. Swanberg
Bridge burning and the postcard view on back cover: Austin McEntee
CNE&W locomotive and crew on back cover: Dudley J. Stickles

5 4 3 2 1
Manufactured in the United States of America on acid-free paper.

CONTENTS

PREFACE   7

CREATION
1. DREAMING OF A BRIDGE (to 1871)   9
2. LAYING A CORNERSTONE (1873)   15
3. BOSTONIANS INVESTIGATE (1875-1876)   21
4. PLACING PIERS IN THE RIVER (1876-1878)   27
5. DESIGN, FINAL PHASE (1886-1887)   33
6. BUILDING, FINAL PHASE (1886-1888)   41
7. THE BRIDGE OPENS (1889)   53
8. CONNECTING THE BRIDGE TO THE WORLD (1887-1893)   57
9. IMMIGRANT LABORERS (1886-1920s)   66

RAILROAD PASSENGER USE
10. THROUGH PASSENGER TRAINS (1890-1916)   71
11. TROLLEY CARS ON THE BRIDGE (1897-1904)   81
12. LOCAL PASSENGER TRAINS (1890-1934)   85
13. THE ROOSEVELTS AND THE BRIDGE (1873-1945)   95
14. WEST POINT FOOTBALL SPECIALS (1921-1955)   103

RAILROAD FREIGHT USE
15. MILK TRAINS (1889-1930s)   108
16. BATTLE OVER THE SPRINGFIELD BRANCH (1871-1922)   112
17. MAYBROOK GATEWAY (1901-1940s)   117
18. FREIGHT TRAINS (1890s-1950s)   127
19. CENTRAL STATES DISPATCH (1952), by Wallace W. Abbey   140
20. CIRCUS TRAINS (1889-1960s)   147

OTHER USE
21. JUMPING OFF THE BRIDGE (1888-1895)   153
22. WALKERS ON THE BRIDGE (1889-1920)   158
23. A BRIDGE FOR AUTOMOBILES? (1919-1930)   162
24. HOBOES (1890s-1940s)   166
25. POUGHKEEPSIE REGATTA (1895-1949)   171

RAILROADERS AT WORK
26. CONDUCTOR BLODGETT, STORYTELLER (1906-1910s)   176
27. STUDYING FOR EXAMS (1909-1946)   181
28. "THE CNE BOOMER," A FOLKSY TALE (ca. 1920)   184
29. WATER BOY DI ROSA (1927-1939)   187
30. BILL FELL, PAINTER ON THE BRIDGE (1940s)   191
31. LANTERN SIGNALING: THE CANSAS FAMILY (1930s-1940s)   195
32. TOWERMAN BEAUJON (1930s-1950s)   197
33. MARY CARMODY, CREW CALLER (1939-1969)   201
34. DISDAIN FOR A PUSHER (1960)   203
35. TOWERWOMAN COOPER (1956)   205
36. DRIVING DIESELS OVER THE BRIDGE, by Engineer Peter McLachlan (1960s)   209
37. CONDUCTOR ALEXANDER, PROTECTOR (1939-1970s)   211

RISKS, ACCIDENTS, SAFETY
38. TELEGRAPHY AWRY: A HEAD-ON COLLISION (1916)   216
39. CARRIE'S COW, Fiction by John K. Jacobs (1920s?)   220
40. SPECTACULAR SPILL (1943)   225
41. FOLLY ON THE BRIDGE (ca.1960)   227
42. BRIDGE SAFETY AND MAINTENANCE (to 1974)   229

TRANSITION
43. DECLINE (1960-1974)   241
44. THE BRIDGE BURNS (1974)   247
45. RESTORE OR DEMOLISH? (1974-1984)   251
46. ALTERNATIVE PROPOSALS: FOR SHOPPERS, TOURISTS, JUMPERS? (1977-1990s)   261
47. WALKWAY (1991-   )   267

APPENDIX:
SAMPLE LOCOMOTIVE ROSTER
by Leroy Y. Beaujon   276

CHAPTER NOTES   275
SOURCES   288
INDEX   291

PREFACE

THE POUGHKEPSIE RAILROAD BRIDGE was the first bridge to be built over the Hudson River from the ocean all the way up to Albany. It was a technological wonder. Opened in 1889 soon after the Brooklyn Bridge opened, it is not only higher above the water than the Brooklyn Bridge, and founded deeper in the water, but also longer. When it opened, its promoters claimed it was the longest bridge in the world.

Over many years, the bridge and its connecting rail lines carried local and long distance passengers. They carried a rich variety of freights, including bulk commodities like coal and oil, and farm products like meat, fruit, and milk. They knitted small and big communities together. They speeded up life.

After trains had slid across the bridge for eighty-five years, a portion of the bridge burned in 1974, closing off its traffic. While since then imaginative proposals have been made for restoring the bridge to trains or using it otherwise, it remains a question what will become of it. After more than a century, the bridge still stands over the mighty Hudson, an awesome black shape, encrusted with memories.

Whatever its future, its history and the history of its connecting rail lines is studded with colorful anecdotes as well as illustrations of the changing nature of technology and transportation, work and play. It is this varied experience, important to the myriads of people affected by it, that this book attempts to document. It includes the story of building the bridge and its connecting lines, of traveling over them, of shipping over them. It includes the story of the conductors and engineers who ran thrashing locomotives over the bridge and on eastward into New England or westward toward Pennsylvania and beyond. It includes hints of what the bridge meant to people who lived near it, raced boats under it, heard trains whistling on its lines. It looks not only at how the F. D. Roosevelt family related to the bridge and the bridge route, but also how hoboes and boomers did. Altogether this book is a comprehensive history of the bridge and its connecting rail lines, the first ever published.

It is a pleasure to thank the many who helped make this book possible. They include all those over more than a century who recorded or otherwise retained memories of what happened, only some of whom can be named here or elsewhere in this book. I thank railroaders, shippers, passengers. I thank families, rail fans, libraries, historical societies, collectors, museums. I thank those, whether now living or dead, who enlarged my understanding of bridges and railroads, including Clark Bonesteel, Willard Rogers, Ferris Davis, Austin McEntee. I thank those who contributed recollections, anecdotes, stories, including Conductor Albert Alexander, Towerman Leroy Beaujon, Engineer Peter McLachlan (his account of running trains over the bridge), and John K. Jacobs (his short story). For their own memories, I thank many persons, including Dorothy Gruner, Arthur McComb, Jack Swanberg, Sam Christian. For making available clippings, letters, notes, photos, and suggesting avenues to explore, I thank many, including Heyward Cohen, Ken Shuker, Bill Sepe, W. Gifford Moore, and Shirley Anson. I thank Wray Rominger of Purple Mountain Press for suggesting the idea of this book and for his long, patient encouragement. I thank those who have criticized portions of the manuscript, including David Krikun, William Schnitzer, Charles Benjamin, and especially persons who have criticized virtually all of it, including Richard Chartier and my daughter Susan Mabee Newhouse. Finally, I wish to thank others of my family, including my wife Norma and granddaughter Elizabeth, for hints and long forbearance.

Carleton Mabee
Gardiner, NY

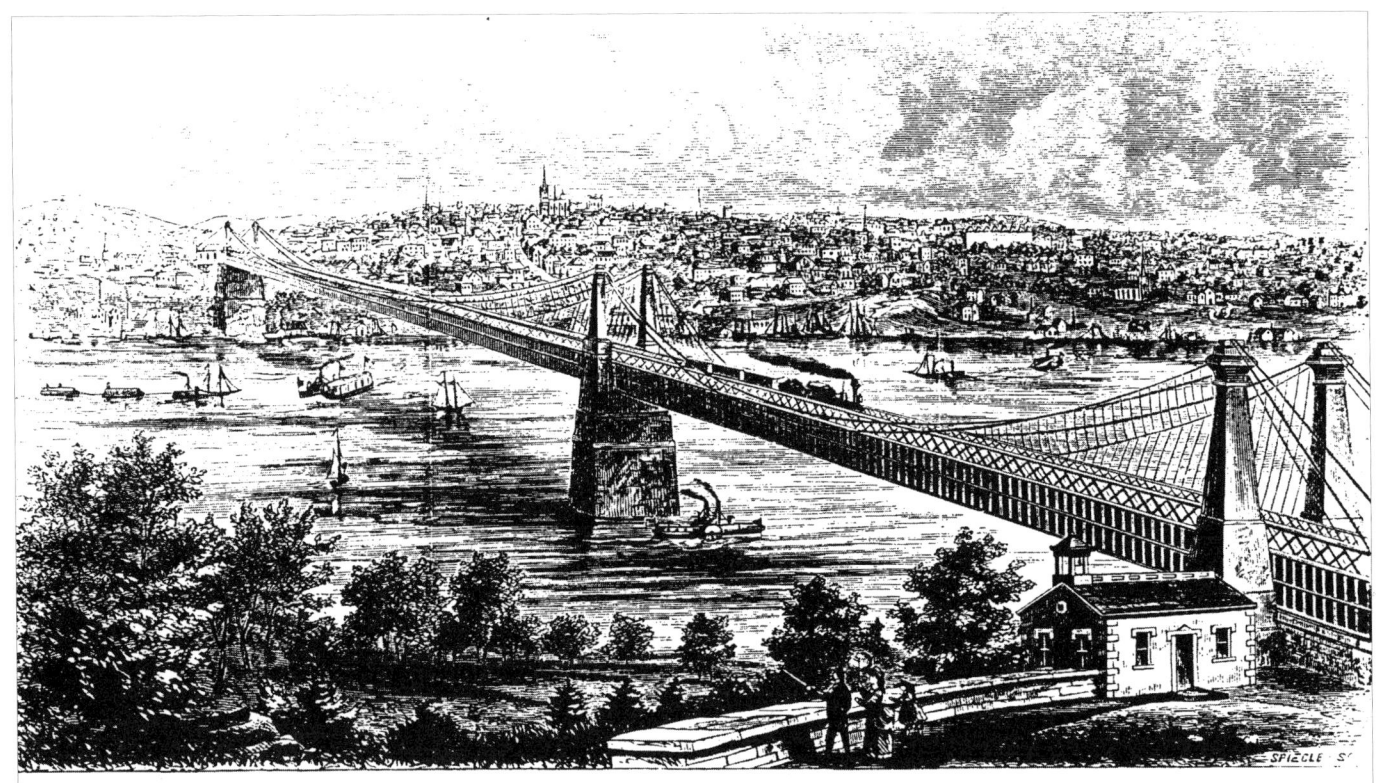

DESIGN FOR A SUSPENSION BRIDGE WITH ONE PIER IN THE WATER, published in a pamphlet by the newly formed Poughkeepsie Bridge Co. in summer 1871 (its title page is opposite). While the company seemed reluctant to concede that any piers in the water would be necessary, it learned that for adequate support, the bridge would need not only one but several piers.

Creation

1. DREAMING OF A BRIDGE (to 1871)[1]

It was a daring dream. While bridges had been built across the narrow, upper Hudson River, no bridge had yet been built over the broader, lower Hudson, from Albany south to the sea. It was doubtful that such a long bridge had ever been built anywhere in the world.

BEFORE THE CIVIL WAR, an engineer proposed that a railroad bridge be built across the Hudson River at Poughkeepsie. He published his proposal in 1855 in the form of a letter to the *Poughkeepsie Eagle*, but the proposal seemed so absurd that the *Eagle* ridiculed it, and soon it seemed to be forgotten.[2]

It was not until after the Civil War in the late 1860s, during a period of feverish railway building across the continent, that the idea of a railroad bridge across the Hudson at Poughkeepsie began to be heard respectfully. At the time no bridge had yet been built over the Hudson River from Albany south to the ocean, about 150 miles.

The Hudson, often being wide and forceful, slowed transportation between significant sections of the nation which needed to reach each other. To the Hudson's east were sections of both New York State and New England in which much of the nation's manufacturing was concentrated. To the Hudson's west were coal-producing Pennsylvania and the grain-producing Midwest.

People and goods moved across the Hudson by ferries, barges, or other vessels which were unreliable because they were dependent on the weather. Also they often required goods to be loaded and reloaded, which could be costly. Even so, it was claimed that more people and freight crossed the Hudson than any other river on the continent, if not in the world.

A railroad bridge came to be proposed for Poughkeepsie especially because the Hudson was not excessively wide there; because bluffs on each side of the river would facilitate raising the bridge high enough to permit river shipping to pass easily under it; and because near there on both sides of the river, railroads were already built or being planned which could easily be extended to connect with the bridge. On the west shore of the Hudson, it was no longer far to the Erie Railway which reached toward Pennsylvania and the Midwest. On the east shore of the Hudson, by 1868, the Poughkeepsie and Eastern Railroad

> BRIDGING THE HUDSON
> AT POUGHKEEPSIE,
> FOR A
> SHORT DIRECT THROUGH LINE,
> FROM
> New England to the Coal Fields
> AND FROM THE
> WEST TO THE EAST,
> FOR
> AN ALL RAIL ROUTE,
> SAVING
> One Hundred Miles in Distance
> OVER PRESENT LINES, AND
> Eighty Miles over any other Proposed Route.
>
> POUGHKEEPSIE:
> EAGLE PRINTING HOUSE.
> 1871.

Company was planning to build a railway from Poughkeepsie to the Connecticut border, and another railway was planning to connect from there on to Hartford.

By 1868, the *Poughkeepsie Eagle* was no longer laughing at proposals for a Poughkeepsie bridge. Eagle editor John I. Platt claimed that especially because of the favorable topography of the Poughkeesie region, "Poughkeepsie is by far the best place between New York and Albany for crossing the river by a bridge."[3]

Poughkeepsie, with a population of twenty thousand, was as large as any city on the Hudson between New York and Albany. Poughkeepsie already had a north-south railroad running through it along the Hudson River, called the Hudson River Railroad (or Hudson Line), which had already become part of the Vanderbilt-controlled New York Central system, linking New York City, Albany, Buffalo, and the Midwest. Poughkeepsie was not only a river port but also a manufacturing city, with John Adriance's factory producing mowers, George Innis's producing dye, and Matthew Vassar's producing beer. Poughkeepsie was already known as the country home of Samuel F. B. Morse whose telegraph invention had led to placing wires along railroad tracks everywhere, enabling trains to run faster and safer. Poughkeepsie was also becoming known as the home of Vassar College, which had recently been founded by the Vassar family with the unusual goal of offering women an education on par with the education offered men.

In January 1871, when the Poughkeepsie and Eastern Railroad opened from Poughkeepsie northeast to near the Connecticut border, that railroad's chief engineer, Pomeroy P. Dickinson, argued for a bridge at Poughkeepsie to connect with his railroad, explaining, "The table lands of Dutchess and Ulster [counties]. . .open the way of approach on both sides of the river better than in any other locality." At about the same time, the *Poughkeepsie Eagle* declared that if there were such a bridge across the Hudson, "only nine miles" of railroad would remain to be built westward "to connect us with the Wallkill Valley [Rail]road at New Paltz, and through that with the Erie and its numerous branches leading into the vast coal fields of Pennsylvania." In February 1871, the aggressive head of Poughkeepsie's Eastman Business College, Harvey G. Eastman, advocated the building of a bridge in a public letter, saying, "If the reader will draw a line on the map from Boston to Pittsburg he will be surprised to find that Poughkeepsie is not only directly in the line," but also "nearly so" are Springfield, Hartford, and the Pennsylvania "coal fields."[4]

Harvey Eastman, who was to become an outstanding advocate of a Poughkeepsie bridge, had taught as a young man in a business school run by his uncle in Rochester, NY. Eventually Eastman had founded his own business college, Eastman College in Poughkeepsie. Harvey Eastman promoted both his college and the proposed Poughkeepsie bridge with the same furious energy that his famous Rochester cousin, George Eastman, was to employ in manufacturing Eastman Kodaks.

A LONG-TIME ADVOCATE OF THE POUGHKEEPSIE BRIDGE, John I. Platt was editor of the *Poughkeepsie Eagle*. He was also a member of the state Assembly (*Poughkeepsie, The Bridge*, 1889)

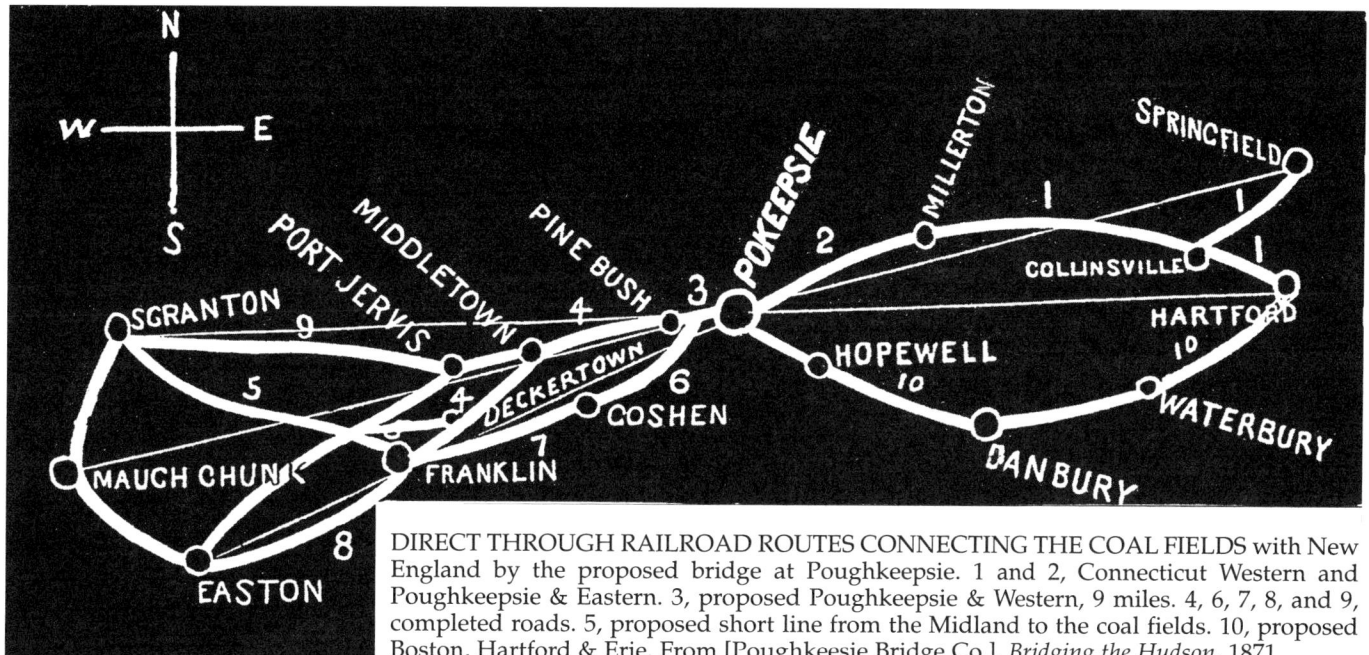

DIRECT THROUGH RAILROAD ROUTES CONNECTING THE COAL FIELDS with New England by the proposed bridge at Poughkeepsie. 1 and 2, Connecticut Western and Poughkeepsie & Eastern. 3, proposed Poughkeepsie & Western, 9 miles. 4, 6, 7, 8, and 9, completed roads. 5, proposed short line from the Midland to the coal fields. 10, proposed Boston, Hartford & Erie. From [Poughkeesie Bridge Co.], *Bridging the Hudson*, 1871

By March 1871, when Harvey Eastman's college had already acquired more students than Vassar College, Eastman became Poughkeepsie's mayor. Within a few days, Mayor Eastman, Eagle editor John I. Platt, and Engineer P. P. Dickinson had drawn up a proposed charter for the Poughkeepsie Bridge Company, and local members had introduced it in the state legislature. The charter would empower the company to build a bridge across the Hudson at Poughkeepsie not only for railroad trains, but also, in provisions likely to garner local support, for pedestrians (called "foot passengers"), and for "teams, vehicles, cattle, horses, sheep, swine."[5]

However, there was fierce opposition to the legislature's adopting the charter. It came from the Vanderbilts' New York Central system which, by this time, had already built a railroad bridge crossing the Hudson at Albany and feared the competition of another such Hudson River bridge. It came from navigation interests that feared a bridge would both reduce the need for shipping across the river and interfere with safe navigation up and down the river. It came from cities along the river which feared a Poughkeepsie bridge would reduce their importance as Hudson River water crossings. It also came from interests that favored

HARVEY G. EASTMAN WAS THE OUTSTANDING ADVOCATE OF THE BRIDGE IN HIS TIME. Informed and articulate, he became mayor of Poughkeepsie while still in his thirties (*Poughkeepsie, The Bridge*, 1889).

JOHN FLACK WINSLOW WAS WELL KNOWN AS A MANUFACTURER OF IRON PRODUCTS, including railroad rails, and as president of Troy's Rensselaer Polytechnic Institute (RPI). He became the first president of the Poughkeepsie Bridge Co. (RPI, *Biographical Record*, 1887)

building a new bridge across the Hudson elsewhere, as at Anthony's Nose, just north of Peekskill, a proposal that Horatio Allen, an early president of the Erie Railway, had helped design. Of course it was impossible to assuage all the opposition interests. But at least to assuage navigation interests, Eastman and his fellow Poughkeepsie bridge promoters, with questionable wisdom, allowed their charter to be written without specifying that any piers could be placed in the river to support the bridge.

On hearing that the legislature had passed the charter, a crowd of Poughkeepsians, scarcely aware of any question about placing piers in the river, gathered to celebrate. On May 13, 1871, they gathered in front of the Dutchess County court house, waving flags and setting off fireworks. They gave "three rousing cheers" for Mayor Eastman.[6]

Before May was out, the new Poughkeepsie Bridge Company had organized themselves. For directors they chose Engineer Dickinson and such Poughkeepsie business figures as John P. Adriance, Matthew Vassar Jr., and George Innis. They chose as vice president the zealous Mayor Eastman. They chose as president the practical John F. Winslow who had become known for his iron foundries at Albany and Troy which were among the largest in the nation. He was known also for having been president of one of the earliest American colleges to educate bridge engineers, Rensselaer Polytechnic Institute in Troy. Winslow had recently moved to Poughkeesie to retire.

To design the bridge, the directors employed Horatio Allen who was about to become the president of the American Society of Civil Engineers. They employed him despite his having done engineering studies for the rival bridge proposed for Anthony's Nose. The directors had in mind, they told Allen, a suspension bridge of one span, without any piers in the river, with its superstructure

HORATIO ALLEN, CIVIL ENGINEER, had wide experience, as on the Delaware and Hudson Canal, on the Croton Aqueduct, and as president of the Erie Railway. In 1871, the Poughkeepsie Bridge Co. asked him to propose a design for its bridge (*Engineering News*, Jan. 11, 1890).

to be of iron—most existing railway bridges at this time had been built of wood, but in the last decade larger ones were being built of iron. However, in July 1871, Allen made clear that, no matter what the directors might prefer, the Hudson at Poughkeepsie was too wide for a bridge to be built across it, even an iron bridge, without placing supporting piers in the river. The directors also asked the advice of another well-known engineer, James B. Eads, who at the time was directing the construction of a great bridge over the Mississippi at St. Louis, an experimental structure being built of both iron and steel. Eads also said that the proposed bridge required placing piers in the river, and he recommended four of them, while, alas, the bridge charter still failed to stipulate that the bridge could have any piers in the water at all.[7]

POUGHKEEPSIE BRIDGE CO., SELECTED INCORPORATORS, 1871
Adapted from [Poughkeepsie Bridge Co.], *Bridging the Hudson*, 1871

In this early period of the bridge project, all its known leaders were from Poughkeepsie or nearby.

Winslow, John F., (Pres.)	Iron and steel manufacturer (retired)	Poughkeepsie
Eastman, Harvey G. (Vice Pres.)	Head, Eastman Business College, and Mayor	Same
Innis, George (Treas.)	Dye manufactuerer	Same
Adriance, John P.	Agricultural machinery manufactuer	Same
Booth, Oliver H. (Vassar nephew)	Vassar Brewery	Same
Collingwood, James	Lumber and coal dealer	Same
Dickinson, Pomeroy P.	Civil engineer; Superindendent, Poughkeepsie & Eastern Railroad	Same
Gaylord, George R.	Shipping merchant	Same
Hasbrouck, Abram	Merchant	Highland
Innis, Aaron	Dye manufactuerer	Poughkeepsie
Nelson, Homer A.	Attorney	Same
Pelton, George P.	Carpet manufacturer	Same
Platt, John I.	Newspaper publisher	Same
Rapelje, Lawrence C.	Merchant	Hopewell Jct.
Vassar, Matthew Jr.	Vassar Brewery	Poughkeepsie

DESIGN FOR THE "MOST DIFFICULT" BRIDGE, PREPARED BEFORE THE BRIDGE CORNERSTONE LAYING, BY THE KEYSTONE BRIDGE CO. (Keystone Bridge Co., *Descriptive Catalogue*, 1874). This design was truss in style rather than suspension like the 1871 one. It provided for four piers in the water, as the bridge charter had been revised to allow.

2. LAYING A CORNERSTONE (1873)[1]

According to the engineers who were designing the bridge, it was "probably the grandest and most difficult that engineering skill has ever been required to undertake."

HARVEY EASTMAN believed that it would be hard to persuade the state legislature to amend the bridge company's charter to permit placing piers in the river. To help persuade the legislature, he ran for membership in it himself, and was elected by a large majority. Once in the legislature, Eastman, along with other members from the nearby region on both sides of the river, pressed for amending the charter to allow placing four piers in the river. Although opponents afterwards claimed that Eastman and his associates had pressed for the amendment "by corrupt means," by May 1872, it passed.[2]

The next major problem was to marshal the funds to build the bridge. Although many railroads and bridges in this period were built with the help of government funds, Eastman and his associates prided themselves on not asking for it. They raised some of the funds locally. But the project was too big for local financing alone.

Among the significant railroads whose lines already reached fairly close to the bridge site, and thus might be expected to benefit from a bridge, two were the Erie and the Ontario & Western (then called the Oswego Midland). But they were both financially troubled and unwilling to invest in the bridge project.

However, the Pennsylvania Railroad, although its lines did not run as close to the bridge site as several other railroads, was looking for better connections to New England. In fact, the Pennsylvania's directors had a committee expressly looking into such connections. Its chairman was

A. L. Dennis, a Newark, NJ, banker, and one of the members was Scottish-born Andrew Carnegie of Pittsburgh, who, though only in his thirties, already controlled extensive interests in railroads, iron manufacture, and bridge building. Negotiating with the Poughkeepsie Bridge Company, Carnegie came to believe that if the Pennsylvania Railroad used the proposed Poughkeepsie bridge to stretch its lines into New England, it could better compete with its ambitious rival, the Vanderbilts' New York Central Railroad. In the spring of 1873, Carnegie tried to persuade the Pennsylvania Railroad's long-time

THE PENNSYLVANIA RAILROAD'S PRESIDENT J. EDGAR THOMSON pushed vigorously for the Poughkeepsie Bridge until disaster struck. He was known as the "great railroad king"(*American Railway*, 1889).

president, J. Edgar Thomson, a Philadelphian of Quaker ancestry, to support the bridge project. Unlike most railway presidents, Thomson was an engineer. Relying especially on his slow, deliberate judgment, he had become an early advocate of steel rather than iron rails. This had helped him build up the Pennsylvania Railroad to become the most extensive railroad in the nation, and he was called "the great railroad king."[3]

Carnegie assured Thomson that the traffic crossing a Poughkeepsie bridge would soon become "enormous," and Thomson had confidence in Carnegie. Moreover, Thomson was aware that often having been ill, he might not live long, and that a bridge could provide an enduring monument of his work. He decided to support the bridge project. Thomson and Dennis together subscribed to $1,100,000 worth of bridge company stock (as in other Pennsylvania Railroad projects, it is not clear to what extent Thomson and Dennis were subscribing personally and to what extent they were subscribing on behalf of their railroad). Their subscription was enough to give them control over the Poughkeepsie Bridge Company.[4]

Responding to its new situation, the Poughkeepsie Bridge Company elected new officers, including Dennis as president, and several other Pennsylvania Railroad figures as directors, including Thomson and Carnegie. Then the Poughkeepsie Bridge Company chose the Keystone Bridge Company, which Thomson had invested in and Carnegie controlled, to build the bridge. It also chose Keystone's president, J. H. Linville, to be its chief engineer.

Like Thomson, Linville was of Quaker ancestry. Though without a formal technical education, Linville worked his way up to be chosen by President Thomson as the Pennsylvania Railroad's engineer in charge of bridges. Like Allen before him, Linville planned the bridge to be essentially iron in material, but unlike Allen, he planned it to be truss in design. Moreover, he designed it to be

THIS VERSION OF THE DESIGN FOR THE BRIDGE WAS PUBLISHED FOR THE CORNERSTONE LAYING, in the *Poughkeepsie Daily Eagle*, Dec. 18, 1873 (McEn). It is presented here only in selected detail. This version, unlike the version illustrated at the front of this chapter, clearly included a roadway under the railroad tracks—the roadway is shown as being used by carriages, wagons, and pedestrians. The difference between these versions may reflect differences of opinion about whether the roadway should be built.

higher, 190 feet over the water, making it easier for ships to pass underneath. He planned to support the bridge with concrete and stone piers in the water, four of them, as the revised charter allowed. Even with four piers, this meant, Linville's Keystone Bridge Company claimed, that the spans between the piers would be "the longest spans of truss-bridge ever attempted in this or any other country." Altogether, the Keystone Bridge Company boasted, because of the bridge's "immense height," the length of its spans, the Hudson's volume of water, and the depth it was necessary to go in the water to reach rock bottom, this bridge was "probably the grandest and most difficult that engineering skill has ever been required to undertake."[5]

How Linville planned to place the pier foundations in the river was a key question. The traditional way to do it was essentially either to dump in stones or to drive in wooden piles. But the depth of the water and the width of the spans planned for this bridge made such methods unsuitable.

A newer method was to sink into the river large cribs (or caissons), constructed of heavy timber and containing hollow chambers which were kept under air pressure to keep water out. The hollow chambers permitted men to work inside to remove river bottom sediment as needed, thus allowing the cribs to settle gradually close to the river's rock bottom. To the fascination of millions of Americans, this method was being used at the time to build the pier foundations for great new bridges at both St. Louis and Brooklyn. But it was a slow and expensive method. Also at both sites, because of the use of this method, a significant number of workmen had suffered the bends, a disease which was then little understood, but which inflicted debilitating pain on some of the workmen, including Washington Roebling, the chief engineer of the Brooklyn Bridge. This method would be even more dangerous if used in the Hudson at Poughkeepsie because the water was deeper there than at either the St. Louis or Brooklyn bridge sites. While in this period builders seldom focused on the safety of their workmen, still Linville and his associates preferred to avoid such threats to their men if they could.

Linville was willing to experiment. He chose instead what the *Poughkeepsie Eagle* called a "comparatively new" method, the use of "pneumatic piles." They were hollow iron tubes, "about three feet in diameter, made in sections so as to be extended to any required length." Each tube, one at a time, was to be driven down through the water into the river sediment until it reached near the river's rock bottom. Meanwhile, the sediment accumulated in the tube was to be pumped out as needed, and eventually the tube was to be filled with concrete. Another tube would then be placed beside it and the process repeated until the area was covered with concrete-filled tubes. Then the tubes were to be "bolted together and fastened by a band passed around the whole." After that, the spaces between the tubes were to be filled with concrete; a riprap of heavy loose stones was to be added around the outside of the whole to bolster its stability; and then the masonry of a bridge pier could be built on top.

Because the charter required the construction work to begin before Jan. 1, 1874, by early the previous December the bridge promoters had arranged for the work to begin. They arranged for it to begin not on one of the difficult piers in the water, but rather on an easy pier on shore, a pier for the bridge approach. This pier site was conveniently located in Poughkeepsie, north of the ferry landing, on a bluff known as Reynolds Hill. Workmen cleared the site down to bedrock, and then blasted the rock enough to level it. Huge blocks of blue stone, weighing more than six hundred pounds each, came in by sloop from a quarry across the river, and workmen set them in place to begin the pier foundation. The company scheduled a cornerstone laying ceremony at this site for Dec. 17, 1873.[6]

For the ceremony, a four-car excursion train arrived from Hartford, coming first over the Connecticut Western to the New York State line, then on the Poughkeepsie and Eastern Railroad the rest of the way, bringing prominent New Englanders. Also a special drawing-room car came up from New York on the Hudson River Railroad, bringing in associates of the

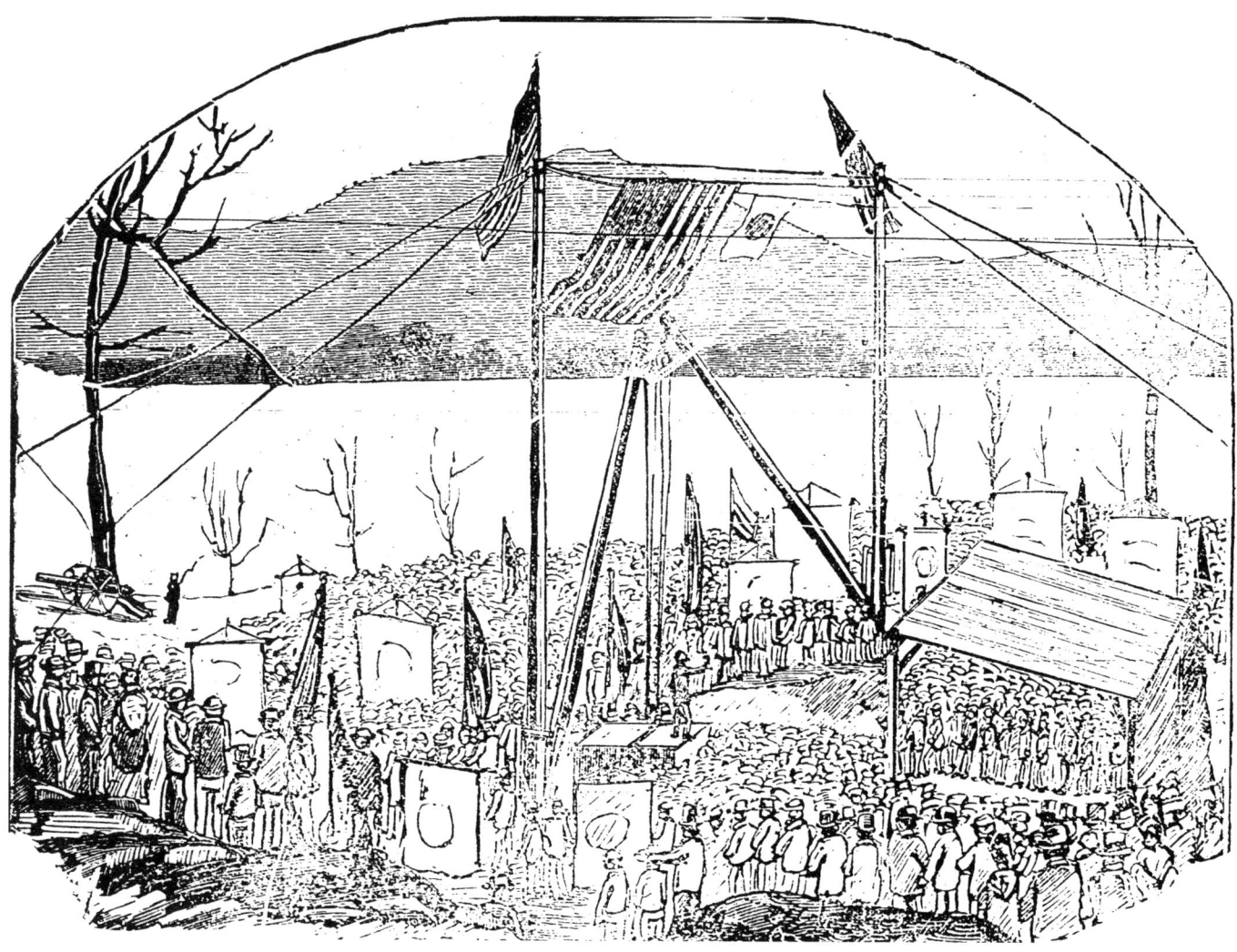

FLAGS AND BANNERS FLEW as a crowd watched the laying of a cornerstone in the base of a bridge pier in Poughkeepsie, overlooking the Hudson River, Dec. 17, 1873. Among those present were top Pennsylvania Railroad officials (*Poughkeepsie News-Telegraph*, Jan. 5, 1889).

Pennsylvania Railroad, including Dennis, Andrew Carnegie, Chief Engineer Linville, the "railroad king" Thomson, and the Hyde Park patrician James Roosevelt (the future father of Franklin D. Roosevelt). A procession moved from downtown Poughkeepsie to the pier site, with fraternal orders in their regalia, 500 Eastman College students, and 50 carriages, amid what the *New York Herald* claimed was "certainly 10,000 people." The stone was laid with Masonic ceremonies, the presiding grand master being James W. Husted, a member of the state legislature from Peekskill.[7]

In the afternoon, bridge promoters gave 300 guests a dinner at the Collingwood Opera House, where they could see on display a "confectionery model of the bridge," and a huge map showing the rail lines that they hoped would converge at the bridge. The dinner, served by Poughkeepsie caterer James Smith (whose sons were beginning to be known for their Smith Brothers cough drops), included green turtle soup, buffalo tongue, and "Bridge Pudding." Mayor Eastman presided. His Eastman College Orchestra played the "Po'keepsie Bridge and Railroad Galop," composed for the occasion, which included hints of a locomotive whistling, puffing, and rumbling over a bridge. Speakers included the Bridge Company's President Dennis; the mayors of Boston and Hartford (the latter insisted that, while it seemed "almost impossible" to build such "a great bridge," it would be done); and Poughkeepsie lawyer Carpenter who predicted that the bridge would bring so much wealth to Poughkeepsie that its "urchins," instead of matching their usual pennies, would match ten dollar gold "eagles."

As for the "urchins" themselves, rather than wait for the bridge to enrich them, some of them decided to take advantage of the immediate bridge excitement. "Being seized with the bridge mania," they "went to some marble yard and stored their capacious pockets with stone chippings," which, according to the *Kingston Freeman*, they sold "at twenty-five cents per specimen" to "bucolic residents" as "pieces of the cornerstone of the great bridge."

The *Albany Argus* expected the bridge to "claim the admiration of the world." The *New York Herald* argued the bridge was an "imperative" for moving Pennsylvania coal to New England. The well-known Dutchess County historian Benson J. Lossing, who served on the committee which arranged the cornerstone laying, predicted extravagantly that while there were already sixty-eight daily trains, passenger and freight, running along the Hudson River line north and south through Poughkeepsie, when the new bridge opened there would be even more trains running across the bridge, and the population of Poughkeepsie would double within five years. The *New York Times* reported that the cornerstone laying was "an imposing incident in a great work," that "Hartford is in ecstasy" over it, "Springfield is jubilant," and "Pennsylvania thinks nothing better could possibly happen."[8]

Even as the cornerstone laying was arousing grand expectations, a major economic recession had already begun which became known as the Panic of 1873. Ironically, one of its causes was the over-expansion of railroads, and many of the railroads to which the Poughkeepsie Bridge seemed likely to connect were soon either bankrupt or in danger of becoming so. Within a few months, the recession had deepened (despite its name, it was to last at least until 1878), and the work on the bridge, which had produced scarcely any visible construction except for the base of one on-shore pier, faltered. In addition, in May 1874, Pennsylvania Railroad President Thomson died. Thomson's estate, along with the Pennsylvania Railroad and its director Dennis, while still friendly to the bridge cause, found themselves without the funds to carry out their plans for the construction of the bridge. Thomson's estate offered to pay all the current debts of the Poughkeepsie Bridge Company, but refused to provide any additional funds to build the bridge, regardless of what Thomson had subscribed, and the company, fearing prolonged litigation if it sued, agreed to accept this arrangement.[9] Bridge construction stopped. Eastman and his associates were grievously disappointed.

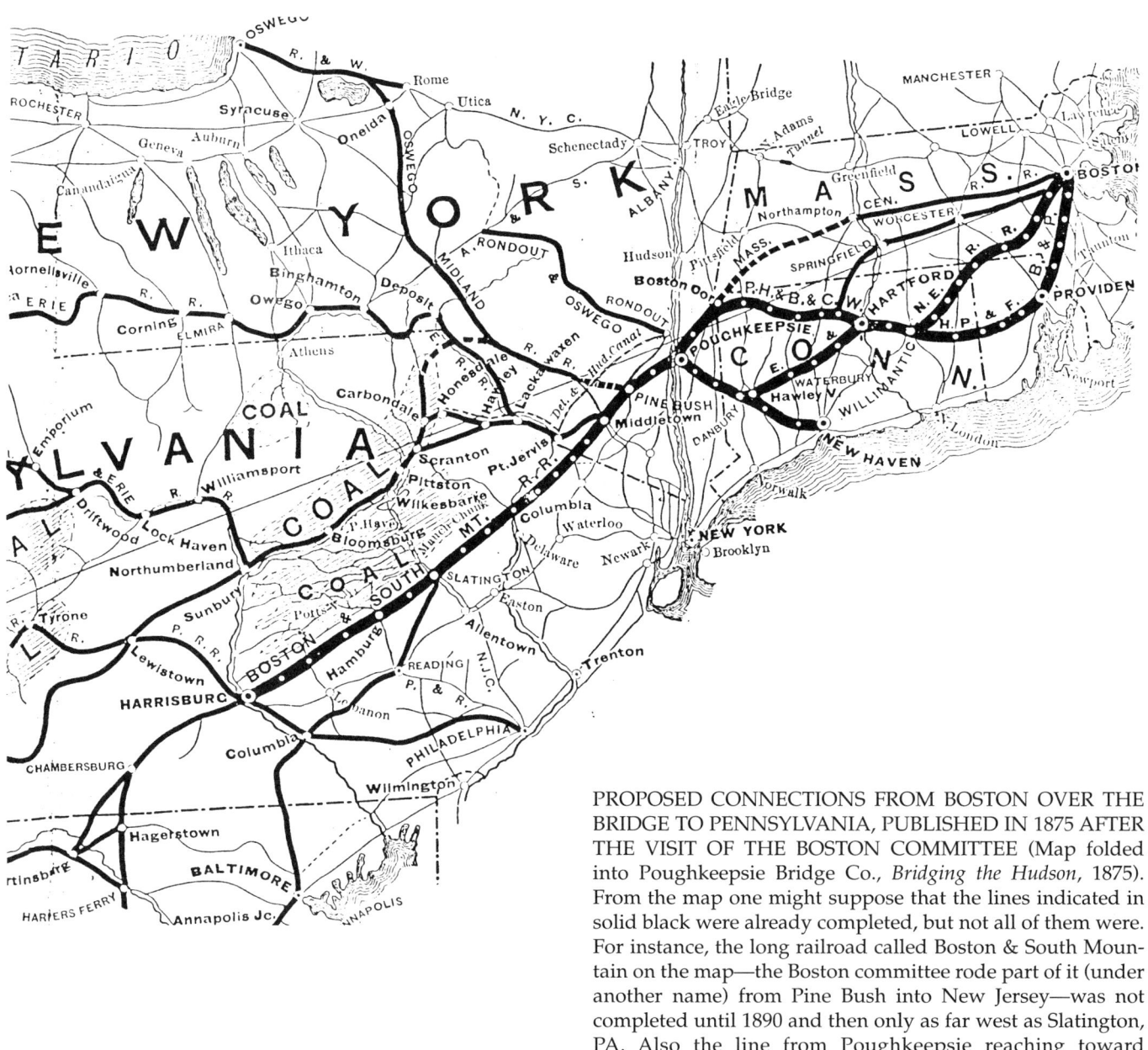

PROPOSED CONNECTIONS FROM BOSTON OVER THE BRIDGE TO PENNSYLVANIA, PUBLISHED IN 1875 AFTER THE VISIT OF THE BOSTON COMMITTEE (Map folded into Poughkeepsie Bridge Co., *Bridging the Hudson*, 1875). From the map one might suppose that the lines indicated in solid black were already completed, but not all of them were. For instance, the long railroad called Boston & South Mountain on the map—the Boston committee rode part of it (under another name) from Pine Bush into New Jersey—was not completed until 1890 and then only as far west as Slatington, PA. Also the line from Poughkeepsie reaching toward Danbury, CT, was not completed until 1892.

3. BOSTONIANS INVESTIGATE (1875-1876)[1]

For New England, what would be the most efficient transportation route, under the least monopolistic control, for bringing in coal from Pennsylvania, pork from the Midwest, and beef from the Southwest?

A GROUP OF LEADING BOSTONIANS decided to investigate whether building a bridge over the Hudson at Poughkeepsie would be in Boston's interest. They decided to go directly to Poughkeepsie to investigate.

The group was a committee created by the Boston Board of Trade in December 1874. The committee feared that Boston was losing out in the race among eastern cities, including New York, Philadelphia, and Baltimore, to develop major rail routes to the rapidly growing West.

Boston already used the Boston & Albany Railroad route which crossed the Hudson by bridge at Albany and continued by the New York Central to Buffalo and the West. This was a northerly route, and another route under construction, a Boston & Maine route, was to be even more northerly. If it ever was to be finished, it was to pass through a five-mile tunnel in the mountains of northwestern Massachusetts, the Hoosac Tunnel, then across the Hudson by bridge north of Troy, and then continue on west via the New York Central. Neither of these routes provided Boston a direct route for the coal of Pennsylvania, the pork of the Ohio River Valley, or the beef of the Southwest. Besides, as many Bostonians perceived it, the Boston & Albany route, if not the new Hoosac Tunnel route too, was under the control of the Vanderbilts' New York Central system which favored New York City, treated New Englanders with "insolence" and delay, and charged them "arbitrary rates."[2]

Harvey Eastman, Poughkeepsie's go-go promoter, had been trying to persuade New Englanders that a Poughkeepsie bridge would benefit them, and he was winning respect in doing so. Eastman has more "practical knowledge" about the proposal than "any other man," said the *Hartford Times*. In November and December, when Eastman and other Poughkeepsians had campaigned for the bridge in Boston, several of its power brokers had listened to him, including the Boston mayor and the Massachusetts governor-elect. Eastman had explained to them that the bridge route could even offer a through Boston-to-Washington overnight passenger service which would avoid New York City, with its customary "disagreeable" transfers from trains to ferries and back to trains again.[3]

At the same time that Eastman was in Boston, a delegation from Newburgh, about fifteen miles south of Poughkeepsie, also arrived in Boston, to urge instead the advantages of an alternate rail route from Boston to cross the Hudson at Newburgh. It would cross the Hudson not by bridge but by ferries which would carry railroad cars across the river to docks where they could roll onto tracks again. Similarly, Kingston, on the Hudson about fifteen miles north of Poughkeepsie, and Peekskill, on the Hudson about thirty miles south of Poughkeepsie, promoted (though without sending delegations to Boston) still other alternative transportation lines across the Hudson. Pennsylvania coal already came cheaply to Kingston by the Delaware & Hudson Canal, and now that a new rail line, the Rhinebeck & Connecticut, was being completed from the Hudson River opposite Kingston to the Connecticut state line, the coal could cross the Hudson by water and then continue by train all the way to Boston. Peekskill was promoting the building of a railroad bridge across the Hudson as part of another all-rail trunk route from Pennsylvania across the Hudson to New England.

How much these alternate proposals—the Newburgh, Kingston, and Peekskill proposals—would divert the Boston committee from its interest in a Poughkeepsie bridge was questionable.

In Boston on the morning of January 19, 1875, a nineteen-man delegation, including merchants, bankers, aldermen, railroad executives. and newsmen, boarded a luxurious "palace car" in which they rode all the way to Poughkeepsie. The car followed a little-known rail route, using several different railroad lines, which Eastman proposed to develop into a new trunk line from Boston via Poughkeepsie to Washington. They left Boston on the New York & New England line. In Willimantic, CT, their car was shunted to the Hartford, Providence, & Fishkill line, on which it continued only as far as Hartford, where it was switched onto the Connecticut Western through Canaan, CT, and finally onto the Poughkeepsie & Eastern to Poughkeepsie. It was a journey altogether of 228 miles, taking ten hours, but all in the same special car.

The next morning, Harvey Eastman and his associates picked up the visitors at their hotel. Loading them into sleighs, they bundled them into "wolf robes," and whisked them to the hill where the bridge cornerstone had been laid so hopefully a little over a year before, in a pier base which remained unfinished. In the evening at a banquet, the Eastman College band played, and the tall, gaunt Eastman told the Boston visitors,"When the world was made, the Almighty marked out. . .the great highway from the east to the west. . .through the city of Poughkeepsie, and on this line the travel of freight and passengers must go."

On the second morning, Eastman and his friends again brought sleighs to pick up the Boston visitors. This time they drove the sleighs across the Hudson River on the ice—in the coldest weather, when navigation on the river ceased, it was customary to cross on ice.[4] Driving westward in Ulster County, they followed approximately the route they believed a rail line would be likely to follow to connect the bridge to railroads running toward Pennsylvania. They skirted hills. They met farmers who said they were anxious to have a railroad which would ship their fruit and milk to market quickly, while it was still fresh. They passed through Modena.

Then in Gardiner, about nine miles from Poughkeepsie, they crossed the rail line on the west side of the Hudson which was closest to the bridge site, that is, the Wallkill Valley Railroad. While earlier the bridge promoters had focused on connecting their proposed bridge rail line to the Wallkill line, now they seemed to ignore it. Perhaps they were now uneasy with the Wallkill because it was related to the big Erie rail system. Though once Eastman had worked for the bridge cause in association with the leaders of the powerful Pennsylvania Railroad, now he was saying, with the Boston committee, that he preferred to have the whole Poughkeepsie Bridge route, from Boston to the Pennsylvania coal fields, be an independent trunk route, not under the control of any monopolistic rail system like the Erie or the Pennsylvania.

When the sleigh riders reached Pine Bush, about eighteen miles from Poughkeepsie on the boundary of Orange County, they were chilled and hungry. To their delight, they found there a special New Jersey Midland train waiting for them, with "a lunch car and two Pullman Palace cars well warmed." Pine Bush was the northern head of a New Jersey Midland rail line which ran southwest through Middletown, NY, to Deckertown, NJ. As their train took them toward Deckertown, it stopped at intermediate stations where they found people were "crazy" to be on a rail route leading to the Poughkeepsie Bridge. In Deckertown they heard assurances that from there the South Mountain Railroad was constructing a line into the Pennsylvania coal region.

Returning by the Midland line to Middletown, they changed there into a special Erie train which they took back to the Hudson River, but this time at Newburgh, where they stayed overnight. The next morning the visitors, accompanied by representatives of freighting and railroad interests, crossed the Hudson by ferry to the Fishkill Landing region, where they inspected various yards, existing or potential, for transferring freight brought by water across the river from Newburgh onto rail lines heading eastward toward New England. Returning to Newburgh, the visitors were entertained at lunch at Homer Ramsdell's "palatial mansion" whose lawns sloped gracefully toward the Hudson River.

Ramsdell, the outstanding transportation spokesman for Newburgh, much as Eastman was for Poughkeepsie, was the proprietor of a large Newburgh-based Hudson River barge system. He had been on the board of the Wallkill Railway. He was now on the board of the Erie Railway, and had been its president. As a negotiator, Ramsdell was so adept that, according to railroad magnate Jim Fisk, he could carry more eels in his arms without spilling them than any other man in America.[5]

Ramsdell admitted to the visitors that while Poughkeepsie was a suitable site to bridge the Hudson, Newburgh was not, the river there being too wide. But Newburgh already handled considerable east-west rail traffic by water, he explained, and he proposed to develop a rail float service there so that rail cars could be ferried over the river without having to unload and reload their cargo.

Ramsdell's view, ever-so-diplomatically expressed to the visitors, was that the time for constructing the Poughkeepsie Bridge had not yet come. What railroad promoters should do first, Ramsdell advised, was to build up the rail traffic from the coal fields across the Hudson at Newburgh, using car floats to cross the river, and then, only if this traffic increased enough to warrant it, build a bridge at Poughkeepsie. But if they did build a bridge, Ramsdell said, they should avoid doing what Eastman urged, connecting it to Pine Bush because from there much of its connection to the Pennsylvania coal fields was still only talk. Instead, he urged, connect the bridge by a rail line from the western end of the bridge down the Hudson's west shore about fifteen miles to Newburgh, from which Erie lines already reached to Pennsylvania and the West. Or if necessary, Ramsdell added, connect the bridge by building a line nine miles to Gardiner from which the Wallkill line already connected to the Erie Railway. Though Ramsdell inevitably represented the interests of the Erie Railway and of his home city of Newburgh, with both of which Eastman and his friends could seem uneasy, Ramsdell was so diplomatic that he was later to be elected to the board of the Poughkeepsie Bridge Company.

The visiting committee returned to Boston, and soon afterward reported unanimously in favor of building the Poughkeepsie Bridge. "We have no hesitation in saying that it is the interest of Boston, and...of all southern New England," they said, "to have this bridge built as speedily as possible." They brushed aside all the alternate proposals, admitting that it would be desirable to have Newburgh's car floats available during the time the bridge was being built, but insisting it would not be wise in the long run for a trunk route to depend on any such unreliable water crossing. Building the proposed bridge at Poughkeepsie seemed feasible despite its "stupendous" size, they said, because of the suitable nature of the terrain, and also because in recent years other great bridges, though not as large, had been or were being successfully built over great rivers. (A steel-arch bridge over the Mississippi at St. Louis, engineered by James B. Eads, had just opened in 1874; it was about two-thirds the length of the proposed Poughkeepsie Bridge.) The committee doubted that the present plan to make the Poughkeepsie Bridge a single-track bridge was wise, considering the traffic it was likely to generate, and recommended instead making it double track from its beginning, even though doing so would substantially increase its cost.[6]

A few months later, in June 1875, the Poughkeepsie Bridge Company reorganized itself. The company's Pennsylvania Railroad-related directors having resigned, the company once again elected its major officers from Poughkeepsie, again including Winslow as president and Eastman as vice president. But it chose other directors from New England, where its new support seemed to be concentrated: it chose three from Boston, as well as others from Providence, Hartford, and New Haven.

Then the Bridge Company campaigned for the stock subscriptions necessary to restart building the bridge. It held meetings in Boston and

Hartford. The Bridge Company's President Winslow himself took part in such meetings. He also campaigned by entertaining at his "princely" estate just north of Poughkeepsie overlooking the Hudson River. Promoters were even brash enough to try to persuade certain interests in generally-hostile New York City to subscribe, arguing that a bridge would provide a shorter route from New York City to the Midwest, that is, a route north via the Hudson River line to Poughkeepsie, then across the bridge and on west by a variety of railroads.

Many newspapers were encouraging. The *New York Times* predicted that "one-half of the capital stock can be raised on the streets of Boston." The *Hartford Courant* said the proposed bridge route gives "promise of cheaper food and cheaper fuel for New England." The *Boston American Union* declared that for Massachusetts this bridge means "trade and profit, and honor and glory."[7]

But pessimists suspected that building the bridge, like many other engineering projects, would cost more than expected. They questioned whether the existing rail lines which were likely to become part of the bridge's connecting routes were of easy enough grade or solidly enough built to be practicable for the heavy use being proposed for them. They questioned whether these rail lines, or others likely to connect with the bridge, winding through hills as they would on both sides of the Hudson, could save as many miles as Eastman argued, regardless of how they looked on simplified maps. They questioned if for coal, saving a few miles en route to market would make

RESIDENCE OF BRIDGE COMPANY PRESIDENT WINSLOW where he entertained prospective bridge supporters. The insert shows the view from his lawn to the Hudson River (Poughkeepsie, *The Bridge*, 1889). This site, on the northern edge of Poughkeepsie, later became Woodcliffe Pleasure Park, and still later Marist College.

much difference anyway. Pennsylvania coal, they said, was already being shipped economically to New England much of the way by water, as by the Delaware & Hudson Canal to Rondout (Kingston), and then along the Hudson to whatever rail connections were convenient for moving it the rest of the way. Also, it was being shipped economically by train to such Atlantic coastal ports as Philadelphia and Hoboken, and then by water along the coast the rest of the way to New England. One civil engineer—J. P. Gould who was the engineer for the D & H Canal-related Rhinebeck & Connecticut Railroad which provided a rival route for shipping Pennsylvania coal into New England—considered the bridge proposal to be incredible in the manner of the proposals of Baron Munchausen, the German teller of incredible tales.[8]

Officers and Directors of the Company

PRESIDENT................JOHN F. WINSLOW, Poughkeepsie, N. Y.
VICE-PRESIDENT,........H. G. EASTMAN,   "   "
TREASURER,.............GEO. INNIS,   "   "
SECRETARY.............GEO. R. GAYLORD,   "   "
CHIEF ENGINEER.......P. P. DICKINSON, New York.
COUNSEL................HOMER A. NELSON, Poughkeepsie, N Y.

DIRECTORS.

JOHN F. WINSLOW, Poughkeepsie.
GEO. INNIS, Poughkeepsie.
OLIVER H. BOOTH, Poughkeepsie.
H. G. EASTMAN, Poughkeepsie.
JOHN O. WHITEHOUSE, Poughkeepsie.
GEO. P. PELTON, Poughkeepsie.
GEO. R. GAYLORD, Poughkeepsie.

JOHN I. CLARK, Boston.
CHARLES H. ALLEN, Boston.
R. M. POMEROY, Boston.
GEO. M. BARTHOLOMEW, Hartford.
GEO. WILSON, Providence.
N. D. SPERRY, New Haven.

Moreover, many railroads, suffering from the long-continuing depression, begun in 1873, were going bankrupt about this time, including railroads potentially related to the Poughkeepsie Bridge such as, on the west side of the Hudson, both the Erie and the Wallkill Valley, and on the east side, the Poughkeepsie & Eastern. Such bankruptcies naturally discouraged further investment in anything related to the region's railroads.

For months, despite Winslow's experience and Eastman's fire, the bridge cause seemed to move forward only sluggishly. However, the long-continued depression gradually brought labor and material costs down, reducing estimates for the cost of bridge construction. Also, several of the nation's great bridge-building companies expressed an interest in building the bridge. Finally, the Poughkeepsie Bridge Company negotiated a contract for building the bridge for $3.4 million with the American Bridge Company of Chicago, an experienced company, which itself undertook to help raise some of the money. While as late as June 1876, it was still uncertain whether enough funds had been raised to allow building to begin again, the readiness of the American Bridge Company to proceed with the building gave the final impetus to lift the subscriptions up to $1 million. This was enough, the weary Poughkeepsie Bridge Company decided, to permit construction to begin again at last, if on a shaky basis, in the summer of 1876.[9]

OFFICIALS OF THE POUGHKEESIE BRIDGE CO., as listed in its *Bridgng the Hudson*, 1875, after the visit of the Boston committee. The names of several New Englanders had recently been added, including the Boston committee's chairman Clark.

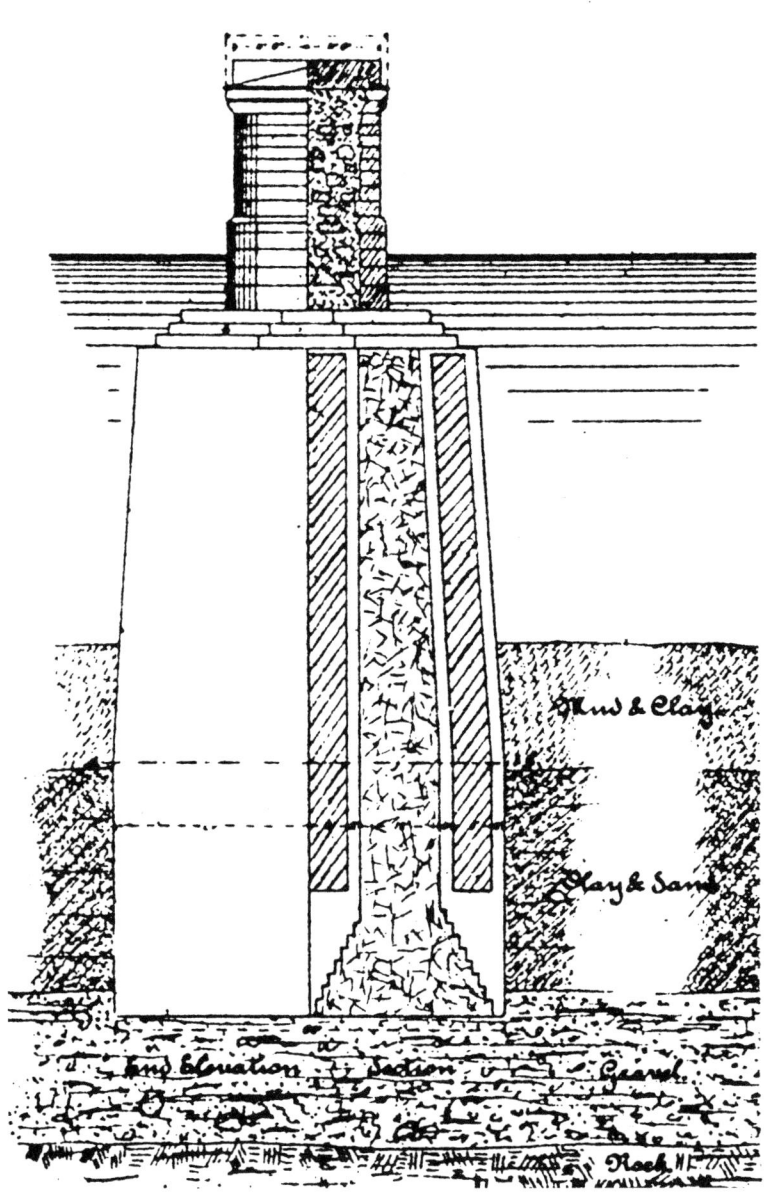

DRAWING OF A CRIB (OR CAISSON) AS PLACED IN THE RIVER, with a stone-faced pier erected on top of it. As indicated, the crib has been sunk down first through water, then through "mud & clay," then "clay & sand," and finally into "gravel." A crib "section" has been exposed to show one of the interior pockets. As shown, the pocket has apparently been filled with concrete (O'Rourke, "Construction of the Poughkeepsie Bridge," ASCE, *Transactions*, 1888).

4. PLACING PIERS IN THE RIVER (1876-1878)[1]

Broken timbers, cracked concrete, delayed wages.

IN 1876, when the work of constructing the bridge began for the second time, the plans still called for the bridge to be a rectangular truss bridge, with four piers in the water. A leading British engineer, Sir Sandford Fleming, who came to visit the site, said it would be "the largest and finest iron truss bridge in the world."

The engineer primarily in charge was the chief engineer for the American Bridge Company, W. G. Coolidge. He reconsidered what method to adopt to place the four piers in the river. He rejected the pneumatic pile method which Engineer Linville had chosen in 1873. He also rejected what the *Poughkeepsie Eagle* called the "enormously expensive and dangerous" method, which had been used in building the St. Louis and Brooklyn bridges, of using timber cribs with underwater air-pressured chambers in them. Instead he chose the relatively new method of using giant timber cribs without air-pressured chambers.

As crowds of people watched, workmen began to construct a giant crib on the Poughkeepsie shore. They bolted timbers together to make the crib about 50 by 100 feet wide and initially about 30 feet high but designed so that, after it was launched into the river, it could be built up much higher. Each crib required over a million feet of timber. Most of the timber, hemlock, came from Pennsylvania, some shipped over the Erie Railroad to Newburgh and then up the Hudson by barge, and some shipped by the Delaware & Hudson Canal to Kingston and then down the Hudson by barge.

Broken stone to fill certain pockets in the cribs arrived from quarries along the Hudson. Cement came up the Hudson by barge from Jersey City, at least 75,000 barrels of it. On Poughkeepsie wharves, as timber, stone, and barrels of cement accumulated, they grew into "colossal piles."

By February 1877, carpenters had built up the crib to the desired size. Then workmen tried to shove the crib, huge and awkward though it was, down the ways. They tried using jacks, pulleys, hawsers, and chains, some of which broke. Groaning and shouting, they improvised. They even used a battering ram. After struggling for five days and torchlit nights, they finally slid the crib into the river. Then they cut away the river ice enough to allow the crib to be floated to a temporary location near the shore, where carpenters built it up higher.

Meanwhile, two massive barges were being specially built in Newburgh, each seventy-five feet long, framed with yellow pine shipped from the Florida Keys, and bolted together with iron. After the barges had been towed upriver to Poughkeepsie, each one was outfitted with two steam-operated derricks which also were built of yellow pine. The *Poughkeepsie Eagle* pronounced the barges "the longest and best derrick barges ever built on the Hudson."[2]

In early spring when the river ice had broken up, tugs towed the two barges to a site about 500 feet off the west shore, the site where the bridge's first river pier was to be erected (the pier later to be known as pier 2). Workmen anchored the barges there. Then a tug, puffing hard, slowly moved the awkward crib from the Poughkeepsie shore out to the pier site, with men riding on top. After about an hour and a half, it reached the site beside the two barges, where men anchored it, to keep it floating in place.

For several months, a swarm of fifty or sixty men worked from the two barges to build the crib higher. They sawed timbers and, with the help of the derricks, hoisted and bolted them in place. At the same time, using the derricks to operate giant buckets, they dumped rocks and gravel into selected crib pockets to weight the crib farther down through the water and on down into the river-bottom sediment. They also let down buckets into certain open crib pockets to dredge out sediment, to clear the way for the crib to settle as directly as possible onto the river's rock bottom.

By July 1877, even though the rock bottom seemed deeper than the contractor expected, the crib had settled close to it. Over a period of several days the engineers sent a diver down through the crib's dredging pockets to test the bottom of each pocket with a crow bar, until he reported that the dredging had approximately reached firm rock. Then, using the derricks to operate buckets, they dropped concrete mix through the water into all the suitable crib pockets, filling them up. The crib had turned into a largely concrete foundation on which to build up a stone pier.

Meanwhile, in accordance with the plan for constructing all the river piers, on top of the crib workmen had been building a temporary cofferdam. It was being built so that as the crib settled

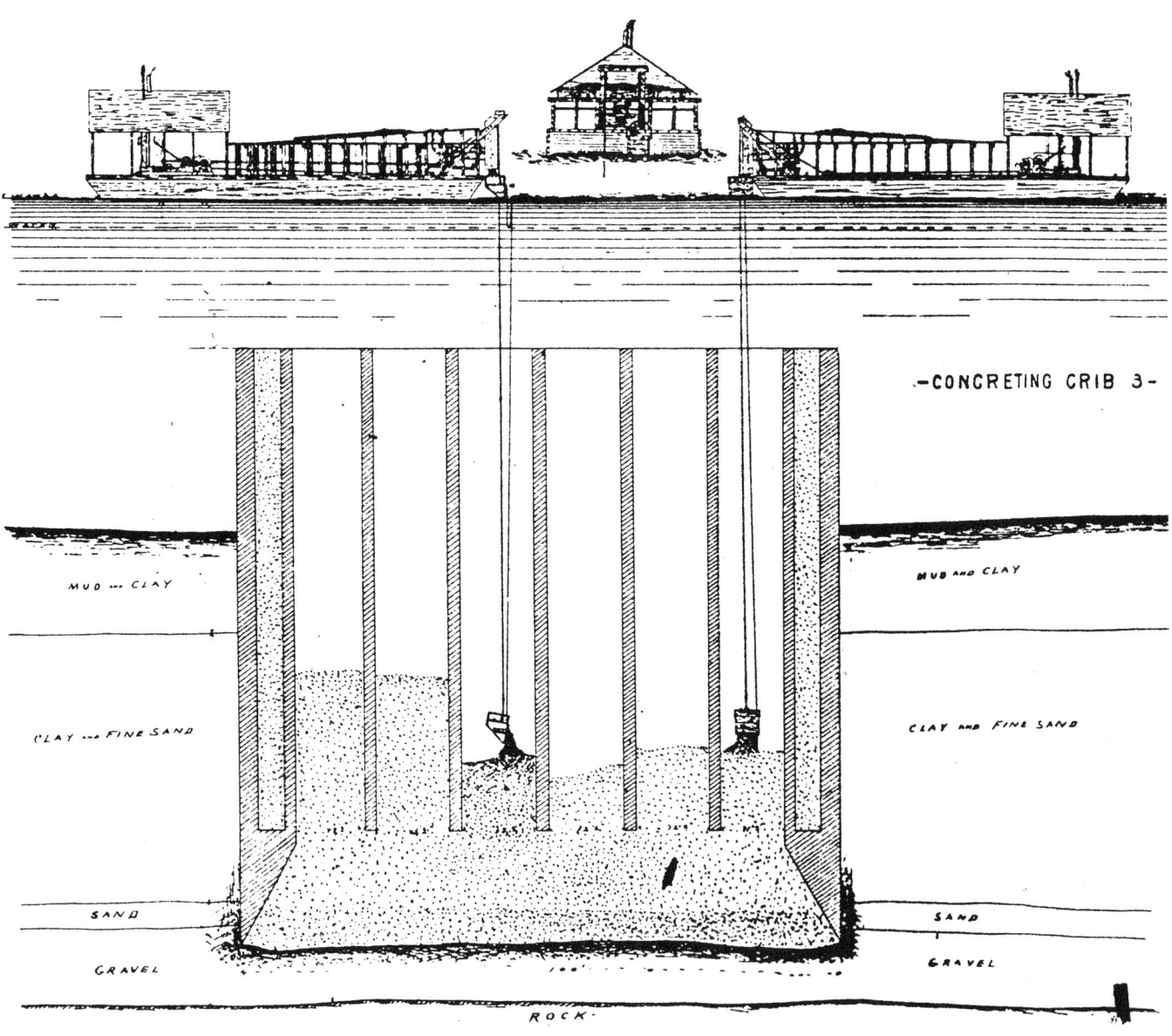

below the level of the water, water could be pumped out from inside the cofferdam to provide masons a dry place to work on top of the crib—the by-then concretized crib. Masons could build there the stone foundation for the pier.

In September, the water was almost all pumped out from inside the cofferdam when suddenly the pressure of the water broke through the cofferdam's walls, and the walls collapsed. Water rushed in. Engineers sent a diver down into the cofferdam and around the edge of the crib to inspect the damage. The diver reported that for several feet down into the crib, the collapse of the dam's walls had pulled out bolts that were supposed to hold the crib's timbers together, had torn timbers, and had cracked concrete.

To repair the damage, the engineers felt forced to construct an underwater air-pressured chamber, "a pneumatic caisson," exactly the expensive and dangerous device they had hoped to avoid. They built the chamber so that it could be placed about twenty feet under the surface of the water, directly over the site of the damage. One shaft gave the men access into and out of the chamber, other shafts sent materials in and out, all shafts being fitted with "air locks." The plan was that when the repair work was completed, this chamber would not be removed but be filled with concrete, becoming like the crib itself a part of the foundation for the pier.

By mid-November 1877, the air-pressured chamber had been put in place, ready for the workmen to enter it to make their repairs. The work began satisfactorily, but as ice formed on the river, most of the bridge construction work stopped, including the underwater work.[3]

Meantime, the American Bridge Company, already weakened by the long continued Panic of 1873, was shaken by the unexpected expense of the crib rupture. It was also hurt, the company claimed, by the piers being required to be sunk deeper into the river than expected. It asked the Poughkeepsie Bridge Company to compensate it for the extra work to the amount of $380,000. The Poughkeepsie Bridge Company's Chief Engineer Dickinson told the American Bridge Company, without consulting his board first, that he thought the compensation should be paid. Embarrassed by Dickinson, the Poughkeepsie Bridge Company, itself under financial stress from many of its subscribers delaying their payments, created a committee to investigate, with Homer Ramsdell of Newburgh as its chairman. The committee found in the contract no provision for contingency payments, and recommended that the Poughkeepsie Bridge Company not pay the extra claim.

Nevertheless, the committee met in a New York hotel with representatives of the American Bridge Company, including its President H. A. Rust and its on-site Chief Engineer Coolidge. They met in a spirit of mutual concern, such as Ramsdell was capable of fostering. They worked out a deal by which the Poughkeepsie Bridge Company, while not paying the extra compensation requested, agreed to pay the American Bridge Company for its work more rapidly than heretofore, to help keep it afloat.[4]

By March 1878, work in the air-pressured chamber began again, along with other bridge work. By April, the work in the chamber was completed. Nevertheless, as the year advanced, the American Bridge Company showed signs of falling apart. It delayed paying its bills for supplies. It delayed paying wages to its employees. Its bridge work slowed. While in mid-summer, there were still 112 men at work on the bridge, by December, the American Bridge Company had been forced into liquidation. Bridge work stopped completely, as it had four years before. The next spring, 1879, there were rumors that the work would soon resume, but it did not. Once again the bridge promoters were devastated.

Facing page: CONCRETING A CRIB. Once a crib was placed in the correct position in the river, steam-operated buckets, working from nearby barges, dredged out selected crib pokets as needed, and then poured conrete into them, filling them up. The result was to transform the crib, pocket by pocket, into a largely concrete foundation for a pier. (While the drawings on pages 26 and 28 were produced to illustrate methods used in constructing the bridge in 1886-7, the same methods were also used in 1876-78. O'Rourke, "Construction of the Poughkeepsie Bridge," ASCE, *Transactions*, 1888).

Meanwhile, Poughkeepsie's preeminent bridge promoter, Harvey Eastman, though only forty-five years old, died, a victim of one of the common diseases of the time, tuberculosis. His disappearance made it harder to mount another drive to resume bridge construction.

When the Poughkeepsie Bridge work stopped this time, one pier had been built up to one foot above high water, another to twenty feet above. Cynics declared that these pier stumps were Eastman's monument.[5]

To warn passing navigators, the bridge forces placed lights on the two pier stumps, and employed a watchman, Captain W. H. Sweetman, to tend the lights. He set himself up in a little hut on the Poughkeepsie shore, from which he boated out to the lights to keep them burning, day after day, year after year. Still no bridge had been built across the implacable Hudson from New York Harbor all the way north to Albany.

HOMER RAMSDELL (Mott, *Between the Ocean*, 1899), the most prominent Newburgh businessman in his time, was a director of the Poughkeepsie Bridge Co. in 1877-78. He owned Newburgh docks and steamers (facing page, above). The famous Hudson River steamer *Mary Powell* was named for his mother-in-law. He entertained promoters of the Poughkeepsie Bridge at his Newburgh residence (below), on the hill above Liberty Street. (Both, Nutt, *Newburgh*, 1891).

THE TWO MAJOR DESIGNERS OF THE POUGHKEEPSIE RAILROAD BRIDGE

CHARLES MACDONALD was born in Gananoque, Ontario, Canada. He was educated as an engineer at Rensselaer Polytechnic Institute, Troy, NY, the earliest American instituion to grant engineering degrees. Macdonald grieved that civil engineers were blindly obedient to the financial interests which dominated railroads (ASCE, *Biog. Dict. of Amer. Civil Engrs.*, 1991, by permission of ASCE).

THOMAS CURTIS CLARKE was born in Newton, MA. He attended Harvard, but received his engineering education, as was common at the time, as an apprentice. Clarke considered that after the Civil War, railways were the engine that fueled the American economy, enabling it to flood the world with American products (*Boston Herald*, June 18, 1901).

5. DESIGN, FINAL PHASE (1886-1887)[1]

While "an ignorant man may invent a safety-pin," an engineer can design a bridge only if he "commands an army of experts."

EIGHT LONG YEARS passed with no further construction of the Poughkeepsie Bridge. During this period, the Bridge Company's directors were so discouraged that they failed to meet for years at a time.[2]

However, a burst of renewed interest occurred in 1886. By this time, the amount of rail traffic crossing the Hudson at Newburgh in rail-car ferries, especially traffic in Pennsylvania coal, had become impressive, strengthening the claim that there would be significant traffic for a Poughkeepsie bridge. Moreover, the presidents of the major New England railroads, meeting at Grand Central Station, threatened that they would push for a new bridge over the Hudson somewhere north of New York City if they were not provided with less extortionate transportation arrangements through New York City. Their threat, whether a bluff or not, helped enliven the Poughkeepsie Bridge cause.

Additionally, promoters of building a rival rail bridge over the Hudson somewhere just north of Peekskill, after nearly twenty years of talk, had finally taken a definite step. With the support of some Boston capitalists, they had drawn up specific plans to build a bridge near Storm King Mountain. The *New York Times* claimed that it was especially the Storm King bridge activity which "led to a movement among the dry bones at Poughkeepsie."

A more decisive stroke, however, was that Brooklyn railroad contractor John Clarke Stanton persuaded Philadelphia utility executive William W. Gibbs that a Poughkeepsie bridge would significantly help to market Pennsylvania coal. Gibbs then decided to lead in completing the construction of the bridge. To carry out his decision, Gibbs and his associates bought most of the Poughkeepsie Bridge Company's stock. With Gibbs in control, a large contingent of Pennsylvanians, rep-

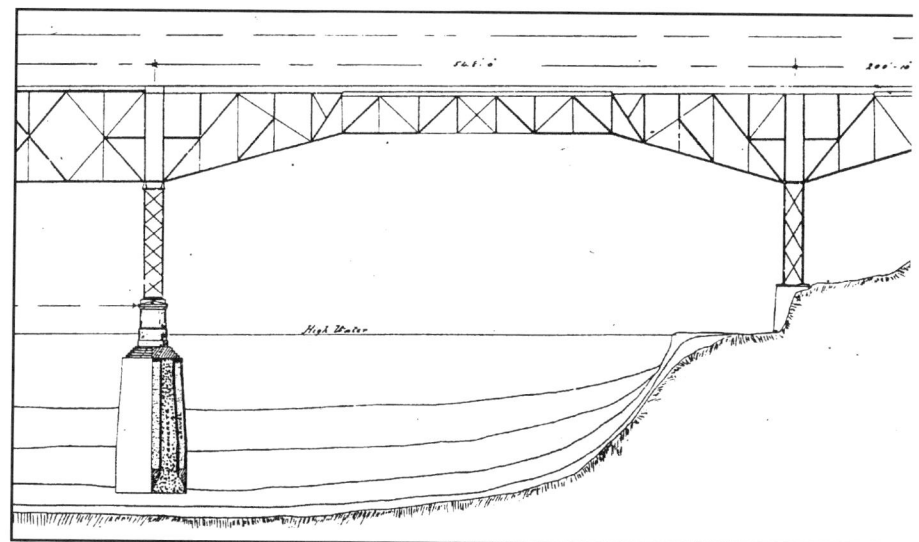

PROFILE DRAWING OF A CANTILEVER SPAN for the Poughkeepsie Bridge
(O'Rourke, "Construction," ASCE, *Transactions*, 1888)

resenting what the *New York Sun* called "the heaviest financial houses in Philadelphia," became Bridge Company directors.[3] The company's new Gibbs-led administration promptly created the Manhattan Bridge Company as its agent to build the bridge. The Manhattan Bridge Company secured a mortgage for the necessary funds, and began to act as general contractor. The third major effort to construct the bridge was underway, the first having been supported especially by Pennsylvanians, the second by New Englanders, and the third by Pennsylvanians again.

DESIGNERS

By August 1886, the Manhattan Bridge Company had contracted for the Union Bridge Company both to design the whole bridge and to construct the bridge proper, while it arranged for other companies to construct the bridge approaches. The Union Bridge Company was what the bridge promoters called "the leading bridge building company of the world." The company, they believed, could marshal expertise and equipment quickly. It promised to complete the Poughkeepsie Bridge by January 1, 1888, the deadline set by the amended bridge charter.

There were five partners in the Union Bridge Company. Three of them were based in the company's shops which were to test and fabricate the steel for the bridge. Of these three, George S. Field and Edmund Hayes were based in the company's smaller shops at Buffalo, NY, while Charles S. Maurice was based in its larger shops at Athens, PA, which were called the largest bridge building shops in the U.S.[4] It was the remaining two partners, however, who took the most responsibility for the design and building of the Poughkeepsie

POUGHKEEPSIE BRIDGE CO., DIRECTORS, Elected Jan. 1887, With Changes in April.

Sources include Poks. Bridge Co., Minutes, Jan 3, April 20, 1887, UCONN

Van Benthuysen, Watson (pres. until April, then v. pres.)	Banker, former pres. of a Louisianna railroad	NY
Gibbs, W. W. (pres from April)	Manager, United Gas Improvement Co.	Phila.
Appleton, Julius H. (resigned April)	Paper Manufacturer	Springfield, MA
Brock, Arthur	Iron Manufacturer	Lebanon and Phil., PA
Cameron, Simon	Former US Senator	Harrisburg, PA
Drayton, J. Coleman	Son-in-law of John J. Astor; director, O & W Railroad	NY
Elsworth, Edward	Mayor	Poughkeepsie
Gallaudet, Peter W.	Banker	NY
Gibson, Henry C.	Dir., Fidelity Insurance Trust	Phila.
Knevals, Sherman W.	Law partner of ex-Pres. Arthur	NY
McCormick, Henry	Furnace manufacturer; banker	Harrisburg, PA
McMichael, Morton	First National Bank	Phila.
Moore, Andrew	Moore & Sinnot, distillery	Phila.
Platt, John I.	Editor of the *Eagle*	Poughkeepsie
Williamson, Isaiah (resigned April)	Cambria Iron Co. (called one of Phila.'s richest men)	Phila.

WILLIAM W. GIBBS, PHILADELHIA UTILITY EXECUTVE, led in financing the final phase of building the bridge (*Poughkeepsie, The Bridge*, 1889).

THE MORTGAGE.—The mortgage covers all the approaches, viaducts, franchise, real estate, and property of every nature. The bridge and approaches will be, from point to point, nearly five miles in length.

AS AN INVESTMENT.—As a matter of investment it is believed that this bridge will be the best paying investment in the United States. For further particulars, write us for information on any point not covered in this circular, and also for prospectus. We offer for sale these bonds at *par* and *interest*, subject to advance in price without notice.

FEBRUARY 1st, 1887.   P. W. GALLAUDET & CO.

Under Gibb's lead, THE GALLAUDETS, NEW YORK BANKERS, RAISED MONEY TO BUILD THE BRIDGE.
Using extravagant claims, they sold bonds backed by a mortgage on the bridge property (Advertising circular, BEAUJ).

Bridge. The two were based in the company's office in New York City. They were Thomas Curtis Clarke, the company's president, and Charles Macdonald.[5]

The Massachusetts-born Clarke, about fifty-nine when he became involved with the Poughkeepsie Bridge, had graduated from Harvard (his older brother James Freeman Clarke, a Boston Unitarian pastor, helped pay his way). Then, following a common practice, Thomas Clarke had studied engineering as an apprentice. Like his brother James, Thomas Clarke became a prolific writer, which helped him to become what the *New York Times* called, "one of the best-known civil engineers in America."

Charles Macdonald, born on the St. Lawrence River in Gananoque, Ontario, Canada, was a little younger, about fifty, when he became related to the Poughkeepsie Bridge. Macdonald had graduated from Rensselaer Polytechnic Institute (RPI), in Troy, NY, which was the earliest American college to grant engineering degrees. While Clarke was better known, both Clarke and Macdonald, before becoming involved in the Poughkeepsie Bridge, already had professional experience with railroading in Ontario, the Midwest, and metropolitan New York. Both Clarke and Macdonald served at various times as president of the Union Bridge Company and of the American Society of Civil Engineers. Both Clarke and Macdonald were pioneers in the building of metal bridges, first of iron and later of steel.

Clarke proudly called the Poughkeepsie Bridge "one of the great bridges of the world," and explained that for a year he was at the bridge site "nearly every day," but he did not claim that he dominated the designing of it, or even that he and Macdonald did. Rather he claimed that he and his company associates designed it together. Clarke was well aware of the collegial professionalism necessary to design a great bridge, remarking once that while "an ignorant man may invent a safety-pin," an engineer can design a bridge only if he "commands an army of experts."[6]

While both Clarke and Macdonald wrote for engineering journals, Clarke wrote also for more general periodicals such as the *Atlantic*, *Scribner's*, and *Christian Examiner*. He often wrote and spoke in the style of a man who was confident without needing to be arrogant, and could allow himself to be light and amused. He sometimes wrote on architecture (he preferred buildings to make their purpose boldly evident in their form, and to avoid ornament other than that which grew out of that form), and he sometimes wrote on subjects as little related to engineering as poetry (he valued poetry especially when it was spontaneous in style, included "expansive sympathies," and said "much in few words").[7] When he wrote on more technological subjects such as bridges, railroads, harbors, and city transit systems (he helped design New York City elevated railways), he often chose to take a broad perspective on them, including their impact on common life.

Both Clarke and Macdonald were critical of both their fellow civil engineers and railroads. Macdonald grieved that engineers were blindly obedient to the financial interests which dominated railroads, and he blasted railroads for charging discriminatory rates without regard to the public interest. Clarke protested that railroad directors were often rich retired businessmen who were ignorant about the practical necessities of railroading, and employed inadequately educated engi-

ONE OF THE UNION BRIDGE CO. PARTNERS, C. S. MAURICE, ran the company's advanced steel fabricating shop, northwest of Scranton, PA (WILLM)

neers. However, both Clarke and Macdonald took pride in America's leadership in building railways. Clarke considered that since the Civil War, railways had become the engine which had fueled the American economy, enabling it to flood the world with American products, and that American railways had progressed far in advance of all other nations, except Canada. A cause of the American advance, Clarke believed, was American engineers' "contempt" for precedent. A related cause, he said, was that American manufacturers, including Andrew Carnegie, had more "courage" to replace antiquated machines with advanced ones, helping Americans to build cheaper railways. Another factor was that in locating rail lines, while the British tended to defy nature, insisting on keeping rail lines almost straight and level no matter what natural objects were in the way, Americans were more willing to adapt to nature, being willing to pass rail lines around mountains rather than tunnel through them. Besides, while European engineers tended to be isolated from builders and suppliers, American engineers work closely with them, enabling bridge builders to shape and drill their bridge parts in the shops where they were made, so that as soon as they reached the bridge site, they were ready to be put into place.[8]

DESIGN

During the two previous starts at building the Poughkeepsie Bridge, in 1873 and 1876-78, the bridge had been designed conventionally as a rectangular truss bridge with, as Engineer Eads had advised, four piers. In this final design phase, Clarke and Macdonald continued to follow Eads' advice to design the bridge to have four piers. But influenced by the experience of their Buffalo partners, Field and Hayes, in helping to build a cantilever bridge over the Niagara River, they designed the Poughkeepsie Bridge as a cantilever bridge. That is, they designed it to have as many cantilever spans as the strength of the piers would allow: three spans. According to the *New York Herald*, it was to be the longest cantilever bridge in the world.

Cantilever spans are constructed essentially like brackets reaching out from the two sides of the space to be spanned until they meet in the middle. To be stable, the brackets must be held in position by strong anchoring. In the case of the Poughkeepsie Bridge, navigation interests, insisting that the bridge should not interfere with navigation any more than necessary, forced the designers to make the piers too narrow to support cantilever brackets all the way across the bridge. So it was necessary to alternate cantilever spans with truss spans which would help anchor the cantilevers.

THE POUGHKEEPSIE BRIDGE: ITS LEADING DESIGN AND CONSTRUCTION ENGINEERS

Date	Name	Work	Born
1869-1889	Pomeroy P. Dickinson	For bridge owners, supervised design and construction	Rochester, NY
1871	Horatio Allen	Design	Schenectady, NY
1873	Jacob H. Linville	Design	Pequea, PA
1876-1878	W. G. Coolidge	Design	[Office in Chicago]
1886-1888	Arthur B. Paine	Directed surveying, acquiring land, contracting	Charleston, SC
1886-1888	John F. O'Rourke	Construction	Tipperary, Ireland
1886-1888	Thoma Curtis Clarke	Design	Newton, MA
1886-1888	Charles Macdonald	Design	Gananoque, Ont., Canada

The principle of cantilever construction had been known for centuries in both Europe and China, but until about this time it had been little applied to modern bridge building. Aware of the long history of cantilever construction though he was, Clarke nevertheless declared cantilever construction, as it was being practiced at the time, to be "the most notable invention" in recent bridge construction. What American engineers were finding particularly appealing about it was that when cantilever bridges were built over water, they had the advantage over rectangular truss bridges that their spans could be built without erecting falsework (scaffolding) in the water to support them, thus reducing both their cost and their interference with navigation.

Later the acerbic bridge designer J. A. L. Waddell, who had taught at RPI, claimed that the Poughkeepsie Bridge was designed as a cantilever bridge because at the time cantilever bridges were popular for their "novelty." While cantilever structures are suitable for bridging a narrow gorge with rocky sides because they reduce the need for costly scaffolding, Waddell explained, there was "no good reason whatsoever for making the great Poughkeepsie Bridge a cantilever structure." For that site, he said, a simple truss span bridge would "probably" have been cheaper.[9] It is a question whether Waddell, concentrating as he did on engineering, took sufficiently into account the opposition of navigation interests to placing any more falsework in the Hudson than necessary. In any case, after the Poughkeepsie Bridge opened, engineers considered it appropriate to build many other cantilever bridges that were also not over narrow gorges, including the Queensboro Bridge over the East River (1909) and the Tappan Zee Bridge over the Hudson (1955).

Clarke knew that pressures to cut costs could weaken bridges so that they would collapse, as they often had. He "shuddered" when he recalled how slender the iron posts of his own early bridges were, and how slight their lateral bracing. As friends of the Poughkeepsie Bridge were likely to know, a few years before on the Connecticut Western rail line which was slated to become one of the Poughkeesie Bridge's major connecting

TWO DRAWINGS FROM DESIGNER CLARKE'S ARTICLE ON RAILWAY BUILDING, published in *Scribner's Magazine*, June 1888, while the Poughkeepsie Bridge was still under construction. The upper one shows the three cantilever spans of the Poughkeepsie Bridge. In the lower one, originally published in 1811, Pope, an innovative ship carpenter, was proposing to build a bridge over the Hudson at New York City, a proposal which has been called the first known plan to build any bridge over the Hudson there. Boldly conceived, it was to consist of one cantilever span, constructed of wood.

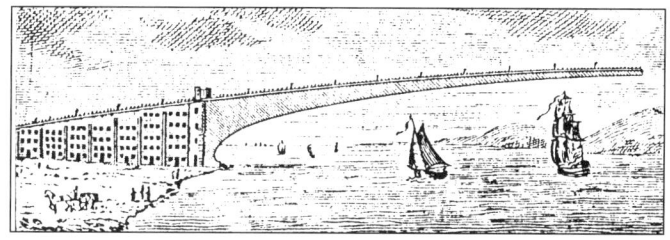

lines, when an excursion train was carrying a crowd of revivalists singing Gospel songs across a wooden bridge near Tariffville, CT, the bridge groaned. Its timbers cracked. The passengers' singing turned to shrieks. Several of the train cars toppled off the bridge and sank into the water. Thirteen passengers died. What caused the collapse was never determined for certain.

Despite safety concerns, the Poughkeepsie Bridge was designed to be built cheaply. Bridge Engineer Gustav Lindenthal, who followed the Poughkeepsie Bridge story over many years, recalled later that "it was well known" among engineers, though not among the general public, that the final push to complete its construction "was rather a desperate effort to get the bridge done on any terms." However, being pushed to build cheaply did not seem to daunt Engineer Clarke. He praised American in contrast to British

EVOLUTION OF THE DESIGN OF THE BRIDGE SUPERSTRUCTURE		
Date	Plan	Primary Material
1871 (Bridge Co. incorporated)	Suspension	Iron
1873 (cornerstone laid)	Rectangular truss	Iron
1876-78 (work resumed)	Rectangular truss	Iron and steel
1886-88 (work completed)	Cantilever	Steel

content; hard steel, he said, while stronger, was brittle and hence "treacherous."[11]

Despite pressure to keep costs down, Clarke and his associates designed the Poughkeepsie Bridge to withstand, he said, the weight of "the heaviest coal trains" expected to cross it, even if there were more than one train on the bridge at a time. They designed it to withstand winds that were strong enough "to blow trains off the bridge." They allowed for the metal of the bridge to withstand changes of temperature by arranging to install at symmetrically spaced junction points what Clarke called "movable links," including sliding bearings and eye bars held together by great pins thrust through their eyes.[12]

Clarke and Macdonald understood how the Brooklyn Bridge was built because they were

railway building for its "almost uniform practice of getting the road open for traffic in the cheapest manner and in the least possible time, and then completing it and enlarging its capacity out of its surplus earnings."[10]

Certainly Clarke and Macdonald designed the Poughkeepsie Bridge to be daring. They designed it to be both longer and higher than the Brooklyn Bridge. While they designed the bridge's approaches to be built of iron, which was adequate for the shorter spans over land, they designed the longer spans over the water to be built of steel. According to the *New York Times*, one of its spans at 525 feet will be "the largest and heaviest steel truss in the world." At this time when most of the railroad bridges in use were still made of wood, and the recently opened Brooklyn Bridge was an early example of a great bridge constructed essentially of steel rather than iron, the methods for manufacturing steel, testing it, and using it in bridge construction were still rapidly evolving. For the Poughkeepsie Bridge, Clarke insisted that the steel be "mild steel," that is open hearth steel having a low carbon content. He refused to use "hard steel," with a high carbon

POUGHKEEPSIE BRIDGE COST (Difference among the costs reported may result from variations in what is included in the count)			
1875	Estimate	By Engineer P. P. Dickinson	$4.2 million
1876 as placing piers in the river began	Contract for construction only	By American Bridge Co.	3.4
1886 as construction was resuming	Estimate	By *Enigineering News*, Sept. 4, 1886	4.2
1887 as construction continued	Estimate	By Poughkeepsie Bridge Co.	5.0
1888 as completed	Report, Jan. 24, 1921	By the New Haven's Engineer of Structures	3.6
1921 with additional strengthening to this date	Same report, Jan. 24, 1921	By the New Haven's Engineer of Structures	5.5

related to it. At RPI, Macdonald had been a classmate of Washington A. Roebling, who, after his father died, became the chief engineer for building the Brooklyn Bridge. Both Clarke and Macdonald served as trustees of the Brooklyn Bridge while it was under construction from at least 1881 until it opened in 1883. In addition, while they were working on the Poughkeepsie Bridge, both served on a three-man committee on how to enable the Brooklyn Bridge to carry more traffic.[13]

Nevertheless, to place the piers for the Poughkeesie Bridge in the water, Clarke and Macdonald avoided the Brooklyn Bridge-style air-pressured underwater chambers. They believed that the extreme depth of water and muck through which it was necessary to place the Poughkeepsie Bridge piers, 130 feet or more, deeper than the depth at the Brooklyn Bridge site, made air-pressured chambers too dangerous. Instead they chose what Clarke called an "American" device, the same which Engineer Coolidge had chosen for the 1876-77 attempt to build the bridge, that is, the sinking of open timber cribs (or caissons) without air-pressured chambers in them, down as far as possible onto the river's hard bottom.[14]

Whatever detailed plans Clarke, Macdonald, and their associates chose, the bridge they designed was high, the river was wide, deep, and powerful. Something could go wrong, with disastrous results. Their responsibility for balancing safety, economy, and the needs of traffic was profound.

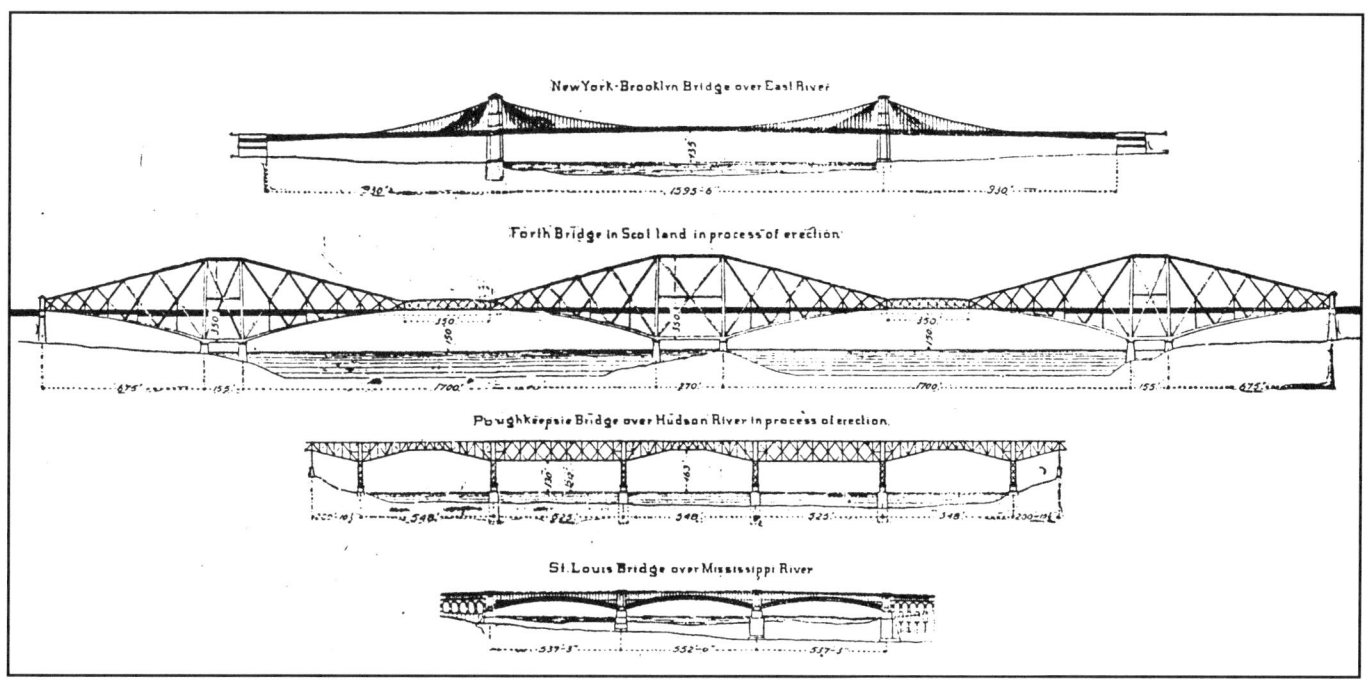

PROMINENT BRIDGES OF THE TIME COMPARED by Drawings to Scale (*Engineering News*, Jan 14, 1888). In style, the Brooklyn Bridge (opened 1883) is a suspension bridge, both the Forth (1890) and Poughkeepsie (1889) bridges are cantilever, the St. Louis Bridge (1874) is arch. In length over water, the Forth Bridge is longest, the Poughkeepsie Bridge second. In clearance for ships to pass underneath, Poughkeepsie Bridge is highest, the Forth second.

AT WORK ON HIGH STEEL, SCARCELY SEEMING TO BE AWARE OF THE DANGER. While this drawing is not necessarily of Poughkeepsie Bridge workers, it was published while the bridge was being built, in an article by one of the bridge's major designers, Clarke, in *Scribner's*, 1888.

6. BUILDING, FINAL PHASE (1886-1888)[1]

"The men become so familiar with labor at high altitudes" that *"they do not really understand the dangers that constantly beset them."*

THE UNION BRIDGE COMPANY, the major construction contractor, appointed John F. O'Rourke to be its chief engineer for the bridge. Under the oversight of Clarke and Macdonald, O'Rourke directly superintended most of the bridge construction. He was only thirty-two years old. The *Poughkeepsie News-Telegraph* called him "perhaps the youngest man that ever constructed a work of the magnitude of the Poughkeepsie bridge."[2]

Born in Ireland, O'Rourke came to New York with his parents as a toddler. He graduated in engineering from New York's Cooper Union, and afterwards taught engineering there. When the West Shore Railroad was being built along the Hudson River, he helped to construct its tunnels through the rocky Hudson Highlands. While working on the bridge, O'Rourke allowed hints of fun to show in his blue eyes. He liked to joke, as by claiming that from the bridge site he could see

JOHN F. O'ROURKE, the popular engineer who was directly in charge of building the Poughkeepsie Bridge in its final phase (O'ROUR)

mosquitoes on Storm King Mountain, and he was popular with bridge workmen. He was also popular in Poughkeepsie homes. When he met Katherine B. Innis, daughter of Aaron Innis and niece of George Innis, who were both dye manufacturers and both promoters of the bridge, he and Katherine became enamored of each other. However, when he wrote a report for an engineering society on his progress in constructing the bridge, his style was strictly conventional, as if in his struggle to rise in his profession, he knew there were occasions to hold down his youthful exuberance.

To fix the precise size and location for the bridge piers in the water, for both the two piers already begun and the other two not yet begun, O'Rourke arranged for surveyors to do the necessary triangulations from shore. He was satisfied when the distances they reported in several independent triangulations checked out with each other, as well as with a direct measurement made across the ice, all identical within a range of .24 feet. *Engineering News* praised this result as "exceedingly close."[3]

Cable to moor the cribs arrived from the Roebling family mill in Trenton, NJ. Cement arrived by train from Buffalo, sometimes at the rate of two car loads per day. As much as 30,000 barrels of the cement was not the more artificial Portland cement but natural cement, made especially from natural limestone, most of it to deposit in the cribs below the water level, for the base of the piers.[4] Sand also arrived, by scow, to mix with the cement in the cribs; the sand was at first unloaded by wheelbarrows, but this being considered too slow, carpenters built gangplanks wide enough to permit unloading by horse and cart instead. A tow dropped off several barges of gravel, high quality gravel, brought up the river from Roa Hook, near Peekskill. Other tows brought in stone, including blue stone for facing the piers, some from Ulster County's Kingston region, some from farther upriver, from Rensselaer County's Sandy Hill Quarry. Some of the facing stone was cut to specifications right at the quarries, so that it was ready for placing in the pier foundations. Timber arrived by train, much of it from Michigan, to be used especially for scaffolding. During the winter, when so much timber came in by West Shore trains that there was not enough room to store it on the nearby shore, some of it was dumped onto the ice on the river and toggled together there to prevent its floating away when the ice moved out in the spring.

LABORERS

From the late 1840s to the 1870s, most of the common laborers who had built railroads in the Mid-Hudson region had been Irish immigrants, who often seemed feisty to the region's longer established residents. By the 1880s, when first the West Shore line and then the Poughkeepsie Bridge and its connecting lines were being built, most pick-and-shovel laborers were milder-mannered Italian immigrants, unable to speak English.

According to the *Poughkeepsie News-Telegraph*, Italian immigrant laborers on the bridge were not taking jobs away from Americans, as was sometimes claimed, because they often did the "hard and mean work" which no one else "could or would do." Most of the Italians received from $1.00 to $1.25 a day. Skilled workmen, however,

CONSTRUCTION ON THE POUGHKEEPSIE SHORE, Oct. 6, 1887, showing scaffolding for the erection of the east anchor tower and its accompanying approach span (ADRI). Visible within the scaffolding for the approach span is a dark supporting girder which is already in its permanent position at an angle. On this east shore, unlike the west shore, extensive docks and yards facilitated the handling and storage of consruction supplies.

DIVER ON TOP OF A COMPLETED CRIB (ADRI). Such divers went down into the crib pockets to check how well the crib had settled to the river bottom. They also went down the outside surface of the crib to check for any cracking in the timbers or concrete.

who were likely to be a mix of many backgrounds, such as "Teutons, Gauls, Britons, Celts, and 'Americans,'" were paid more, about $3.00 a day.

To speed up their work, contractors kept some men at work in winter. According to the *News-Telegraph*, in the winter of 1887, in preparation for constructing bridge piers, "the musical clink of stone cutters tools" could "be heard from early morning until sunset."[5] Also during the winter, at the sites of the two pier foundations already built years before and left unfinished, divers plunged into the water to inspect the foundations, staying there an hour at a time, using hoes to scrape away debris so they could better check for defects. From what the drivers reported, engineers decided that parts of those two piers had been significantly cracked. For this reason and also because cantilever spans required thicker masonry in their supporting piers than the earlier-planned truss spans required, engineers ordered that some of this pier work be removed and redone.

As early as January 1887, besides the engineering staff, there were 200 men busy at the bridge site. About that time, workmen were taking advantage of the ice on the river, where it was thick enough, to walk across it to work sites, including walking all the way from one shore to the other. By April about 1000 men were working at the site and by July, 1500.

Sightseers watched workmen unload and deposit supplies. They watched work boats take workmen out to the pier sites and come back again. According to the *Poughkeepsie Eagle*, passengers on trains running along the Hudson River shore gawked "with astonishment" at the bridge work, and talked of "nothing but the bridge." For closer views of the work out in the water, sightseers could ride out daily on the "little steamer Gypsy." Or they could ride the regular Poughkeepsie-Highland ferry, which, although it had altered its usual course because of the bridge work, continued to run, with its flags flying.[6]

PLACING FOUNDATIONS IN THE WATER

In the second attempt to build the bridge in 1876-78, difficulty in placing the foundations for one pier in the water had been a factor in bringing that construction attempt to an end. In this third attempt to build the bridge, difficulty in placing the foundation for another pier again threatened construction.

Much as in the previous attempt to build the bridge, O'Rourke arranged to begin to build two huge cribs on shore. As sightseers watched with delight, carpenters built them of heavy hemlock timbers, with the timbers interlaced, and with pockets in them open to the air. Early in 1887, with O'Rourke hovering about "in an enormous buffalo skin coat,"[7] workmen launched these two cribs into the river, even if they had to launch them through the ice. Then workmen built them up higher. Eventually after the weather warmed, they floated them out to the pier sites, way out in the river, and tried to anchor them in the precisely correct position.

One day in the spring of 1887, after the steamer *Miller* had towed a crib out into the river into approximately the correct position for pier 4, a formidable problem developed. The crib had already been attached to one anchor—it was a huge wooden box anchor, called a Chinese anchor, put together in log-house fashion, and filled with broken stone—when complications occurred with the anchor cables, causing delay. By the time the *Miller* was ready to begin pulling the crib into precise position, the tide was running out briskly (the tide reaches up the Hudson past Poughkeepsie), and by chance a spring run-off—what O'Rourke called a surprise "freshet" was also flowing in the river. The freshet was simply "unexampled," he explained. When the freshet "was at its worst," to keep the crib in position, the steamer *Miller* turned on as much steam as its "boiler could stand." The *Miller*'s engine throbbed. Its "screw revolved with terrific force." The whole steamer shook. But the *Miller* could not prevent the crib, the "huge mass of timber" that it was, from slipping downriver, and dragging the steamer with it. Engineers arranged for the tug *Hart* to come to the *Miller*'s assistance, pushing her nose against the crib, but still the crib kept drifting downstream. Around six o'clock in the evening, even though a corner of the crib ran aground for a bit, it still continued to drift. O'Rourke was embarrassed, to say the least. A rumor arose that the crib's floating away meant that the whole idea of building the bridge was floating away with it.

The crib had drifted three miles downriver when the tide turned and the crib finally stopped drifting. Then O'Rourke marshaled three steamboats, including one of the regular Poughkeepsie-Highland ferries, to head the crib back upstream. By midnight, with the help of the tide and "an entire absence of wind," the three steamers had pulled the crib back into its proper position. Thereafter engineers increased the number and variety of anchors to hold the crib in place. They drove piles—of unusual length—into the river bed to support the anchors. Although various anchor cables broke, the workmen gradually improved their placing of the cables and their adjustment of the cable tension. When several days later, crib 4 seemed finally secured exactly where it was supposed to be, steam whistles gave out "ear-piercing blasts," and "a hundred throats" gave forth "resounding cheers and a tiger," a traditional yell. Later Chief Engineer O'Rourke, reviewing the whole construction of the bridge, reported coolly that placing the crib for pier 4 "was, perhaps, the most difficult feat." *Engineering News* later declared the builders had "triumphed over many difficulties," and called O'Rourke "able."[8]

After each crib was anchored at its assigned site, it was gradually sunk by dropping rocks into some of its open pockets. As the crib sank, carpenters kept building up the crib to make it higher, that is, they kept laying additional courses of timber on top of it. At about the same time, derricks, mounted on a nearby scow, lowered their huge buckets through some of the cribs' empty pockets and hauled up bites of the mud, clay, sand, and gravel accumulated over countless years on the river bed, gradually removing enough bites to allow the crib to settle lower, closer to the firm rock of the river bottom. Gradually, also, derricks dropped buckets of concrete, mixed from natural cement by concrete mixers on the accompanying scow, into selected water-filled crib pockets, where it hardened. The concrete's added weight further helped to lower the crib into the hard gravel near the river bottom.

When the crib had settled well, and all the remaining pockets had been filled with gravel or concrete as appropriate, a pile driver was towed out to the crib, and a few taps from its 6,500 pound hammer helped to settle the crib still more firmly. The timbers which originally formed the crib remained with it permanently under water, where away from air, they would not rot any more than stone would. Thus each crib, once open to the air, gradually acquired a central core of concrete which was to be the foundation for a pier.

As the crib placing continued in the summer of 1887, the American Society of Civil Engineers, while they were holding a conference at the pillared Hotel Kaaterskill high in the Catskill Mountains, arranged to visit the bridge site. About 150 of them took a train to Kingston where a steamer picked them up and, while they partook of a lobster lunch, brought them down the Hudson. The steamer passed by the bridge pier work and docked at Poughkeepsie. There, five tugs picked up the visitors and took them out on the water where, under the guidance of Chief Engineer O'Rourke and others, they inspected work sites close up. At one pier site, they watched a great bucket "with its monster ugly jaws. . .delving into a six-foot [wide crib] pocket and bringing up black pasty deposits from the bed of the Hudson. . .eight or ten bushels at each dive, and dumping them into the river outside of the crib frame." The visitors were awed.[9]

ERECTING SPANS

For the bridge towers and superstructure, engineers ordered steel. Much of it was poured at Carnegie-related factories in the Pittsburgh area. Then it was sent to the Union Bridge Company's shops at Athens or Buffalo, where it was tested (the *Scientific American* said the Union Bridge Company's new steel-testing machine at Athens was "the most powerful testing machine in the world"), and then it was cut, holed, and riveted into the requisite shapes, ready for use. Only then was the steel shipped by train to the bridge site. So much steel was needed, however, that American factories could not produce it quickly enough. The Union Bridge Company did not thereupon settle for using the more available iron, but rather sent company partner Field to England to secure the additional steel. When the English steel arrived in the U.S., it was sent, like the American steel, to the shops in Athens or Buffalo, to be tested and fabricated, and only then shipped to the bridge site. Near the bridge site, much of the river shore

CONSRUCTION OF THE WEST ANCHOR TOWER (ADRI). The West Shore Railroad normally had two tracks, but here it has a third track used to handle construction supplies. However, as the picture indicates, because this Highland shore had steep cliffs and railroad tracks close to the river's edge, it provided less space than the Poughkeepsie shore for the docks and yards desirable for handling construction supplies.

seemed to become cluttered with steel, machinery, timber, stone, sand, and gravel, but *Engineering News* considered their arrangement "admirable."[10]

The original charter for the bridge had provided for it to carry not only railroad tracks but also a roadway for carriages, as many local drivers were well aware. In early 1887, some of the bridge company officers were still pushing to install a carriageway in a story underneath the railroad tracks. But Arthur B. Paine, who had been appointed by the over-all contractor, the Manhattan Bridge Company, as its chief engineer for the bridge, questioned the wisdom of installing such a roadway. Carolina-born and educated in Europe, Paine had recently been chief engineer of one of the railroads which was expected to connect to the bridge, the railroad being slowly built from Pine Bush, NY, to Slatington, PA. Called "very gentlemanly," Paine reminded the bridge forces that a convenient ferry was already available to carry ordinary vehicles like carriages and wagons across the river, and persuaded the bridge forces that for such traffic to drive on the bridge would require building long, expensive approaches, and that the amount of this traffic would likely be too small to justify the expense. When an attempt was made in the state legislature to require the Poughkeepsie Bridge Company to build the carriageway, not only did the company oppose it but so did interests which favored the continuation of the ferry, and it failed to pass. By June 1887, to the resentment of many Poughkeepsie area residents, the company had pushed through the legislature an amendment to its charter which omitted providing for a carriageway.[11]

TRAVELING DERRICK AT WORK ON TOP OF THE WEST ANCHOR TOWER (McEN). Men are standing on the derrick at various levels, some of them in seemingly risky positions. Steel work has already been erected inside the tower's scaffolding. At right lower down is the scaffolding for the erection of a rectangular truss span.

The bridge was designed to have two major steel spans which were not cantilever spans but were rectangular truss spans which the cantilevers needed for support. To construct these two truss spans, temporary "false-work" (scaffolding) had to be built up from the river bed. Though it inevitably slowed construction, the bridge builders agreed they would erect false-work in the water only between one set of piers at a time, thus reducing annoyance to navigators.

To provide a foundation for the false-work, workmen drove piles into the river bed. The water and the mud were so deep that these piles had to be an astonishing 130 feet long or more. They were made of yellow pine and spruce, bolted together. Workmen drove them in with a three-ton steam-powered hammer. They drove them in clusters, hundreds of clusters, in row after row. Above the water level, workmen topped the piles with square timbers, and then began to erect on top of the timbers interlaced timbers, especially hemlock. Eventually spectators on the river bank could see massive scaffolding rising, with "workmen walking here and there through the mazy interlacings of timber," appearing in the distance as small as ants. Engineer Clarke was proud of this "timber trestle work," calling it "one of the largest and most successful" ever to be constructed in the United States.[12]

Once a section of the false-work was completed, a temporary deck was built on its top, and then four sets of temporary tracks were laid on the deck. Two of the sets of tracks were usually occupied by a traveling hydraulic riveter and the other sets by a traveling derrick, both steam-operated. The traveling derrick was put together at the site, of timbers joined by plates and ties of iron. This and other derricks used were larger than the ones which had been used in building the Niagara cantilever bridge; *Engineering News* called them the "most complete" such derricks ever built. The traveling riveter, similarly put together of timber and iron at the site, set rivets "like a pair of scissors with a small ram between the rear jaws." Steel pieces on the bridge were normally not welded together, but riveted, some riveted by the hydraulic riveter, some by hand. Rivets were up to eight inches long, which, according to *Engineering News*, were "probably the longest ever used in this country."

Once a reporter was allowed to go up into a maze of timber false-work. Up there he saw "charcoal furnaces" scattered about, heating rivets so hot that they became white. He watched "great broad-shouldered workmen," hammering hot rivets into steel trusses with "huge sledgehammers." They were standing on the center of boards which were held up only at each end, and every blow the men delivered on "the resounding steel" made the boards "spring to and fro."

One day workmen who were building up false-work, story by story, noticed that it was shaking. They discovered that several "urchins" had climbed a guy rope and were "performing gymnastic evolutions on it." According to a newspaper report, after "several of the youngsters had smart applications of leather laid on hard and rubbed in well," it was thought they would not continue a game so dangerous to the workmen up above.[13]

When the two truss spans which linked certain pier towers were about completed, giving the towers the strength to brace the cantilever arms, then it was time to erect the three cantilever spans. To do so, the cantilever arms were extended boldly, steel piece by steel piece, out into the void, high over the water. In a dramatic spectacle, a traveling derrick moved each piece out on top of the completed part. Then the derrick lowered each piece into the appropriate position. Men standing on a platform suspended in the air by rope from above connected each piece in place.

As construction progressed, navigation interests, continuing their fierce hostility to the bridge, tried to persuade Congress to require that the piers already built for the bridge be removed as obstructions to navigation. But the truth was, friends of the bridge argued, that the width available for shipping to pass between the piers of the new bridge was larger than the natural width available for shipping in some heavily used sections of the Hudson River, such as near Storm King Mountain. Moreover, while construction proceeded, as Engineer Clarke observed on his frequent visits to the bridge site, tows consisting of steam tugs pulling seventy to eighty barges tied together managed to pass without the "slightest difficulty." When a few minor shipping accidents did occur at the site during construction, *Engineering News* refused to take them seriously, reporting that at least one of them might have been deliberate.[14]

Besides being obliged to beat back obstructionist legislation, the bridge forces were obliged to go to the state legislature with an embarrassing request of their own. By early 1887, it had become clear that bridge construction was not moving ahead fast enough to be completed comfortably and safely by the time the revised charter required, which was January 1, 1888. The Poughkeepsie Bridge Company was forced to ask the legislature for an extension of time, as the company already had done four times before. Once again the company ran into opposition from both navigation and rival bridge interests. However, the bridge forces struck a deal with one key rival-bridge advocate, the assembly's speaker, James William Husted. He had supported the Poughkeepsie Bridge when he had been the Masonic grand master in charge of laying its cornerstone in 1873. But since then Husted had identified himself with the rival Storm King Bridge proposal, to build a railway bridge over the Hudson just north of his home town of Peekskill.

As Husted himself explained, the contractor John Clarke Stanton of Brooklyn, an associate of Gibbs, had come to Husted, asking him on behalf of the Poughkeepsie Bridge cause, to join forces with them. Stanton "told me that he came to me not in my capacity as a legislator," Husted said, "but simply as a business man interested in an opposition company. I consulted with our [Storm King Bridge] people on the matter. . . . I felt pretty sure that while there were two such projects in the field there would be constant fights which would cripple and retard both. The Poughkeepsie Bridge seemed well planned; its railroad connections. . . .appeared to be adequate, and, for these reasons alone and purely from a business point of view, I consented to abandon the Storm King Bridge enterprise and join with its rivals." Apparently as a reward for his switch, Husted secured a financial interest in the Poughkeepsie Bridge Company and in its associated rail lines, and became president of at least two such lines, one being the Hartford & Connecticut Western. According to the *New York Times*, Husted made a "corrupt bargain:" he "virtually sold his influence as Speaker. . .in behalf of a private concern." It was with the help of this bargain that the Storm King Bridge cause withered, and Poughkeepsie Bridge Company won from the legislature the extension of time it needed to complete its bridge, an extension to January 1, 1889.[15]

Despite construction running behind its original schedule, in early 1888 the *Scientific American* reported that construction was proceeding efficiently. Americans were constructing the Poughkeepsie Bridge more efficiently, the *Scientific American* and others explained, than the British at the same time were constructing a similarly long cantilever railroad bridge over the Firth of Forth in Scotland. The Americans used relatively small, slender members for their truss work, while the British used comparatively large, ponderous ones. The Americans did much of the necessary cutting, riveting, painting, and numbering of truss members in workshops so that when they were delivered at the site, they were ready to be put up at once, while the British did much of this work only after the truss members had been delivered at the site.[16]

As the bridge was nearing completion, on one occasion O'Rourke was up on it watching two gangs of "brawny" men working in rivalry to be the first to get certain bolts in, when by chance a workman in one gang was crowded out of his position. He lost his foothold, and dangled by his arms from a slender steel piece, in the air, 200 feet above the water. According to O'Rourke, the dangling workman, "in the most matter of fact way," asked one of his gang to move a bit, and then, "concentrating his strength in one effort, he regained a footing on the iron. . . . The men become so familiar with labor at high altitudes," O'Rourke explained, that "they do not really understand the dangers that constantly beset them."

On another occasion at about the same time, a derrick, high up on the false-work near pier 5, was

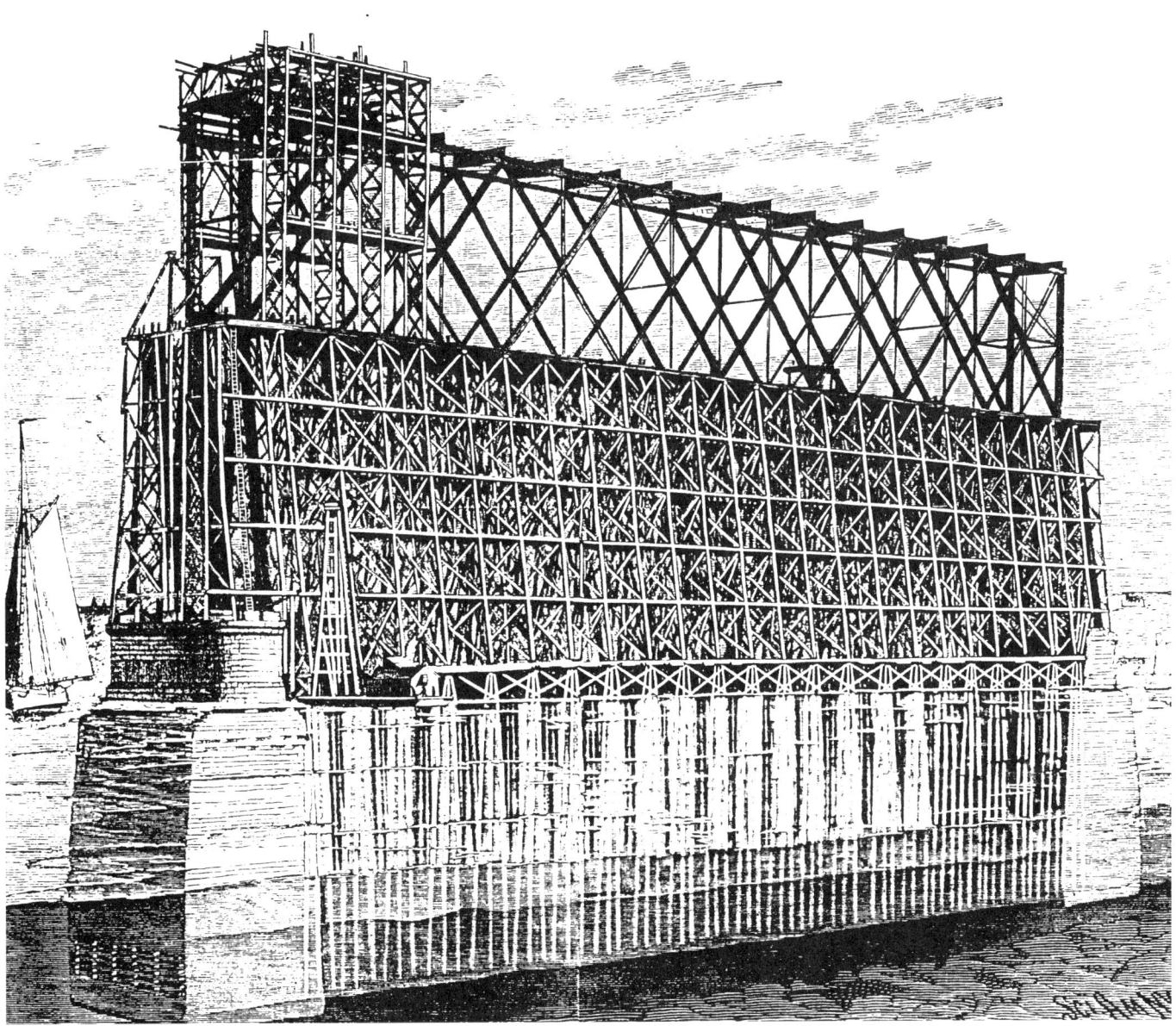

TEMPORARY SCAFFOLDING ERECTED BETWEEN TWO PIERS, WITH A PERMANENT TRUSS SPAN BEING BUILT ON TOP OF IT (from engineer Clarke's article on the Poughkeepsie Bridge in *Scienctific American Supplement*, 1888). The sailboat to the left indicates the water level. The two piers and the piling in between extend not only down into water, as indicated by light shading, but also, as indicated by darker shading, reach far down into river sediment.

hoisting a huge metal piece by rope from a scow far below on the water, with men on the scow guiding the ascent. In a squall of wind, the metal piece caught in some beams, and because the derrick was still pulling on the rope, the rope broke. The metal piece began to fall. Workmen shouted a warning. Two men on the scow jumped out of the way into the water. One man was pulled back into the scow safely, but the other, Austin Merritt, of Highland, aged thirty-five, who had long been working on the bridge, disappeared. Engineer O'Rourke, who happened to be nearby in a rowboat at the time, kept watching for him to rise to the surface, but never saw him. He suspected that Merritt must have been caught in an eddy that carried him down in the water. Merritt was the eighth workman to die in the bridge construction.

On August 30, 1888, near the eastern shore,

ONLY A SMALL OPEN SPACE REMAINS TO BE CROSSED TO MAKE A CONTINUOUS BRIDGE, Aug. 28.1888. Looking northwest from the Poughkeepsie shore (McEN). The massive scaffolding which shows in the center as still remaining in place will probably soon be removed since the truss span for which it was erected appears to be complete.

the last pin was driven to connect the two arms of the bridge's last cantilever span, so that a continuous bridge was at last in place. A cannon had been hauled up to the top of the bridge for the occasion. Present along with the cannon, as it was fired to celebrate the event, were several of the major figures in the creation of the bridge, including Chief Engineer O'Rourke, the Manhattan Bridge Company's gentlemanly Engineer Paine, the Poughkeesie Bridge Company's long-term engineer Dickinson, as well as the bridge designer Macdonald, the bridge financier Gibbs, and at least two women guests.

Women did not work constructing the bridge. Evidence is not available that even in the neighborhood of women's rights citadel Vassar College, it was even proposed that women should work constructing the bridge. But a few days after the celebration, a woman, Mrs. John Kindlen, accompanied by her bridge-employed husband, climbed up the bridge approach in Poughkeepsie, and then walked out on the bridge over the water. By this time, wooden walkways were already in place on both sides of the bridge deck, but most of the iron railing which was intended to protect pedestrians on the walkways was not yet in place. In addition, along the center of the bridge deck, since railroad ties had not yet been put down, there were large openings through which one could easily see through the bridge trestle work down to the water. However, according to the *Poughkeepsie News-Telegraph*, John Kindlen gave his wife "no assistance except at rare intervals holding her arm."

HARTFORD & CONNECTICUT WESTERN RAILROAD (H&CW). A Poughkeepsie Bridge affiliate: Selected Officers, 1887-1888, while the Poughkeepsie Bridge was being built. Office: Hartford. (From State of CT, Railr'd. Comm's. An'l. Reps., 1887, 1888)		
Husted, Jas. W., pres.	Speaker, NY Assembly	Peekskill, NY
Barnum, Wm. H., vice pres.	Railroad investor	Lime Rock, CT
Appleton, Julius H.	Paper manufacturer, pres. of the Poughkeepsie Bridge Co. in 1886	Springfield, MA
Brock, John W.	Iron manufcturer; pres., Manhattan Bridge Co., the gen'l contractor for building the Poks. Bridge	Phila. and Lebanon, PA
Cornell, Thos.	Promoter of Hudson River shipping and such railroads as the Rhinebeck & CT (already absorbed by the H&CW) and the Wallkill Valley	Rondout [Kingston], NY
Gibbs, Wm. W.	Leading financier for building the Poks. Bridge	Phila.
Holly, Alexander H.	Banker; ex-governor of CT	Lakeville, CT

She walked with her husband "unhesitatingly," all the way across the bridge, and, added the newspaper, reflecting the expectations of the times, she "didn't once scream."[17]

By November 1888, the false-work had been taken down, and the piles that had been driven into the river bed to support the false-work were being pulled up, making it easier for vessels to pass by. Thousands of ties on which to lay the rails on top of the bridge were arriving—they had been cut in the vicinity of Ellenville, NY, and shipped by boat on the Delaware and Hudson Canal to Kingston and then down the Hudson River. Along side the wooden walkways on top of the bridge, protective iron railing was going up. Construction was almost complete.

"DRIVING THE LAST PIN," CREATING A CONTINUOUS BRIDGE ACROSS THE HUDSON, Aug. 29, 1888. ((*Poks., The Bridge, 1889*) Looking northwest from the Poughkeepsie shore. People gawk. People show on the traveling derrick, even on top of it, as well as on the top of the bridge span opposite, on the work platform suspended in the air, and on nearby girders.

FIRST TRAIN ON THE POUGHKEEPSIE BRIDGE, AN INFORMAL TEST TRIP, DEC. 29, 1888 (McEN). Conspicuous among the celebrants is a show-off standing on top of the locomotive cab. If the celebrants seem too many to have all arrived on this one-car train, some may have walked out from Highland—the train is shown at the Highland end of the bridge.

7. THE BRIDGE OPENS (1889)

A "colossal" bridge, "a monument of engineering," providing views "unrivalled in this or any other country," and built to "last forever."

ACCORDING to the Bridge Company's amended charter, the bridge was required to be "opened for use" by January 1, 1889. To accomplish this, the bridge managers planned, before that date, to bring a train from the bridge-affiliated Hartford & Connecticut Western Railroad (H&CW) onto the bridge via the tracks of the New York & Massachusetts line (as the reorganized Poughkeepsie & Eastern was now called). For this purpose, late in December 1888, the bridge managers built a temporary connection between the bridge tracks and the New York & Massachusetts tracks in Poughkeepsie. They built it without adequate planning, with furious, wasteful speed.

On Saturday December 29, the expected train had arrived in Poughkeepsie, and the bridge contractors arranged to send its locomotive and one coach out onto the bridge informally, as a test. The locomotive was H&CW's #10, named the Tariffville. Providing unfortunate symbolism for this occasion, the locomotive was well known as the same one which had fallen off the bridge at Tariffville, CT, when that bridge had collapsed.[1]

One of the construction engineers, playfully questioning the soundness of the new bridge, contrived symbolic safety devices to protect the Tariffville locomotive as it tested the new bridge. The devices consisted of small parachutes and balloons to be fastened to the locomotive so that if the bridge collapsed under the locomotive's weight, supposedly the locomotive would fall gently.

For the test, the Poughkeepsie Bridge Company's Engineer Dickinson, who had been promoting the building of the bridge for at least seventeen years, rode in the locomotive's cab with his hand on the throttle. Dickinson was anxious to discover if the train would shake the bridge.

Starting from Poughkeepsie, the train moved slowly at first. As it passed along Parker Avenue, track workers dropped their tools, and climbed onto the train wherever they could. They climbed onto the cowcatcher, platforms, coal tender, even on top of the coach; they seemed to be confident the bridge would hold up for the trip. As the train moved onto the bridge's approach viaduct, Dickinson tooted his whistle, passengers in the train waved their handkerchiefs, factories blew their whistles, onlookers gawked. When the train crossed high over the New York Central's Hudson Line tracks, according to the *Poughkeepsie Eagle*, locomotives below "shrieked forth an enthusiastic salute." Then, as many people on both sides of the river marvelled to see, the train rumbled "at a live-

DRIVING THE FIRST TRAIN ACROSS THE BRIDGE, POMEROY P. DICKINSON BLEW HIS WHISTLE CHEERFULLY. But he was anxious. (*Poks., The Bridge*, 1889)

ly rate" out over the Hudson. When Dickinson decided the train was not shaking the bridge, not even giving it "a tremor," it "was probably the proudest moment of his life."[2]

Two days later, December 31, 1888, one day before the charter required that the bridge be "opened for use," the bridge promoters held a modest ceremony. First the Manhattan Bridge Company formally turned the bridge over to the Poughkeepsie Bridge Company. Receiving the bridge for the Poughkeepsie Bridge Company was the *Eagle* editor John I. Platt, who had been an advocate for the company since its inception in 1871, and was now its vice president. Other company officials participating included the financial powerhouse who as much as anyone had made completion of construction possible, the Philadelphian W. W. Gibbs. Later in the day, company and construction officials, continuing their ceremony, boarded a train for an official demonstration trip over the bridge. Headed by the engine Tariffville again, this train pulled two cars, both crowded with officials and guests, including such long-time Poughkeepsie promoters of the bridge as John P. Adriance and George Innis. The train carried them "at good speed" across the bridge, high over the historic river.

The next day, January 1, 1889, the Bridge Company officially opened the bridge. The company did so especially by inaugurating a regular passenger service over it, open to the public. To the Bridge Company's embarrassment, however, the service ran only back and forth over the bridge. When the company had tried to arrange regular rail connections toward the east, it had found that the nearest railroad, the New York & Massachusetts, which was not affiliated with the bridge managers, asked impossible terms. The company's failure to make connections on the east side of the river apparently helped to delay arrangements for connections on the west side too. Meantime, the bridge itself was still being tidied up. A second train track was being laid on it, and for months yet a crew of twenty-five men would be crawling about the bridge's steel network, painting it.[3]

The bridge's major designers, Macdonald and Clarke, were proud of their new bridge, but at various times they seemed to differ on how they regarded it and bridges at large. Macdonald, taking a visionary stance, told a convention of engineers that at their best, engineers are artists. If they design bridges to use no more material than necessary, as presumably he had in helping to design the Poughkeepsie Bridge, then they attain "artistic excellence." More down to earth, Clarke declared in a lecture at Cornell University, that although the new Poughkeepsie Bridge was a "considerable piece of engineering," it was not "a thing of beauty." In America where so many railway bridges are being built as quickly as possible, he explained, "aesthetic considerations are little regarded. Utility alone governs their design."

Other observers responded to the bridge variously. While the bridge was still being built, the *New York Herald* said that it looked forward to seeing its "graceful latticework against the sky." When the bridge was taking its final shape, *Engineering News*, though little inclined to aesthetic judgment, declared that while in the spans of this new bridge, "the absence of verticals of any mass looks odd," yet it "gives to the structure a very airy but pleasing appearance." After the bridge opened, a passenger on a steamer passing under the bridge on its regular route on the Hudson, a Boston brunette, "sweeping her eyes from cliff to cliff," called the bridge a "frail framework," and asked, could it really "sustain the weight of a train?" "A grinning youth of seventeen" wondered if it was even strong enough to "hold up a Vassar girl."[4]

When 800 excursionists came down the Hudson by steamer from Albany to visit the new bridge, according to an Albany reporter who accompanied them, as they saw the "great bridge "in the distance, they also felt it to be insubstantial: "its open iron work" looked to them "like the web of a great spider suspended between heaven and earth." The steamer docked on the Highland side of the river. From there, the excursionists "patiently toiled up the long and steep hill that leads to the beginning of the [bridge's] western approach," and then, using passes the bridge

The Bridge Opens

AN ADVERTISING CARD CELEBRATING THE NEW BRIDGE (Lithograph, about 1890, SHUKR). Adriance, Platt, & Co. made mowers and reapers. Its large factory shows to the right on the Poughkeepsie shore. River shipping served the factory at its front, while the Hudson Line railroad (which connected to the bridge) served it at its rear.

management had already arranged for them, walked out on the bridge.

They found the top of the bridge to be thirty-five feet across, with "wide-boarded walks on each side," and double railway tracks in the middle. Walking 212 feet above the water, higher than if they were walking across the Brooklyn Bridge, according to the reporter they were so jittery that they felt "only two inches" separated them from "eternity." Looking into the distance as steadily as they could, they saw "a scene of incomparable beauty, unrivaled in this or any other country," including "verdant woodland, grassy meadows and fields of waving grain," and the "blue river," winding "its serpentine course through a succession of lofty hills until it is lost in the towering highlands."

When the bridge opened, its promoters called it the longest bridge in the world, but the next year, 1890, when a new bridge opened over the Firth of Forth in Scotland, they conceded that the Poughkeepsie Bridge had become the second longest. Anyway, *Engineering News* called it "colossal," and claimed that, among all the structures in the world, it was "one of the grandest."

Engineer Clarke said that its spans were "the largest cantilever spans yet built." *Scientific American* called it "a monument of engineering." Its promoters celebrated it as "one of the wonders of the world," providing "the grandest" views "to be seen from any railroad line in the world," and built to "last forever."[5]

POUGHKEEPSIE RAILROAD BRIDGE VITAL STATISTICS	
Length over water, anchorage pier to anchorage pier	3094 ft.
Total length of bridge, including approach viaducts	6768 ft (1-1/4 mi.)
Height from water to base of rail track	212 ft.
Vertical clearance for shipping– under connecting truss spans under center cantilever span	130 ft.; 160 ft.
Depth of water and sediment to the hard gravel on the rock bottom	130 to 140 ft.
Width of top of bridge	35 ft.
Usual speed limit on bridge	12 mph.
Steel: "mild" with a strength in thousands of pounds per square inch	60-65 psi.
Weight of one yard of track rail originally by 1907 by 1954	70 lbs. 100 lbs. 131 lbs.

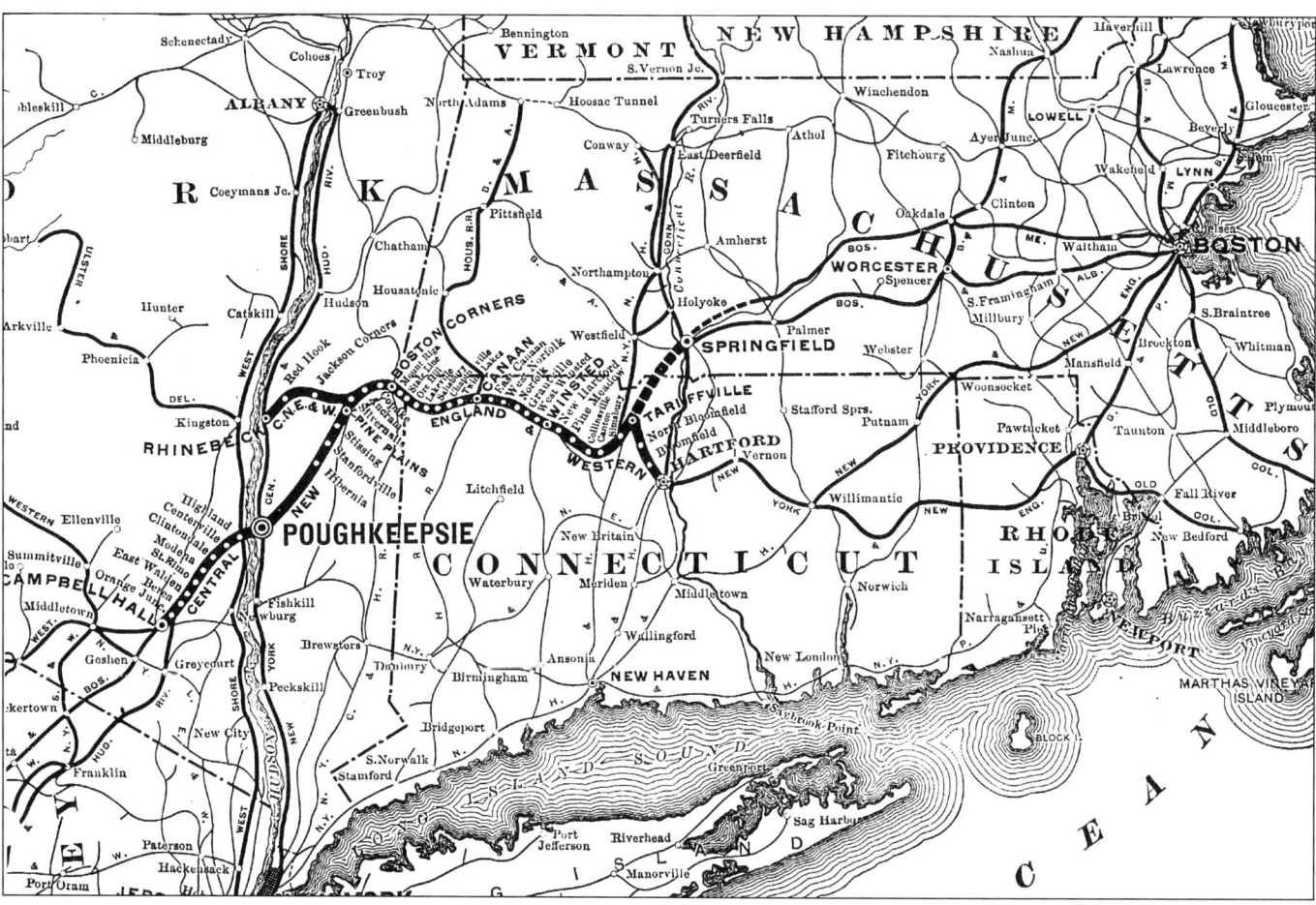

EASTERN RAIL CONNECTIONS FOR THE BRIDGE (*Poks, The Bridge*, Oct., 1889). The main line of the Central New England & Western (the "Poughkeepsie Bridge Route") shows as reaching from Campbell Hall, NY, over the bridge to Hartford, CT. A branch shows from Silvernails to "Rhinebeck" (later called "Rhinecliff") on the Hudson. A proposed branch, marked in a broken line, shows from Tariffville, CT, to Springfield, MA. Not showing as even proposed is the later line from Poughkeepsie southeast to Hopewell Jct. to connect there with the NY&NE to Danbury, CT.

8. CONNECTING THE BRIDGE TO THE WORLD (1887-1893)

Creating the "Poughkeepsie Bridge Route."

WHEN THE BRIDGE was ready for train service on Jan. 1, 1889, as we have seen, there was no place for the trains to go except back and forth across the bridge. The bridge promoters had allowed what *Engineering News* called a "strange delay" in completing rail connections for the bridge.

The bridge promoters had always been aware that efficient east-west rail connections were essential to bring the traffic they wanted onto their bridge. Indeed, when the bridge opened, the bridge forces already had nearly 3,000 men at work on those rail connections.[1] Still, how could the bridge promoters have been so slow as to have no connections completed when the bridge opened?

In regard to eastern connections for the bridge, the Poughkeepsie & Eastern (P&E), a railroad friendly to the bridge forces, had long been running from Poughkeepsie northeast to the neighborhood of Millerton, NY, near the Connecticut state line, where it connected with the Hartford & Connecticut Western on to Hartford. The bridge's promoters had expected to use this route to connect the bridge to much of New England. But the P&E, to the disgust of many Dutchess County residents whose taxes had helped to build it, had fallen into bankruptcy, and had been sold. It had been sold in 1886 to the "shrewd yankee" Henry D. Cone, a paper manufacturer of Stockbridge, MA, who renamed it the New York & Massachusetts. While at first Cone was considered a friendly backer of the Poughkeepsie Bridge, he developed his own ideas on how to make his railroad a gateway for the bridge into New England, with the connivance, it was suspected, of enemies of the bridge; he refused to make his railroad available to the Poughkeepsie Bridge forces on what they considered reasonable terms. So during 1887, the bridge forces felt forced to begin to build a new railroad (called the Poughkeepsie & Connecticut) parallel to Cone's railroad. They began building it from the Poughkeepsie Bridge northeasterly about thirty miles to Silvernails, in Columbia County, NY, where a convenient connection could be made with the Hartford & Connecticut Western via Canaan, CT, to Hartford. They built their new railroad so close to Cone's that the tracks were often within sight of each other. But they were slowed both by lack of funds and by the lingering hope that they could come to terms with Cone. While at one time they had 1,500 Italian laborers working on this new line, by June 1889, five months after the Poughkeepsie Bridge had been opened, "hundreds of track layers" were still laying rail for this new line, and it was not yet available to connect the bridge to Hartford.[2]

Also for eastward connections for the bridge, as early as 1886, a new bridge-friendly railway

> The Newburgh Journal says the building of the Poughkeepsie bridge means the early construction of a railroad from the west end to Newburgh.
>
> The Goshen Republican favors the Poughkeepsie Bridge, and wants to see "trains passing through Goshen from the coal fields to Boston, via the Wallkill Valley Road and bridge."

SPECULATION ON HOW THE BRIDGE WOULD BE CONNECTED TO THE WEST WAS AN OLD STORY BY THE TIME THE BRIDGE OPENED (Both items from *New Paltz Independent*, Feb. 8, 1877)

company had been incorporated to build a twelve-mile line from the bridge southeasterly to Hopewell Junction. There it would make a significant connection for the bridge with the existing New York & New England line (NY&NE) which ran from the Hudson River opposite Newburgh through Danbury and Waterbury, CT, toward Boston. However, it was not until three years later, in the summer of 1889, well after the bridge had been completed, that this new railroad company announced that it was about to start constructing its line to Hopewell Junction. Then as gradually became apparent, not only had this company fallen into the unfriendly hands of Cone and his New York & Massachusetts Railroad, but it lacked the funds to build its new line. Eventually the bridge forces, under the lead of Gibbs and the Brooklyn contractor John Clarke Stanton, took over this stalled enterprise, and reorganized it as the Dutchess County Railroad. Even then they found themselves slowed by the depressed financial climate of the times, and by landowners asking what seemed to them to be exorbitant prices for the necessary rights-of-way. They were also slowed by their squabbling with the ND&C over their delay in picking up cars which had arrived with their construction supplies, clogging the ND&C's yards, leading the ND&C superintendent to call them "a lot of thieves." It was not until more than three years after the bridge had been opened, in May 1892, that the bridge forces completed their new Poughkeepsie-to-Hopewell line, painfully late for the interests of the bridge.

In regard to western connections for the bridge, the bridge promoters found it difficult to choose among the various possibilities which they had long considered. One possibility was to connect to the West Shore line which ran along the west bank of the Hudson from Albany south to metropolitan New York. However, such a connection was scarcely practical because where the bridge tracks crossed over the West Shore tracks there was a severe difference in grade. Other possibilities were to connect the bridge line either to the small Wallkill Valley line (by this time New York Central-controlled) at Gardiner or New Paltz, or, as long had been favored, to another small line—once called the New Jersey Midland—at

SURVEYING NEAR STANFORDVILLE, DUTCHESS COUNTY, NY, FOR THE BRIDGE'S NEW CONNECTING LINE, CALLED THE POUGHKEEPSIE AND CONNECTICUT to run northeastward from the bridge. Photo about 1888, while the bridge was still under construction (LILL). Custom called for surveyors to wear elegant clothing even for tramping in the mud.

DIGGING OUT A BANK BY HAND, REMOVING THE SOIL BY HORSE AND CART, FOR THE BRIDGE'S NEW EASTWARD CONNECTING LINE. Near Stanfordville, about 1888-89. Above, open fields show on a hill. Below, where the horses and carts are, is probably the site for the new rail bed (McEN).

Pine Bush. However, connecting westward through any one of these three lines alone would make the bridge route dependent on a rail line which it did not control.

It was only in September 1888, that the bridge promoters decided instead to build their own line (at first called the Hudson Connecting Railroad) from the bridge twenty-seven miles southwest to Campbell Hall in Orange County, NY. Campbell Hall offered inexpensive, level land for terminal facilities. It also offered multiple rail connections, not only to the little Wallkill Valley Railroad, but also to two more significant railroads, the Ontario and Western (O&W) which reached northwest to Lake Ontario, and the Erie which reached westward to Pennsylvania and the mid-West. Still other rail lines were also planning to connect to Campbell Hall or nearby.

For this new bridge-to-Campbell Hall line, the bridge forces put "thousands" of men to work. The men cut through hills. They plowed and shoveled. They pressed steam drills into rock and set off dynamite. They used horses and wagons to haul earth and rocks. They built up embankments. They built bridges and laid ties. According to the *Poughkeepsie Eagle*, they included a "gang of negroes" from Maryland and Virginia who drove spikes to fasten rails to the ties. The blacks "understand the business," the newspaper explained, "and command fair wages." They "learn more readily than the foreigners generally employed."[3]

On May 22, 1889, more than four months after the bridge had opened, the bridge forces at last began to run the first regularly scheduled train to run over the bridge and any distance beyond. They ran it from Poughkeepsie over the bridge to Campbell Hall as a daily train, each way, for both freight and passengers. By June, the *Poughkeepsie Eagle* was pleased that this new train allowed residents of Orange and Ulster Counties to take the train to Poughkeepsie in the morning, spend several hours doing business there, and return home

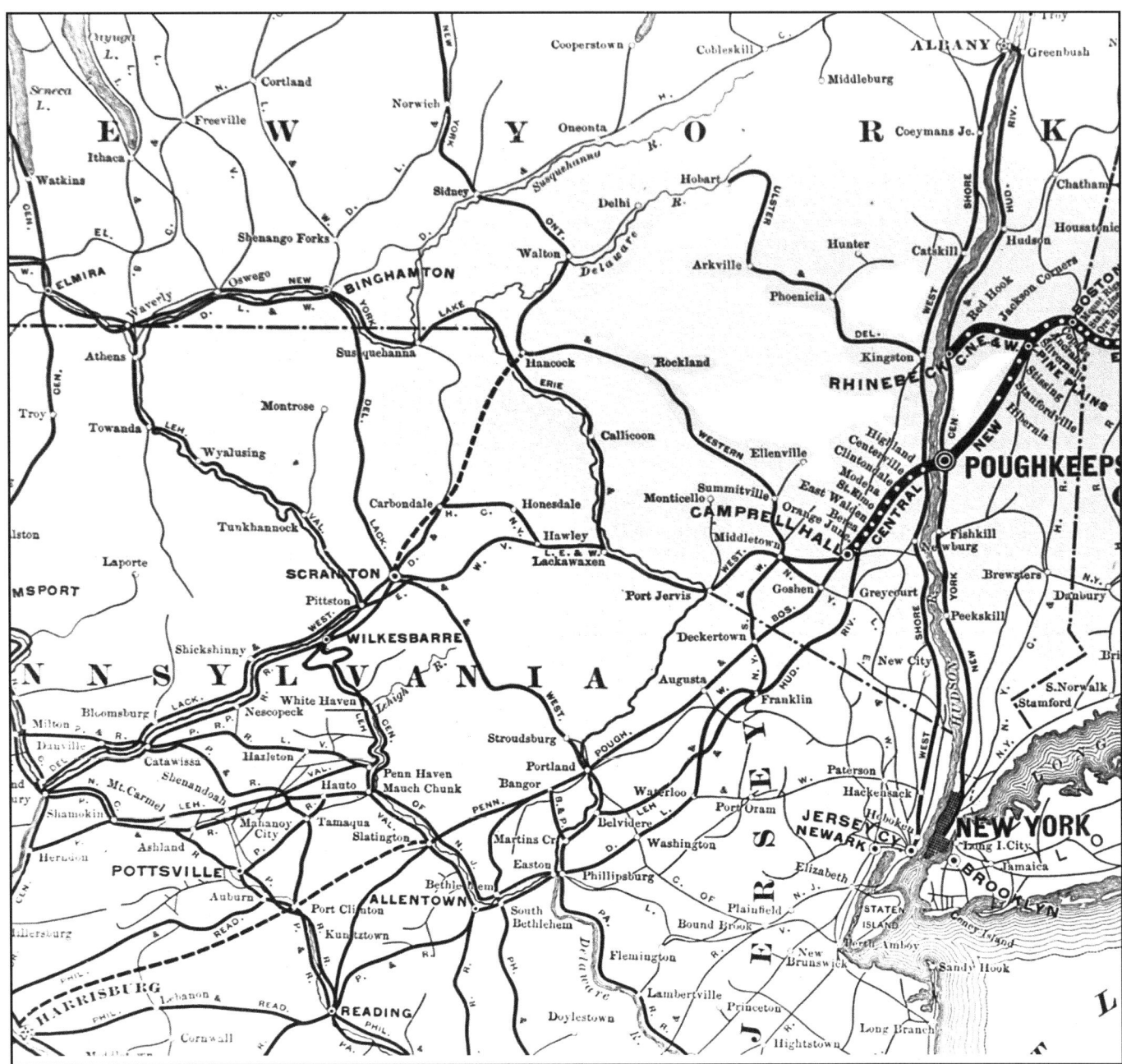

WESTERN RAIL CONNECTIONS FOR THE BRIDGE (*Poks., The Bridge*, Oct., 1889). From Campbell Hall, NY, the western terminus of the Bridge Route (the Central New England & Western), many connections show on the map, including north by the short Wallkill Valley line (marked but not named) to Kingston; northwest by the NY, Ontario & Western through Middletown, Hancock, Sidney, and Norwich toward Niagara Falls and Chicago; west by the NY, Lake Erie & Western (the Erie) through Middletown, Port Jervis, and Binghamton to the Midwest; southwest by the Pennsylvania, Poughkeepsie, & Boston via Goshen, NY, Deckertown, NJ, and Portland, PA, to the coal region near Slatington; southwest by the Lehigh & Hudson River (which connected to the Bridge Route literally not at Campbell Hall but nearby at what is called on the map "Orange Junc.," later called Maybrook) through Greycourt, NY, Franklin and Belvidere, NJ, and Easton and Allentown, PA, into the coal fields; or south by the Erie through Goshen and Greycourt to metropolitan New York (Jersey City).

in the afternoon. However, the *New Paltz Independent* growled that the fifty-seven cent fare from Clintondale to Poughkeepsie seemed as high as the bridge![4]

On June 18, a party of bridge officials boarded a special train at Campbell Hall to decide if the whole route from there over the bridge and on via Canaan to Hartford was ready for regular train service. The whole route, Campbell Hall to Hartford, was already being called part of the Central New England & Western Railroad (CNE&W) which was the name of a new railroad into which the bridge-affiliated rail lines were being merged. Most of the traveling officials were from Philadelphia, including W. W. Gibbs himself. They also included the impressive John S. Wilson, who had been the general traffic manager of the Pennsylvania Railroad and was about to become the new CNE&W's president. The train included a coach already marked with the new initials CNE&W. It also included the legendary wood-paneled observation car of President Thomas P. Fowler of the O&W Railway, which was a major connecting line for the new bridge route.

The officials found that the roadbed from Campbell Hall to the bridge was still being surfaced and aligned, so their train could run at only a moderate speed. They learned that Western Union had not yet installed along the tracks the telegraph wires essential for efficient train dispatching, but was soon to do so. After crossing the bridge into Poughkeepsie, they found that the former Lown residence on Parker Avenue, which the railroad was remodeling into its Poughkeepsie station, was not yet ready, although at least its station platform was in place. From there on east, the track was so new it was "not yet ready for business," but the train clattered on over it "quite steadily" anyway. At Stanfordville, they found the station did not yet have its intended slate roof. At Silvernails, the train switched onto the main track of the already long-functioning Hartford & Connecticut Western. The train hissed on through Canaan, CT, and then labored up into Norfolk, an elegant mountain resort, where the party stayed overnight. The next morning they continued on to Hartford.

It was not until more than a month later, on July 29, 1889, that CNE&W officials at last felt able to begin regularly scheduled service on their whole line, from Campbell Hall over the bridge to Hartford. A few days later, they decided, to the delight of Poughkeepsians, to publicize their new rail system as the "Poughkeepsie Bridge Route," a term which was to continue to be used for many years, even though the names of the railroads using it were to change.

On August 26, 1889, more than seven months after the opening of the bridge, the first coal train direct from Pennsylvania mines passed over the Poughkeepsie Bridge Route. It came from Honesdale by the Erie Railroad to Campbell Hall. Its engine carried flags in celebration. At station after station, crowds cheered, and as the train crossed the Poughkeepsie Bridge, tugs on the water below blew their steam whistles. Although two factory towns, Holyoke and Chicopee, MA, had applied for the honor of receiving this first load of coal, it was all delivered in Poughkeepsie.[5]

In January 1890, when the ability of the bridge to carry the heavy trains that were crossing it seemed well established, bolstering the reputation of Engineer John F. O'Rourke, he came from his home in New York to visit Poughkeepsie again. He came to marry the young lady he had met while working on the bridge, Katherine B. Innis. They held two weddings, one at the Innis home, a private Catholic one, because O'Rourke was Catholic, at 5:30 PM, very short; and then at 6:00 PM, a longer one, a public Protestant one, because the Innises were Protestant. The latter was held at the Reformed Church on the corner of Main and Washington Streets, the church of the early Dutch settlers, and it was crowded. The *Poughkeepsie News-Telegraph* called it "the most royal" wedding ceremony "that Poughkeepsie has known for many years." Afterwards many of the guests took carriages to the Innis home where a "collation" was catered by the Smith Brothers, and according to the *Poughkeepsie Eagle*, "nearly every person known in society was among the guests." Signifying that this was a bridge-made wedding, among the guests were at least ten of the original incorporators of the bridge company, including

Engineer Dickinson and Editor Platt. Later in the evening, when the newlyweds left Poughkeepsie by train, they were apparently unmindful of the symbolism they missed. They took a train not across the bridge, but a train from the Hudson Line station almost in the shadow of the bridge, a train down the Hudson River's eastern shore to New York.[6]

Meantime, local critics complained that the Bridge Route lines on both sides of the bridge seemed to have been laid out not so much to meet the needs of local residents, as to meet the needs of through traffic. In particular, westward from the bridge, the *Poughkeepsie Eagle* pointed out that the Bridge Route had located its line so that it did not go inside several of the villages it might have been expected to serve, such as Clintondale, Walden, and Montgomery, instead placing what should have been major stations in "open fields." The Bridge Route seemed not to have followed what Engineer Clarke considered a basic principle of laying out a rail line, that is, that to accommodate the public, a rail line should be laid through "the important towns, even if the line is made more crooked and longer thereby." The so-called Clintondale station was built about two miles outside of Clintondale village, in what a critic called "the mudhole of Peepertown" from which no roads led to the village. To protest this location, some area residents boycotted the Clintondale station, making a point of using other stations even if they were farther away. In return, those who supported locating the station where it was boycotted the boycotters.

STEPS IN OPENING THE BRIDGE	
1886-1888	Major bridge construction
1888, Dec. 29	First train crossed the bridge, a test only
1889, Jan. 1	Bridge officially opened. First regular train service crossed the bridge, but no farther than back and forth across it
1889, May 22	First significant regular local service crossed the bridge, Poughkeepsie to Campbell Hall, NY
1889, July 29	First regular long-distance train service crossed the bridge, Hartford, CT, to Campbell Hall, NY

Gradually, however, the ground around the seemingly isolated stations was graded and graveled. Watering facilities for horses were constructed. Roads were built to connect the stations to nearby localities. Stores, lumber yards, and coal yards appeared around the stations.

Still, by the end of 1890, traffic on the new Poughkeepsie to Campbell Hall line seemed thin to the *Poughkeepsie Eagle*, especially considering that the line passed through "prosperous farming country." But the *Eagle* saw hope. "At Lloyd there is a little life, two or three houses have been built, and a good store has been built, also a fruit package factory, indicating that the road has a little drawing power at that point.... At Modena a mill is going up at the depot, which may mean the beginning of something in the building line.... At Maybrook a start for a settlement has been made. ... Streets have been laid out, a hotel built, a store opened and quite a number of handsome houses built. We understand that lots are given away to anyone who will build a house worth a certain moderate sum, and that the offer is being availed of."[7]

Meantime, bridge officials were pushing for more rail connections from Campbell Hall toward the Pennsylvania coal mines, connections which they wanted to be sure would be friendly to them. Utilizing the branch of the Erie which ran from Campbell Hall through Goshen to Pine Island, NY, they worked from there to complete a railroad which had long been sporadically under construction as the South Mountain Railroad (later called the Pennsylvania, Poughkeepsie, & Boston). It passed through Augusta, NJ, crossed the Delaware River at Portland, PA, and went on to Slatington near the Lehigh River coal mining region. In 1887, Gibbs reorganized this railroad and himself became its president (it was eventually to be called the Lehigh and New England). In September 1889, it opened for regular trains from Campbell Hall as far as Augusta.

Richard H. Tingley recalled that for several years he did surveying for this Campbell Hall-Slatington route until it was completed in 1890. Tingley remembered working particularly under the direction of John Clarke Stanton who he con-

TRAIN FROM ACROSS THE BRIDGE DISGORGING PASSENGERS IN HIGHLAND, perhaps Spring, 1889 (LLOYD). This photo has been called "the First Train Across the New Bridge, Highland." But the first train to cross the bridge did so on Dec. 29, 1888, which was winter while the photo does not depict winter. Moreover, the first train had only one car and it was a Hartford & Connecticut Western car. While the second train to cross, on Dec. 31, had two cars, it seems unlikely that one of them would be labelled a Central New England & Western car. It was only in May, 1889, that the name for this new railroad (which consolidated all the railroads of the Bridge Route from Campbell Hall to Hartford) was becoming known. It was only in July, 1889 that its incorporation was filed.

THE "STRENUOUS" JOHN CLARKE STANTON, OF BROOKLYN, "put the Poughkeepsie Bridge Route on the railroad map." (*Poks., The Bridge*, 1889)

sidered to be "the chief promoter and financier" of the struggle to connect the Poughkeepsie Bridge to the world. Stanton was "strenuous." "We were all afraid of him. Fortunately he didn't often come to see us in the field; but when he did come — look out." He built the line to Slatington as a no-frills single-track line, Tingley recalled. Stanton was "a wonder on economy." He was resourceful. He had "unbounded energy." He "succeeded in accomplishing what others had failed to do. He put the 'Poughkeepsie Bridge Route' on the railroad map."[8]

Another connection for the Poughkeepsie Bridge to Pennsylvania was the Lehigh and Hudson River Railroad (L&HR). In the early 1880s, this railroad had built a sixty-three mile line from Greycourt, Orange County, NY, southeast through Warwick to Belvidere, NJ, on the

Delaware River. Before the Poughkeepsie Bridge was finished, this line had already brought considerable coal from Pennsylvania and forwarded it via an Erie branch from Greycourt to Newburgh and thence by float over the Hudson to railroads connecting to New England. With the advent of the Poughkeepsie Bridge, the L&HR decided to extend its line from Greycourt the ten miles north to Maybrook to connect there with the Bridge Route. At its southwestern end, by 1889 the L&HR had arranged to lease tracks from the Pennsylvania Railroad along the Delaware River from Belvidere to Phillipsburg, NJ. Then from there, in cooperation with other railroads, the L&HR built a bridge over the Delaware River to Easton, PA, where connections were available with the Lehigh Valley and the Central Railroad of New Jersey. In later 1890, the L&HR was already running coal trains from Easton to Maybrook to be sent on across the Poughkeepsie Bridge.[9]

However, by the next year, it was clear that the Poughkeesie Bridge Route was not prospering. Handicapped by often being single tracked, mountainous, curving, with steep grades, and little populated, the route had not yet been able to pay any interest to its bondholders. Certain bondholders became so frustrated that they forced the Central New England & Western, the owner of the bridge and its immediate connecting lines, into receivership. This led to rumors that several different railroads were interested in buying the bridge. "One day the Pennsylvania was after the property," said the *New York Times* in October 1891; "next day the Erie was doing the purchasing; then the Vanderbilts were going to make the West Shore a freight side track for the New York Central by the exclusive use of it [the bridge]; on some other days other roads were tumbling over one another in a mad plunge to get control of it. On every day that perennial Wall Street kite, the New York and New England, was right on the eve of appropriating the whole outfit." In sum, said the Times, posturing over the bridge and the lines related to it is "one of the liveliest and most elusive schemes that Wall Street has ever known."

Before 1891 was out, a group of Philadelphia capitalists bought the CNE&W. They were a somewhat different set of capitalists from those who had owned it before, many of the new ones being affiliated with the Reading Railroad. In early 1892, they reorganized it as the Philadelphia, Reading, and New England Railroad (PR&NE). They considered it part of the "Reading Railroad System," and chose the system's head, A. Archibald McLeod, as the PR&NE president.

McLeod said that controlling "the Poughkeepsie Bridge Route"—he continued to use that term—gave the Reading Company "just what we

SELECTED BRIDGE-RELATED INCORPORATIONS	
1871	Poughkeepsie Bridge Co.
1887	Hudson Connecting Railroad (to run from the bridge southwest to Campbell Hall, NY, to connect south, west, and north)
1888	Poughkeepsie & Connecticut Railroad (to run from the bridge northeast to Silvernails, NY, to connect with the HC&W to Hartford)
1889	Central New England & Western (to absorb the above companies and the HC&W, and to operate them together as the Poughkeepsie Bridge Route, reaching from Hartford to Campbell Hall)
1890	Dutchess County Railroad (a bridge-affiliated line, to run from the bridge southeast to Hopewell Jct., NY, to connect to Danbury, CT)
1892	Philadelphia, Reading & New England (to take over all the above railroads, that is, the whole Bridge Route)

PHILADELPHIA, READING, & NEW ENGLAND RAILROAD (PR&NE), THE POUGHKEEPSIE BRIDGE ROUTE: SELECTED DIRECTORS, 1893		
Office: Reading Terminal, Phila. (From NYS Bd. of Railr'd Com'rs, *An'l Rep.*, 1893)		
McLeod, A. A., Pres.	Pres., Reading Co. (Philadelphia & Reading Railroad)	Phila.
Brock, John W. (brother of Arthur)	Iron manufacturer; by 1894, replaced McLeod as PR&NE president	Phila. & Lebanon, PA
Brock, Arthur	Iron manufacturer	Phila. & Lebanon
Gibbs, Wm. W.	Utility executive; led in financing Poughkeepsie Bridge	Phila.
Sinnott, Joseph	Distiller	Phila.
Tower, Charlemagne Jr.	Banker	Phila.

have needed—a direct route to New England." Its importance for "our coal trade," he explained, is "incalculable." The *Poughkeepsie News-Telegraph* cheered. "From now on," it said, "the visions" of our bridge "projectors" will be "realized."[10]

By about this time McLeod, for the Reading System, controlled not only such minor lines as the just-completed Campbell Hall to Slatington line, but also such major coal-carrying railroads as the Central of New Jersey and the Lehigh Valley. Moreover, McLeod was intending to facilitate the distribution of Pennsylvania coal by extending his railroad system further into New England. Borrowing recklessly, in 1892 he bought control not only of the New York & New England but also of the Boston & Maine, thus associating the Poughkeepsie Bridge altogether with a huge railroad system.

The New Haven Railroad, which already controlled much of New England railroading and sought to control more, was upset by McLeod's inroads. J. P. Morgan, the banker who was increasing his stake in the New Haven, considered McLeod an irrational destabilizer. As McLeod expanded into New England, Morgan and his New Haven allies retaliated by manipulating Reading securities and by freezing McLeod's lines out of favorable connections with other lines. By 1893, Morgan interests had crushed McLeod and his Reading system, throwing into receivership not only several Bridge Route-affiliated railroads but also the principal Bridge Route railroad, the Philadelphia, Reading, and New England.[11] For the next six years, the Bridge Route, under the bankrupt PR&NE's control, could only limp.

BRIDGE AT SLATINGTON, PA, OVER THE LEHIGH RIVER, BUILT UNDER THE LEAD OF GIBBS AND STANTON, for the Pennsylvania, Poughkeepsie, & Boston (Poks., *The Bridge*, 1889). By 1890, this bridge was being used, in connection with the Poughkeepsie Bridge, both to bring coal from Pennsylvania to New England and to send passengers from New England to Philadelphia and Washington.

9. IMMIGRANT LABORERS (1886-1920s)[1]

Should the immigrants who worked on the bridge be put in boxes and "sunk to the bottom of the river"?

IMMIGRANTS helped to build the Poughkeepsie Railroad Bridge and its connecting rail lines on many levels. The Canadian-born Charles Macdonald figured in designing the bridge, the German-born John A. Roebling's factory supplied cable, and the Irish-born engineer John F. O'Rourke directed construction. Immigrants were especially conspicuous, however, as unskilled laborers.

The immigrants who were destined to provide unskilled labor for building the bridge were primarily Italian males who arrived in America by ship in New York Harbor. They often did so within sight of the Statue of Liberty which was installed there in 1886, the same year the final push to complete the construction of the bridge began.

The Italian immigrants who worked on the bridge were usually recruited in New York City, and came up the Hudson to the Poughkeepsie region without their families, having left them behind in Italy, intending some day to return there. Usually unable to speak English, they were customarily both housed and put to work separately from other workers, in groups of their own.

By the fall of 1886, a gang of fifty Italian immigrants were excavating the foundations of the anchor pier to be placed on the river's west shore. By the next spring, a group of thirty ready to work on the bridge arrived by the West Shore Railroad at its Highland station, and walked from there to a shanty erected to house them. About the same time, another group of 200 arrived by boat, and were housed in a Poughkeepsie warehouse.[2]

The settled people of the Mid-Hudson region were already a mixed people, many of them the descendants of early Dutch, British, French, or German immigrants. Nevertheless, they could be riled by the arrival of new immigrants whose appearance, food, religion, and habits were different from theirs. Around 1850, many of them had been disturbed by the Irish immigrants who had flooded into the region to help build the Hudson Line, the rail line along the Hudson's east shore. More recently, in the early 1880s, many had also been disturbed by the Italian immigrants who had poured in to help build the West Shore Railroad, the line along the Hudson's other shore. With the arrival of more Italian immigrants to help build the Poughkeepsie Bridge and its related rail lines, many again felt threatened.

The Italians' lack of English made them seem unapproachable. Moreover, they seemed, to many American eyes, dark. They were usually southern Italian peasants, poor, unskilled, illiterate, and unfamiliar with modern industry. They seemed obsessively frugal, being given, a local newspaper reported, to eating inexpensive macaroni seasoned with lard, and sending home to their families a large share of their earnings. Although the Italians were said to be more good natured than the contentious Irish, and less demanding than the order-loving Germans, for some Americans who

"Track Laying."
FROM BRIDGE DESIGNER CLARKE'S
"Building of a Railway," *Scribner's*, June 1888

had scarcely known any Italians before, it was difficult to get over their stereotypical image of Italians as either organ grinders or brigands. While Franklin D. Roosevelt was growing up in Hyde Park at the time the bridge was being built nearby, his parents shared the common upper-class scorn toward both Irish and Italian immigrants, preferring to employ better educated British or Scandinavian immigrants.[3]

During bridge construction, the *Poughkeepsie Courier* found opportunity to belittle the Italian laborers, echoing the uneasiness of the settled population toward them. As the *Courier* told the story, on a summer day of oppressive heat, a foreman working on the bridge requested a gang of his workmen to work overtime, and to do what was really not their job, unload gravel from scows. They refused. The boss then fired all forty of them, and, according to the *Courier*, replaced them with immigrant Italians — at the time it was a common practice of American employers when faced with a strike to employ immigrants as strike breakers. The *Courier* considered that the men who were fired were "honest, intelligent workingmen" willing "to do a fair day's work." It called their stand "manly." It reported them all to be residents of Poughkeepsie and some of them taxpayers, while it called the Italians who took their places "the scum of Italy."

On another occasion, as we have already seen, Chief Engineer O'Rourke was trying to use a so-called "Chinese anchor—a great wooden box filled with stone—to hold a crib in a certain position in the river, when the Chinese anchor proved inadequate, allowing the crib to drift downriver. At that time, the *Courier*, with morbid humor, reported it had been proposed to replace the Chinese anchors with "Italian anchors" which it described as large boxes "filled with Italians" which would be "sunk to the bottom of the river."

A little less caustically, the *Poughkeepsie News-Telegraph* depicted the Italians at work on bridge-related rail lines as being slow. They have "stolid faces" indicating "no anxiety" about hurrying the work, it said. "American or Irish laborers could, without doubt, accomplish double the amount of work as is done by the Italians, to whom railroad building is a thing as foreign as the English language."

On the outskirts of Poughkeepsie in the summer of 1888, a gang of Italians were cutting through rocks, preparing the roadbed for the rail line to connect the bridge toward Connecticut. They all worked under the same foreman, an Irishman. All forty of them lived near their work site, in the neighborhood of Saint Peter's Cemetery, jammed together in one single-family house. They cooked with tin pails, over "crude" fireplaces scattered about the yard, and, according to the *News-Telegraph*, ate from a table which was "a pine plank nailed across a stump." They bought their food from an Italian who kept a store in the cellar of their house, food such as bread, macaroni, bologna, and salt pork. This storekeeper, a padrone who spoke "fair" English, was in a posi-

A "CHINESE ANCHOR" (*Engineering News*, Oct. 29, 1887). While cribs were being transformed into foundations for the bridge piers, they were held in place in the river by cables attached to such anchors as these. Each anchor was a giant timber box held together with iron rods and bolts, and filled with broken stone.

tion both to help and exploit the laborers. He kept the records of what they bought from him, and settled with them only monthly. In accordance with the custom of the time, he evidently had recruited this gang of workers himself. When more men were wanted, he arranged for them to be "shipped from New York on his guarantee that they will be employed."4

When the demand for labor on the bridge and some of its connecting lines was beginning to ease, the *News-Telegraph* reported that the night boats "nearly every evening" were "loaded" with Italian laborers bound down the Hudson, returning to New York City. As Italian workers departed, the newspaper offered an ambivalent assessment of them. It reported that these Italians were "sunny," a word commonly used at the time to describe Italians. They had given their labor with "strong arms and hands," the newspaper said, but neither the bridge nor the connecting railroads can "be said to be in any way a monument to the Italian laborer. So far as his knowledge of the work went, he might as well have been digging a sewer or opening a sand bank."

Nevertheless the newspaper added grudgingly, "If the Italian laborer has done nothing for Poughkeepsie by way of contributing to its growth or wealth, he has done a great deal for himself. During his stay here he has overcome that mountain of prejudice that greeted him. He has proved himself anything but the dangerous brigand that so many of our people supposed him to be. At the police court he has been a stranger. The saloons have received little of his wages. Merchants have few if any accounts against him. The Italians have passed through our streets night and day, alone and in groups, and a breach of the peace has not been known. Ladies have gone out to see them work and have passed their cabins unprotected, and not a single insult has ever been offered a woman in Poughkeepsie by any of the Italian contingent. They have held aloof from city life. They have built no homes, paid no taxes, save those who paid rent, they contributed little to either church or school. They have simply lived apart from society, apart from everything but labor, and in so doing helped to convince hardshell Poughkeepsians that there is room enough in this great land for all, as well [as] having supplied an illustration of how perfectly practicable it is for people to mind their own business."

"Poughkeepsie has no reason to regret the coming of the Italian laborer," the newspaper concluded, "nor good cause to shed tears for his departure. Had the money paid to Italian laborers been paid to residents of Poughkeepsie, if they could have been found to work on the bridge and railroad, it would have been a blessing to the city, it is true; but the poor Italian was in no way to blame for the fact that it was not so done."5

After the Italians finished their immediate bridge or bridge-related railroad work in the Mid-Hudson region, all of them did not return to New York City or Italy. Some of them stayed in the region, and more Italians, friends and family, came to join them. According to an Italian immigrant who grew up in the 1890s in Poughkeepsie, most of its Italian men worked on railroads, and especially on the Bridge Route. Along the Bridge Route, as elsewhere in the Northeast, many Italians settled close to railroad tracks. Some long continued to work for railroads and at the same time put the farm experience they brought from Italy to use by doing seasonal labor on local farms or by cultivating little patches of ground for themselves. Since in the social structure of the Italy these immigrants came from, the ownership of land was the defining element, it is not surprising that by about 1900, some Italians, having brought their families from Italy, were buying farmland on the rolling hillsides near the Hudson River. According to an Italian who grew up in the 1910s near the western end of the bridge, when Italians were able to secure even a tiny piece of land, they customarily expressed their gratitude by planting it with grapevines.6

Italians kept arriving along the Bridge Route, many of them drawn by railroad jobs. In 1892, when a freight train on the Bridge Route near Winsted, CT, was blown apart by an explosion of dynamite it was carrying, according to a Winsted newspaper, a gang of "pig-headed" Italians was brought in from Hartford with a wrecking car to clear away the wreckage, but they "proved about as intelligent and useful as a tribe of cigar store wooden Indians." Newspapers might continue to disparage Italian laborers, but railroads kept hiring them. In the late 1890s, when a trolley line was built from New Paltz to Highland where it connected with trains which crossed the Poughkeepsie Bridge, many of the laborers were Italian immigrants. In 1898, when the ND&C wanted to take up some track it no longer used, its general manager sent a foreman to New York to recruit "20 good Italians." He wanted them to be "hardy, strong men" who would work for one dollar and thirty cents for a "heavy" ten-hour day. To house them, he provided a railroad boxcar fitted up with "bunks." Even after 1900, when Italians were arriving in the Poughkeepsie region in volume, most of them were still working for railroads. In 1904, for the improvement of the Bridge Route, 150 Italian immigrants were digging gravel out of a gravel bank at East Walden, in Orange County, and "three large shanties" were erected nearby for them to live in. In 1910, Franklin D. Roosevelt was campaigning for the New York State Senate in Putnam County near Brewster, where the Bridge Route crossed the Harlem line, when he came upon a group of Italian track workers. As a good politician seeking everyone's vote, he jumped out of his car, grasped their hands, and astonished them by talking to them in his version of Italian—he had picked up a little Italian on a visit to Italy. By that same year, there were so many Italians living in Poughkeepsie that they built a church of their own, Mt. Carmel Catholic Church, in the "Little Italy" neighborhood just south of the bridge.[7]

In the early 1900s, when the Bridge Route was expanding its rail yard in Maybrook, NY, Italian immigrants were conspicuous in providing labor for it—they were brought from New York City for that purpose and housed by themselves at first in a colony built for them on the east side of the tracks. Many Italians stayed on in Maybrook to do railroad work, even if for some years they might be segregated in where they lived and slow to be chosen for the better jobs. A boy growing up in the 1920s and 1930s in one of Maybrook's Italian railroad families recalls that some Italians were still so poor that they scrounged in the rail yard in empty box cars for any left over vegetables, and that to him, everyone in Maybrook was either Italian or American. He was not aware of any other ethnic group.[8]

ITALIAN IMMIGRANTS WHO PROVIDED LABOR FOR BUIDING THE BRIDGE AND RELATED RAILROADS HELPED TO CREATE OUR LADY OF MT. CARMEL CHURCH as their community church. It opened 1910, just south of the bridge in Poughkeepsie (Mt. Carmel Church).

In 1890, the number of Italian-born in the population of the five counties most closely identified with the Bridge Route was still small. In the three New York counties, Dutchess (including Poughkeepsie), Ulster (Highland), and Orange (Maybrook), the foreign born from Italy were only the fifth largest foreign-born group, and in the two Connecticut counties, Litchfield (Canaan) and Fairfield (Danbury), only the seventh and eighth largest (in all five counties the foreign born from Ireland were the largest group and from Germany the second largest). By 1910, however, in a dramatic shift, with the help of railroads, the foreign born from Italy had become the largest foreign-born group in both Ulster and Litchfield counties, second largest in the others. By 1930, those from Italy had become the largest in all five counties.

On the Bridge Route through the 1920s, both station agents and train crew were overwhelmingly of north European descent, including many Germans and Irish. From World War II, however, when the demand for labor soared, significant numbers of the children and grandchildren of Italian immigrants, like the descendants of earlier immigrants, were moving up to become regular, long-term Bridge Route employees on a wide range of levels.

PASSENGER TRAIN CROSSING THE BRIDGE IN MOONLIGHT with four lighted cars. Romanticized. Looking north. Postcard, postmarked 1908 (McEN). What looks like a long, low structure crosswise near the center of the river is a mystery.

Railroad Passenger Use

10. THROUGH PASSENGER TRAINS (1890-1916)[1]

Elegant passenger trains slid back and forth across the great bridge, some of them offering parlor or sleeping cars, along with steam heat, gas lamps, and polite service.

MANY AMERICANS were fascinated by railroads. Walt Whitman wrote that locomotives "pant and roar." It seemed to John Frederick Nims that trains "vaunt their way across the continent," shrinking time. When buggy drivers were stopped by trains at grade crossings, they might be intrigued by the names on the train cars passing by, especially names like Grand Trunk or Santa Fe which hinted of distant landscapes where they might escape and reinvent themselves. Not a few train watchers felt with Edna St. Vincent Millay, "there isn't a train I wouldn't take, no matter where it's going," and the farther it was going the better.

In the spring of 1890, the Central New England & Western was negotiating for long distance trains to cross the Poughkeepsie Bridge. They planned one express to go from Boston over the bridge to Harrisburg, PA. They planned another to go from Boston over the bridge to Washington, DC. Both of these trains would have the advantage of avoiding the bottleneck of New York City. Because there were no bridges or tunnels allowing trains to pass through Manhattan Island, train passengers from Boston heading through New York to Philadelphia or Washington were customarily obliged to take ferries from the Bronx mainland around Manhattan to the New Jersey shore.

The express to Harrisburg, which was conceived first, was planned to originate in Boston, with the Boston & Maine supplying two cars to go all the way through to Harrisburg. The correspon-

THE PULLMAN CAR "VERNIA" RAN OVER THE POUGHKEESIE BRIDGE AS PART OF THE WASHINGTON-BOSTON EXPRESS (SMITH). The Vernia was one of nine Pullman cars which the B&O made available for use on the express in June 1893, and the Vernia made ten Washington-Boston round trips over the bridge that month. According to the B&O's bill, Sept. 1893, the B&O charged the PR&NE $2.35 for the car's use on each trip (Burr Collection, Box 4, HARV).

ding train in the other direction would originate in Harrisburg, with the Philadelphia & Reading providing two cars which were to go all the way through to Boston. From both points of origin, the two cars were to be, as first planned, a "day coach" and "a combination baggage and smoking car."

From Boston, the express for Harrisburg would first run west on the Boston & Maine to Northampton, MA, then south on a New Haven branch to Simsbury, CT. From there it would go west again on the CNE&W (the Bridge Route) through Canaan across the Poughkeepsie Bridge to Maybrook, NY, then on the Lehigh & Hudson River and related lines to Phillipsburg, NJ, where it would cross the Delaware River. From there it would run on the Central of New Jersey through Bethlehem to Allentown, PA, and complete its journey on the Philadelphia & Reading tracks to Harrisburg.

The CNE&W's president, John S. Wilson, who led in arranging for these express trains, was already familiar with complex railroad interchanges from his earlier experience with the Pennsylvania Railroad. Wilson's office was in New York, but he negotiated by telegram, letters, and meetings with other CNE&W officers, including Vice President Arthur Brock in Philadelphia, Superintendent S. B. Opdyke in Hartford, and General Freight and Passenger Agent N. R. Turner in Poughkeepsie. He also negotiated with the other railroads concerned, the most stubborn being the New Haven. At one time, Wilson felt only "faint hope" of securing the New Haven's participation because it saw itself as competing with the Bridge Route.[2]

TWO EXPRESSES BEGIN RUNNING

However, all the necessary railroads agreed to participate, including the New Haven, and on May 26, 1890, the Boston-Harrisburg express began running daily, all the way without change of cars. By about that time, the Boston to Washington express was also being planned, and by June 30, it also began running also daily, also without change of cars.[3]

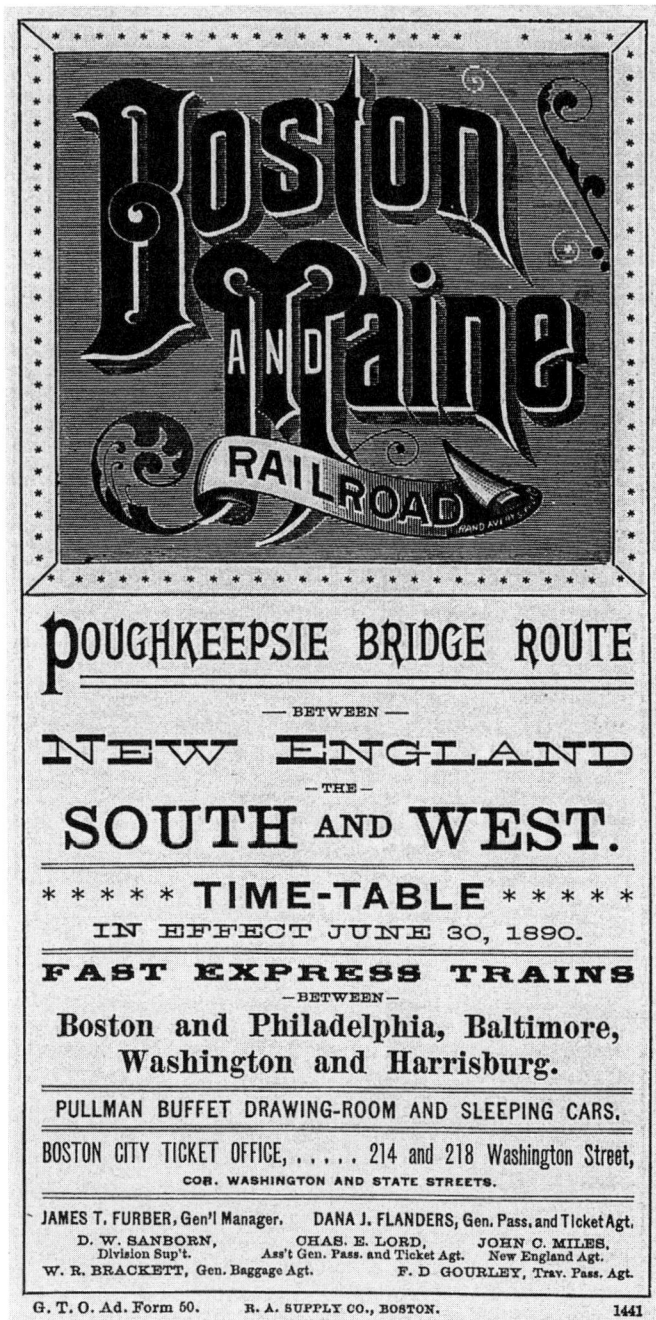

THIS AND THE TWO FOLLOWING ILLUSTRATIONS ARE FROM A BOSTON & MAINE TIME TABLE, JUNE 30, 1890. As the time table indicates, from the Poughkeepsie Bridge west, the expresses to Harrisburg and Washington used different routes to reach Bethlehem, PA.

When the express to Washington first began to run, it followed the same route as the Harrisburg express across New England and over the bridge to Maybrook. From there, however, it continued on to Campbell Hall, and then followed a more northerly route in New Jersey and Pennsylvania, much of it on the Bridge Route-affiliated Pennsylvania, Poughkeepsie, & Boston (PP&B). It passed through Goshen, NY, Augusta, NJ, and Slatington, PA, to Bethlehem. From Bethlehem, whether it reached there by the PP&B route, or, as it often did later, by the more southerly L&HR route which the Harrisburg express also took, it went south on the Philadelphia & Reading to Philadelphia, and then continued on the Baltimore & Ohio line through Baltimore to Washington.

As it turned out, the Boston-Harrisburg express, a day train, carried not just a coach and combination car, as originally intended, but also

BOSTON & MAINE RAILROAD.
CONDENSED TIME-TABLE
BOSTON - TO - HARRISBURG, PA.
Poughkeepsie Bridge Route.

Mls				Week-Days
0	BOSTON......Mass., {Boston & Maine R.R., Boston & Lowell Station,}		Lve.	8.00 a.m.
10	Waltham...... "	........B.&M.R.R..	"	8.19 "
20	South Sudbury, "	.... " ..	"	8.35 "
28	Hudson .... "	.... " ..	"	8.48 "
42	Oakdale...... "	.... " ..	"	9.11 "
54	Rutland ...... "	.... " ..	"	9.37 "
62	Barre .... "	.... " ..	"	f9.49 "
75	Ware......... "	.... " ..	"	10.10 "
83	Bondsville..... "	.... " ..	"	v10.22 "
88	Belchertown... "	.... " ..	"	10.31 "
97	Amherst ...... "	.... " ..	"	10.47 "
105	Northampton.. "	.... " ..	Arr.	11.00 "
	Northampton.. "	N.Y.,N.H.&H.R.R.	Lve.	11.00 "
	Easthampton.. "	" " ..	Arr.	11.18 "
	Westfield..... "	" " ..	"	11.32 "
138	Simsbury, Conn..... "	" " ..	Arr.	12.10 p.m.
	Simsbury, "	........C.,N.E.&W.R.R.	Lve.	12.10 p.m.
159	Winsted "	.... " ..	Arr.	12.51 "
168	Norfolk .. "	.... " ..	"	1.11 "
177	Canaan .. "	.... " ..	"	1.32 "
187	Lakeville, "	.... " ..	"	1.53 "
197	Boston Corners..N.Y.. "	" " ..	"	2.10 "
238	Poughkeepsie ... "	" " ..	"	3.25 "
268	Maybrook Junct.. " ..	" " ..	"	4.25 "
	Maybrook Junct.. "	........O.C.R.R.	Lve.	4.35 "
280	Greycourt .... "	" "	Arr.	5.15 "
290	Warwick........ "	......L.&H.R.R..	"	5.40 "
343	Belvidere ....N.J........... "	" "	"	7.23 "
357	Phillipsburg, "	...........Penn.R.R..	Arr.	7.50 "
	Phillipsburg, "	........C.R.R.of N.J..	Lve.	8.11 "
358	Easton......... Pa........ "	" "	Arr.	8.17 "
369	BETHLEHEM.. "	.... " "	"	8.39 "
374	Allentown..... "	.... " "	"	8.52 "
	Allentown..... "	........P.&R.R.R..	"	9.05 "
410	READING..... "	.... " "	Arr.	10.25 "
438	Lebanon....... "	.... " "	"	11.20 "
464	HARRISBURG "	.... " "	"	12.10 a.m.

f Stops on signal, or on notice to Conductor to leave.
v Stops to take only.

CONDENSED TIME-TABLE
BOSTON TO PHILADELPHIA, BALTIMORE AND WASHINGTON.
Poughkeepsie Bridge Route.

				DAILY.
BOSTON......Mass., {Boston & Maine R.R., Boston & Lowell Station,}			Lve.	5.45 p.m.
Oakdale....... "	........B.&M.R.R..	"		6.56 "
Ware ... . "	.... " "	"		7.55 "
Amherst...... "	.... " " ..	"		8.32 "
Northampton.. "	.... " "	Arr.		8.45 "
Northampton, Mass..N.Y.,N.H.&H.R.R...Lve.				8.50 "
Easthampton, "	" " "	"		9.00 "
Westfield, "	... " " "	Arr.		9.20 "
Simsbury, Conn........C.,N.E.& W.R.R..Lve.				9.52 "
Winsted, "	.... " " "	"		10.30 "
Norfolk, "	.... " " "	"		10.49 "
Canaan, "	.... " " "	"		11.08 "
Lakeville, "	.... " " "	"		11.31 "
Boston Corners, N.Y...	" " "	"		11.48 "
Poughkeepsie, "	.. " " "	"		12.54 a.m.
Campbell Hall, "	.. " " "	Arr.		1.57 "
Campbell Hall........P.P.&B.R.R...Lve.				2.00 "
Goshen................	" " .. "			2.10 "
Deckertown.............	" " .. "			2.50 "
Blairtown..............	" " .. "			3.38 "
Portland...............	" " .. "			3.58 "
Pen Argyl..............	" " ..	Arr.		4.17 "
Slatington..............Lehigh V.....Lve.				6.30 "
Mauch Chunk...........	" " .... "			6.55 "
Penn Haven............	" " .... "			7.09 "
Wilkes Barre...........	" "	Arr.		8.50 "
Pen Argyl..............L. & L. R.R...Lve.				4.20 "
Bethlehem.............	" " ..	Arr.		5.10 "
BETHLEHEM............P.&R. R.R....Lve.				5.15 "
Quakertown............	" " .. "			5.45 "
Lansdale..............	" " .. "			6.15 "
Jenkintown............	" " .. "			6.43 "
Wayne Junction	" " .. "			6.53 "
PHILADELPHIA	" " ..	Arr.		7.04 "
Philadelphia.............B. & O. R.R...Lve.				8.15 "
Chester...............	" " "			8.30 "
Wilmington............	" " "			8.45 "
Newark, Del...........	" " "			9.00 "
BALTIMORE...........	" " .. "			10.35 "
Annapolis.............	" " ..	Arr.		1.45 p.m.
WASHINGTON.........	" " .. "			11.20 a.m.

luxurious "Pullman Buffet Parlor Cars." The Bridge Route, being anxious to have this new train succeed, staffed it, according to the *Poughkeepsie Eagle*, with "polite employees." Because the Boston-Washington express was a night train, it carried Pullman sleeping cars, and, as the CNE&W bragged, they were "magnificent vestibuled" cars. They were provided not only the customary steam heat, piped in from the locomotive, but also buffet dining, separate toilets for ladies, and, in place of the usual oil lamps which tended to drip oil on passengers, gas lamps.

Although many Americans were drawn to the excitement of riding a Pullman car while it hurtled through the darkness, the *New York Times* claimed that in reality, sleeping cars at about this time were so poorly ventilated and their bed sheets so "clammy" that spending a night in them could be "horrible." According to Boston's William Dean Howells, in one of his popular farces, in sleeping cars, black porters might be reassuring, and passengers might find opportunities to repair broken romances, but the cars were coupled together with disconcerting jolts, passengers talked too loudly, and at any moment passengers who forgot which berths were theirs might barge through the curtains into your berth.[4]

The day express for Harrisburg left Boston at 8:00 AM and was scheduled to arrive in Harrisburg at midnight, 16 hours and 464 miles later. The night express for Washington left Boston at 5:45 PM and was scheduled to arrive in Washington the next morning at 11:20, taking 17 hours 35 minutes for its run of about 527 miles. Each run averaged about 30 miles per hour.

After running for five months, the Boston-Harrisburg day express had attracted few passengers and was abandoned. However, in 1892, when the Reading Railroad system took over the Bridge Route (the CNE&W), reorganizing it as the Philadelphia, Reading & New England (PR&NE), it revived the day express, this time with the destination no longer being Harrisburg but Philadelphia (it was called the "Quaker City Day Express"). This time it left Boston on the New York & New England, and switched onto the PR&NE's tracks at Hartford. After crossing the Poughkeepsie Bridge to Maybrook, it ran on the L&HR and Reading toward Philadelphia. Unfortunately this new day express carried few passengers, and lasted only six months.

The Boston-Washington night express, however, attracted more passengers. On March 3, 1893, the day before Grover Cleveland was inaugurated president for the second time, the Washington-bound express carried seven sleeping cars. A few

THIS AND THE FOLLOWING TWO ILLUSTRATIONS are from a Reading Railroad Time Table of Feb. 26, 1893.

THROUGH Pullman Buffet Sleeping Cars.
Boston to Philadelphia, Baltimore and Washington.
POUGHKEEPSIE BRIDGE ROUTE.

			P. M.
Boston, Mass.	{Boston & Maine R. R. Causeway Street. Sta.}	Lv.	*5.50
Hudson			6.36
Worcester			†6.25
Oakdale			6.58
Ware			7.57
Amherst			8.32
Northampton		Arr.	8.45
Springfield Mass., Conn. Riv. R. R.		Lv.	*8.10
Holyoke			8.25
Northampton, N. Y., N. H & H. R. R.			8.45
Easthampton			8.55
Westfield			9.15
Boston, N. Y. & N. E. R. R.		Lv.	
Franklin			
Putnam			†6.00
Willimantic			6.55
Hartford		Arr.	8.00
Hartford, P. R. & N. E. R. R.		Lv.	*9.10
Simsbury			9.52
Winsted			10.33
Norfolk			s10.55
Canaan			11.15
Lakeville			s11.35
Poughkeepsie			1.01
Maybrook		Arr.	1.50
Maybrook, L. & H. R.R'y		Lv.	2.00
Greycourt. N. Y.			2.18
Warwick, N. Y.			2.40
Andover, N. J.			3.40
Belvidere			4.18
Easton, Pa.		Arr.	4.45
Bethlehem, P. & R. R. R.		Lv.	5.30
Quakertown			5.56
Lansdale			6.24
Jenkintown			6.49
Wayne Junction			6.57
Philadelphia, 9th and Green Sts.		Arr.	7.10
Philadelphia, 24th and Chestnut Sts.			8.05
Philadelphia, " " B.&O.R.R.		Lv.	8.15
Chester			8.30
Wilmington			8.47
Baltimore		Arr.	10.30
Washington		Arr.	11.20
			A. M.

s Indicates that trains stop only on signal, or when Conductor is notified. τ Indicates special stop to take or leave through passengers.
* Daily. † Daily, except Sunday.

READING RAILROAD SYSTEM.
NEW ENGLAND DIVISION.

CONNECTIONS.

At Hartford with N. Y., N. H. & H. R. R., (Hartford & Valley Divisions,) N. Y. & N. E. R. R

At Simsbury with N. Y., N. H. & H. R. R., (Northampton Division).

At Pine Meadow with N. Y., N. H. & H. R. R. (Northampton Division).

At Winsted (Naug. Station) with N. Y., N. H. & H. R. R. (Naugatuck Division).

At Canaan with N. Y., N. H. & H. R. R.—Berkshire Division.

At Boston Corners, Millerton and Mt. Riga, with N. Y. C. & H. R. R. R. (Harlem Division),

At Rhinecliff with N. Y. C. & H. R. R. R.

At Poughkeepsie with New York Central & Hudson River, and New York & Massachusetts Railroads.

At Maybrook with Lehigh & Hudson River Ry.

At Campbell Hall with P. P. & B. R. R., N. Y., O. & W. Ry., N. Y., L. E. & W. R. R., Wallkill Valley R. R.

At Hopewell with New York & New England and Newburgh, Dutchess and Connecticut Railroads.

days later, a Boston-bound express brought back from Washington a crowd of inauguration celebrants, including Governor William E. Russell of Massachusetts. Enlarged for the occasion, the train traveled in two sections. The second section alone included seven Pullmans.[5]

During the 1893 Chicago World's Fair, the Bridge Route offered round trip excursion fares, by coach or Pullman, from Boston to Chicago, using the Boston-Washington express for part of the way. Excursionists could chose from two different sections of the Boston-Washington express originating in Boston, one going via the B&M through Northampton, the other going via the NY&NE through Hartford. The two sections joined at Simsbury, CT, and then continued on

CNE LOCOMOTIVE #14, TEN WHEELER, 4-6-0, MADE BY BALDWIN, WITH CREW, AT STATE LINE, NY-CT. Postcard, mailed 1906 (FULLR). Such engines were often used on the Boston-Washington express.

across the Poughkeepsie Bridge, and through Maybrook, NY, Bethlehem, PA, and Philadelphia to Washington. There excursionists could change to the B&O for Chicago. Alternatively, excursionists could leave the Boston-Washington express in Bethlehem, from there taking the Lehigh Valley to Niagara Falls and the Grand Trunk on to Chicago. One September day, as the Boston-Washington express passed over the Poughkeepsie Bridge, it was so crowded with World's Fair passengers that it consisted of two sections of twelve cars each.[6]

However, the Bridge Route's Boston-Washington express competed with popular trains. For Mid-Hudson River residents on their way to Chicago, the express competed with the New York Central's faster and more frequent service via Albany. For Bostonians on their way to Washington, the Bridge Route's express competed especially with the prestigious Federal Express which was operated jointly by the New Haven and Pennsylvania railroads. It ran from Boston along the relatively level sea shore through Providence and New Haven to the Harlem River terminal in the Bronx, from which, because there was still—despite much talk—no rail connection either by bridge or tunnel through Manhattan Island to the New Jersey shore, its rail cars were ferried via the East River around Manhattan in a voyage of about 14 miles to Jersey City. From there its cars were routed on the Pennsylvania's tracks through Philadelphia to Washington. Despite its awkward water bypass around New York City, the Federal Express reached Washington about two hours faster than the Bridge Route's Boston-Washington express.

By 1893, the Philadelphia, Reading, & New England had fallen into financial distress, which, along with the general financial panic of 1893, weakened the Bridge Route's through passenger service. In November 1893, after four glamorous years, the Bridge Route stopped running its Boston-Washington express. When the Bridge Route attempted to revive it in 1896, the revival lasted only one month.[7]

SUMMER TOURING

Meanwhile the promoters of the Poughkeepsie Bridge Route, like promoters of other railway routes, wooed passengers by issuing an illustrated summer travel guide. The PR&NE first issued the guide in 1895, printing 10,000 copies, and continued to issue it annually. After the Central New England (CNE) took control of the Bridge Route in 1899, it issued a similar guide.

The guide noted the scenery along the route, places to stay, and local side trips. It listed train connections to such resorts areas as the Berkshire and Catskill Mountains, and to such great cities as Boston (via Hartford), New York City (via the New York Central's Harlem or Hudson lines), and Chicago (via Campbell Hall). Without modesty, the CNE's guide claimed that the "Poughkeepsie Bridge Route" offered summer travelers "a perfect train service."

More skeptically, at about the same time, the *New York Times* was warning that it is difficult to enjoy railway travel "no matter how fine the scenery." When you are riding "ordinary railroad cars," the *Times* explained, you are likely to be among some passengers who are not "polite and refined," and who therefore, may follow their "purely animal instincts." For instance, when it is warm, they may open windows to let in the air even though it "is black with smoke and cinders."[8]

The guide focused on one through passenger

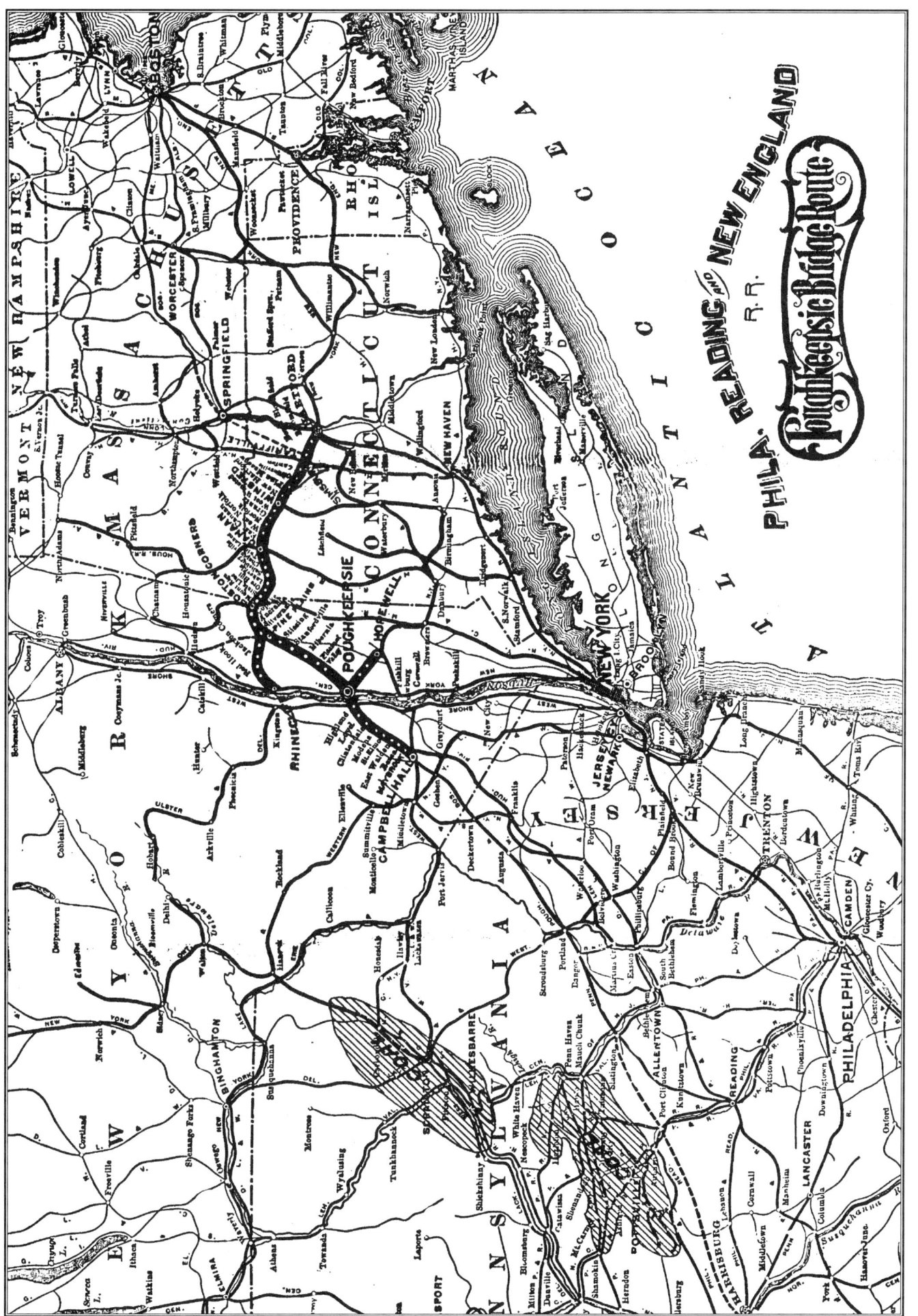

FROM THE PR&NE'S TIME TABLE, JUNE 2, 1895. The PR&NE's own line (the Poughkeepsie Bridge Route) is marked in heavy black in the center. Connections from the bridge to Boston show as by the Boston & Maine (via Northampton and Oakdale) and the NY & New England (via Hartford and Willimantic). Connections from the bridge to the Midwest show as by the Ontario & Western (via Summitville and Sidney), and the Erie (via Binghamton). Connections into Pennsylvania show as by the Penn., Poughkeepsie, & Boston (via Deckertown and Portland), and the Lehigh & Hudson River (via Franklin, Belvidere, and Easton).

train called the "Day Express," which traveled the CNE's main line from its eastern terminus at Hartford, CT, over the Poughkeepsie Bridge to its western terminus at Campbell Hall, NY, a total of 145 miles. Going west, this Day Express left Hartford, "Connecticut's historic capital," daily at 12:30 PM. It rolled east through "level farming districts" where on both sides of the track passengers could see "extensive" fields of "quality" tobacco "which, experts claim, is superior to the famous Virginia brands." The train stopped in Simsbury, a town with a "broad avenue, lined with grand old elms" (an inn, only three minutes walk from the station, prefers "gentlemen guests," at $10 to $12 a week, no children). Steaming up into the Litchfield Hills, the train passed through Winsted, a town whose river-powered factories made wood and leather products, and climbed into Norfolk, Connecticut's "highest railroad resort" (about 1,250 feet "above tide"). There travelers could stay at old farmhouses for $6 per week, or the fancy Hillhurst Inn for $15 to $25 per week, with meals comparable to New York's Delmonico's.

From Norfolk, the tracks twist down around a "perfect horseshoe" into the more open country near Canaan, where New Haven Railroad connections were available north into the Berkshires. Leaving the Connecticut highlands, the train skirted the edge of the Berkshire Hills, and entered New York State's Columbia County at the tiny mountain village of Boston Corners where passengers could change to the Harlem line. Then the express dropped down southwest into Dutchess County's "rich grazing land." It passed through Pine Plains ("excellent roads for wheeling") where passengers could change to the ND&C line to reach Millbrook's Halcyon Hotel (which the ND&C encouraged to exclude Jews as "undesirable.")[9] Continuing on, the train passed through Salt Point (the Van Wagner farm offered boarders a "veranda 100 feet long"), until about four and a half hours from Hartford, at 5:05 PM, it reached Poughkeepsie.

Poughkeepsie, crowed the guide, "is the most important city" on the Hudson River between Albany and New York. Here travelers, with the help of local trolley cars, could visit Vassar College, "the greatest female educational institution in the world." Or also with the help of trolley cars, they could change to New York Central's Hudson rail line, or to luxurious Hudson River steamers, both available only about a mile away from the Bridge Route station. On the steamers they could sail up or down the Hudson ("the most charming inland water trip on the American continent"), with meals "served at all hours."

If passengers stayed on the "Day Express," as they pulled out of Poughkeepsie they would glide out slowly, as safety regulations required, onto "the great cantilever bridge," which was, the guide boasted, the second "largest in the world." The bridge seems "frail," but in fact it "is so strong in its steel frame that it supports long trains as safely as the solid bed of the rocked-ribbed mountain." Below "flows the broad historic Hudson," the "Rhine of America," which the passengers can see from their train "for twenty miles north and south." In the background to the northwest is the "grandeur of the Catskills."

Climbing up from the Hudson River into Ulster County, the train curved first through "deep rock cuts" into Highland where travelers could stay at Mrs. Wisemiller's (with a "beautiful view of Poughkeepsie bridge"), or at Oliver Smith's (we "have our own milk, butter, and eggs—no stale stuff"). Skirting along "great bluffs," the train came out into "the greatest fruit section in New York State." Each year, from such stations as Highland, Clintondale, and Modena, the CNE, according to its guide, ships to New York, Boston, and Philadelphia, "nearly ten million pounds of grapes," as well as lesser quantities of apples, pears, and peaches. Continuing southwest, the train entered Orange County, NY, which, the guide reported, is known "far and wide," for its dairying. "To encourage this important industry not only here, but all along the line," the railroad has selected "twenty sites" for creameries, of which, according to the 1903 guide, twelve had been built.

The Day Express concluded its journey by "thundering" into Campbell Hall, at 6:15 PM, 5–3/4 hours after leaving Hartford. From Campbell Hall, railroad crossroad that it was, travelers could take trains heading toward metropolitan New York, the Catskills, Pennsylvania, Chicago, or, boasted the guide, "almost any part of the world."[10]

After the CNE came under the control of the New Haven Railroad in 1904, it gave up issuing summer travel guides. While it continued to offer a full range of passenger services, the CNE came to emphasize them less, whether they were long distance or local.

THE FEDERAL EXPRESS

Nevertheless, several years later a famous long-distance passenger train began running over the Bridge Route. This happened because both the Pennsylvania Railroad and the New Haven, the CNE's parent, found they wanted it.

By 1911, the Pennsylvania Railroad, with the help of Engineer O"Rourke, had built tunnels under both the Hudson and East Rivers. This allowed the Pennsylvania and New Haven Railroads to run their famous Washington-Boston Federal Express—a train which President Taft often rode—from Washington through Manhattan's Penn Station to Long Island. But from there, since no bridge or tunnel extended to the mainland, the express was obliged to cross by ferry to the Bronx from which it ran on the New Haven Railroad's Shore Line to Boston. However, by October 1912, the two railroads had decided that sending it by ferry even a short distance was subject to too much risk, as from winter ice and congested water traffic. They decided instead to send the Federal Express by the no-water Poughkeepsie Bridge Route. In effect they were reviving the Bridge Route's earlier Washington-Boston express, though the route they chose for it was somewhat different. They chose to route the Federal Express from Washington by Pennsylvania Railroad tracks to Trenton, NJ, by a Pennsylvania branch line north along the Delaware River to Belvidere, NJ, by the Lehigh & Hudson River line northeast to Maybrook, NY, then by the New Haven-controlled CNE Bridge Route over the Poughkeepsie Bridge through Danbury, CT, and finally from Devon by the New Haven's own Shore Line tracks through New Haven to Boston. By about this time, all the tracks from Maybrook over the bridge to Danbury had been double tracked, and on some sections of this track the speed limit for passenger trains had gone up to 50 miles per hour. Still, on this new route the Federal Express took three hours longer than on its former route through New York City.

The Federal Express was an elegant overnight train. It was often pulled by impressive engines, as on the L&HR and over the bridge, by Baldwin-built 123,100-pound locomotives. It carried coaches, drawing room cars, sleeping cars, dining cars. Its conductors wore smart blue coats with brass buttons. Its passengers were nearly all through passengers, and it was a profitable train. It was the most famous passenger train ever to run regularly on the Poughkeepsie Bridge Route.

By 1916, the Pennsylvania and New Haven railroads together had almost completed building a bridge from Long Island to the Bronx mainland, the great steel-arch Hell Gate Bridge. As soon as this bridge opened, through trains from Washington to Boston could pass through metropolitan New York without any water transfer, and the main reason for routing the Federal Express over the Poughkeepsie Bridge vanished. The Federal Express stopped running on the Poughkeepsie Bridge Route in 1916, and by the next year resumed running through New York City.[11]

At various times between 1890 and 1916, many privileged Americans rode long distance trains across the Poughkeepsie Bridge. As they rumbled, day or night, high over the Hudson's deep waters, on their way to and from Boston, Washington, Chicago, or elsewhere, some passengers slept, some were charmed by the view of the river and mountains, others dreamed of the infinite possibilities, legal or illegal, likely or unlikely, which awaited them when they reached their destinations.

STRANGE HYBRID: A COMBINED RAILROAD-TROLLEY TRAIN, IN HIGHLAND, ON ITS WAY TO CROSS THE POUGH-KEESIE RAILROAD BRIDGE. Photo by Charles Collin, 1898 (McEN). The bridge under which the train is passing has traditionally been identified as the highway 9W bridge, but doubters wonder if it could be another bridge slightly to the east, the Little Italy Road bridge. The car in the middle, a railroad coach, had a special coupler to allow it to be coupled to a trolley car. The last car, a trolley car open at the sides, was customarily used only in summer. The small railroad locomotive is PR&NE's "Dinkey" or "Dummy" steam engine #1. According to Leroy Beaujon, it was an 0-4-2T (the "T" means it was a tank engine, that is, the engine instead of having a separate tender to carry its coal and water supply, carried its own coal and water supply, the latter in a tank). Also according to Beaujon, the locomotive was not as previously thought acquired by the PR&NE from a New York City elevated line, but from the North Hudson County Railway, in New Jersey, which had run it on a Weehawken line until that line was electrified.

11. TROLLEY CARS ON THE BRIDGE (1897-1904)[1]

Passengers could ride across the bridge in trolley cars, sometimes even open trolley cars, providing them a dramatic if scary way to observe "the grandest scenery in America."

AN unusual local passenger service began to run over the bridge in 1897. In a period when electric trolley lines as well as railroads were rapidly spreading across America, the new service was a combined railroad and trolley service. Called "suburban" or "rapid transit," it ran from New Paltz eastward through Highland and across the bridge to Poughkeepsie, a total of nine miles. It was operated jointly by a trolley line and the Poughkeepsie Bridge Route, the latter being at first the PR&NE, later the CNE.[2]

The cars for this service were often trolley cars whose wheels had been adjusted to allow them to run on both trolley and railroad tracks. From New Paltz, for the first six miles of the trip, the cars were run as trolley cars, powered by electricity obtained from overhead wires. For the last three miles of the trip, from the Pratt's Mills station over the bridge to Poughkeesie, the cars were pulled by a steam locomotive, a low, short one, called a "dummy."

The cars took thirty minutes for the electric-powered portion of the trip from New Paltz to Pratt's Mills, and fifteen minutes for the steam-powered portion the rest of the way over the railroad bridge to Poughkeepsie. At Pratt's Mills, passengers had a choice. They could continue on by the steam-powered rail line over the railroad bridge directly to Poughkeepsie. Or they could continue by the electric-powered trolley line to the Highland ferry dock from which they could ferry to Poughkeepsie. If they preferred to ride over the bridge, they sometimes had to change cars, but often they could stay on the same trolley car right on through, without change, all the way from New Paltz over the bridge to Poughkeepsie.

From at least 1898 to 1903, such rapid-transit trains often ran from New Paltz over the bridge

THE 'DUMMY' STEAM ENGINE PULLING A TROLLEY CAR ACROSS THE BRIDGE
on its way to New Paltz. Postcard, published 1904 (LLOYD).

CENTRAL NEW ENGLAND RAILWAY,
Poughkeepsie Bridge Route.
RAPID TRANSIT TIME TABLE,
IN CONNECTION WITH THE
New Paltz and Poughkeepsie Traction Co.,
Effective, June 3rd, 1900.

Rapid Transit Trains between POUGHKEEPSIE, HIGHLAND and NEW PALTZ.

Trains for Highland and New Paltz Daily, except Saturday and Sunday.			Trains for Highland & Poughkeepsie Daily, except Saturday and Sunday.		
Trains Leave Poughkeepsie.	Arrive Highland.	Arrive New Paltz.	Trains Leave New Paltz.	Leave Highland.	Arrive Poughkeepsie.
7.25 A.M.	7.33 A.M.	8 15 A.M.	7.10 A.M.	7.45 A.M.	7.55 A.M.
8.30 "	8.45 "	—	—	9.10 "	9.25 "
9.10 "	9.18 "	10.00 "	9.00 "	9.33 "	9.43 "
10.15 "	10.23 "	11.00 "	—	9.34 "	9.52 "
11.25 "	11.40 "	—	10.00 "	10.35 "	10.45 "
12 00 M.	12 08 P.M.	12.45 P.M.	11.45 "	12 20 P.M.	12.30 P.M.
1.15 P.M.	1.23 "	2.00 "	—	12.01 "	12 20 "
2.15 "	2.23 "	3 00 "	1 00 P.M.	1.35 "	1.45 "
3.15 "	3.23 "	4.00 "	2.00 "	2.35 "	2.45 "
4.15 "	4.23 "	5.00 "	3.00 "	3.35 "	3.45 "
5.08 "	5.16 "	6.00 "	4.00 "	4 35 "	4.45 "
5.13 "	5.28 "	—	5.00 "	5.35 "	5.45 "
6.15 "	6.23 "	7.00 "	6.00 "	6 35 "	6.45 "
7.45 "	7.53 "	8.30 "	7.30 "	8.05 "	8.15 "

Saturday and Sunday only.			Saturday and Sunday only.		
Trains Leave Poughkeepsie.	Arrive Highland.	Arrive New Paltz.	Trains Leave New Paltz.	Leave Highland.	Arrive Poughkeepsie.
†7.25 A.M.	†7.33 A.M.	†8.15 A.M.	†7.10 A.M.	†7.45 A.M.	†7.55 A.M.
†8.30 "	†8.45 "	—	—	†9 10 "	†9 25 "
9.10 "	9.18 "	10 00 "	9.00 "	9 33 "	9 43 "
10 15 "	10.23 "	11.00 "	—	†9.34 "	†9.52 "
†11.25 "	†11.40 "	—	10 00 "	10 35 "	10.45 "
12.00 M.	12.08 P.M.	12.45 P.M.	—	†12.01 P.M.	†12.20 P.M.
1.15 P.M.	1.23 "	2.00 "	11.45 "	12.20 "	12 30 "
2.00 "	2.08 "	2.55 "	1.00 P.M.	1.35 "	1.45 "
2.45 "	2.53 "	3.40 "	1.35 "	2 20 "	2 30 "
3.30 "	3.38 "	4.25 "	2.15 "	3 05 "	3.15 "
4.15 "	4.23 "	5 00 "	3.05 "	3 50 "	4.00 "
5.00 "	5.08 "	5.55 "	3 50 "	4.35 "	4 45 "
†5.13 "	†5.28 "	—	4.35 "	5.20 "	5.30 "
5.40 "	5.48 "	6 40 "	5.15 "	6 00 "	6.10 "
6.30 "	6.38 "	7.25 "	6.05 "	6 50 "	7.00 "
7.15 "	7.23 "	8.00 "	6 50 "	7 35 "	7.45 "
8.15 "	8.23 "	9.00 "	8 00 "	8.35 "	8 45 "

†Saturday only.   †Saturday only.

Rates of Fare.—Fare between Poughkeepsie and Highland or Pratts Mills, 10 cents. Excursion tickets, 15 cents. Package of ten tickets, 75 cents. Between Poughkeepsie and Centreville, 15 cents. Between Poughkeepsie and Loyds, 15 cents. Between Poughkeepsie and Ohioville, 20 cents. Between Poughkeepsie and New Paltz, 25 cents. Excursion, 40 cents.

Passengers from New Paltz and stations on the Electric road paying fare to Poughkeepsie, will be furnished by the conductor a ticket good for the ride from Pratts Mills to Poughkeepsie on the C. N. E. Ry.

Purchase Tickets at Office.—Double Fare will be charged if paid on Trains between Poughkeepsie, Highland and Pratts Mills, and conductor will issue to such passengers Duplex Train Tickets.

W. J. MARTIN, Sup't C. N. E. Ry.

directly to Poughkeepsie ten to fourteen times each way daily, more in summer. During the busiest parts of the day, they often ran less than an hour apart.

The trolley cars, varying in size, could carry forty to one hundred passengers each, and the rapid transit service advertised them as "the most luxurious cars obtainable." Some were equipped with air brakes, oil headlights, water closets, and ice-water tanks. In summer, some of the cars were open at the sides, offering a dramatic, if scary, way to cross the bridge. The dummy steam engine was fitted to burn hard rather than the usual soft coal so that even if the cars were open at the sides, their passengers would be little affected by the smoke.

When the rapid-transit cars were crossing the bridge, their promoters promised to run them "slowly," to "give passengers an opportunity to observe the scenery." Passengers riding high up above the Hudson, on "America's greatest bridge," the promoters declared, are "suspended as it were between heaven and earth, with the grandest scenery in America at their feet."[3]

To encourage students to commute by this combined railroad-trolley line from Poughkeepsie to the State Normal School at New Paltz, the line offered student fares at twenty-five cents per day, round trip. To entice Poughkeepsians to attend races in New Paltz—there was a horse racing track on the flats along the Wallkill River within walking distance of the trolley terminal—it offered thirty cent round trip rates. For those who wished to travel in the opposite direction, to attend theater in Poughkeepsie, the trolley line itself sold theater tickets and ran special theater cars from New Paltz. But most of the line's traffic was provided by summer visitors. They might be on their way to hike, swim, or sail. Or they might want to stay at boardinghouses along the trolley line or at resort hotels west of New Paltz in the Shawangunk Mountains, as at Lakes Mohonk or Minnewaska. To associate the trolley line with the resorts, one of its cars was named Mohonk, another Minnewaska.

In 1897, when the rapid transit service was new, on several summer days over 500 people used the service each day. In that year, two parties of Vassar girls took the trip, one a party of 100, the other 125. In 1899, to stimulate traffic on the fourth of July, the Bridge Route—that is, the CNE—made a donation to the New Paltz Independence Day celebration. On that day, more than 1,387 people used the service, many of them riding from Poughkeepsie over the bridge all the way to New Paltz.

As the years passed, however, the number of passengers riding on the rapid transit declined. The Highland-Poughkeepsie ferry service offered strong competition. The ferry had the advantage that it docked in Poughkeepsie close to both the Hudson Line's rail station and Hudson River steamer docks, while the rapid transit rail service took passengers to the Bridge Route's Poughkeepsie station on Parker Avenue about a mile away from them. Although street cars connected the Bridge Route station to the Hudson River front, they did not do so directly; they required an inconvenient change from one street car line to another.

By 1904, the New Haven Railroad, as it took control of the Bridge Route, was concentrating on increasing the bridge's freight traffic. The New Haven found the low-paying rapid-transit service on the bridge a nuisance. While the trolley service from New Paltz to the Highland ferry dock continued, the New Haven discontinued the combined railroad-trolley service over the bridge.[4]

Above. OPEN TROLLEY CAR WHICH MAY HAVE CROSSED THE POUGHKEEPSIE BRIDGE. Car #7, AT ELTING'S CORNERS, between New Paltz and Highland (LLOYD)

Facing page. TIME TABLE, from an advertising card (BENJ)

TWO BRIDGE ROUTE PASSENGER TRAINS (both from STICK). Top, A TRAIN DECORATED FOR THE PRESIDENTIAL CAMPAIGN OF 1896, IN SUPPORT OF MCKINLEY AND "SOUND MONEY," Oct. 24, probably in Millerton, Dutchess County, NY. The engine is PR&NE's # 19 (4-4-0). The engineer, dressed in white, is George Thurston. Below, TRAIN AT SPEED IN NORFOLK, Litchfield County, CT, heading west, June 23, 1924. The engine is CNE's #45 (4-4-0).

12. LOCAL PASSENGER TRAINS (1890-1934)[1]

Off to work, school, or shopping, enveloped in steam, hisses, and smoke with a chance that you might meet someone on the way.

AFTER THE BRIDGE ROUTE had stopped its railroad-trolley service over the bridge, the Bridge Route's regular local passenger service continued to be significant. It long remained significant for the Andrew L. F. Deyo family, a family of French Huguenot descent who lived on their old family farm in Gardiner, NY, about twelve miles west of the Poughkeepsie Bridge, on the Gardiner-Modena road.

According to Deyo family diaries, in the 1890s and early 1900s the Deyos often drove about a mile and a half to the Bridge Route's Modena station. They usually drove by wagon or buggy, or in the winter occasionally by sleigh, and they might do errands in Modena at the same time, buying feed or having their horses shod. Several members of the family were likely to drive to the station, including the father Andrew Deyo, the mother Agnes, and the three children: Joe, who was helping his father on the farm; Bertha, who had become a teacher in Queens, but often visited at home; and Lizzie, even though she had a disturbing nervous twitch.

When the Deyos took a train from Modena, they were usually heading for Poughkeepsie, a 45 to 50 minute trip. Once Agnes, according to her diary, herself "went to Pokeepsie on train." There she "met Edd and Ella," her New Paltz relatives. "Went shopping." Another time, Agnes and her daughter Lizzie "went to Pokeepsie on morning train. [Lizzie bought] a ready made dress at Lucky & Platts." On another occasion, in 1913, Andrew and his wife Agnes took a morning train to Poughkeepsie to buy bed furnishings to send to their daughter Bertha who had recently married and was living in New Jersey. As Agnes often did, she carefully noted the price they paid for each item: springs $7.75; mattress $17.50; "wool comfortable" $5.00.

Less often, the Deyos took a train from Modena in order to change in Poughkeepsie to travel on elsewhere, by one of the many available links of the public transportation network in the Mid-Hudson region. Once Lizzie and a relative named Ella "went from Modena to go on the day boat," that is, in Poughkeepsie they changed to a Hudson River steamer. At another time, Bertha took a CNE train from Modena to Poughkeepsie, and changed there to take a New York Central train heading north: Bertha "went to Syracuse via Modena & Pokeepsie." On another occasion when an elderly aunt, Maggie, was ill in New York, the diarist, Agnes Deyo, went to New York to see her,

THE STATION AT MODENA, ULSTER COUNTY, NY, 1931 (SHUKR). Like many Bridge Route Stations, it was small, low, rectangular, and while passenger trains were still running, used for both passengers and freight. It was on Plate Rd., just off the Modena-Newburgh Highway (Rt. 32). The station had burned down in both 1890 and 1928, but had been rebuilt. In 1931 it was a New Haven Railroad station, and its station agent was Rufus Jenkins who was known for riding to work on an old bicycle.

traveling as she often did with a friend, in this case, her friend Mamie. They took the Bridge Route train to Poughkeepsie, and then changed to a New York Central train heading south. As Agnes explained in her diary, "Mamie got off at 125th St. [in upper Manhattan]. I went on to 42nd St. Went right to see Maggie." Two days later when Agnes headed back home, Andrew's brother Solomon Deyo, an engineer, saw her off in New York: "came with me to the car."

Solomon (Sol) Deyo had been the chief engineer for constructing the first major New York City subway, the Interborough Rapid Transit, and was still its chief engineer. Sol lived with his wife and daughter in New York City but on weekends and holidays often returned to the big Deyo farmhouse in Gardiner where he had grown up, surrounded by locust trees, orchards, and cattle. Once after Sol had been visiting the Deyos in Gardiner, along with his daughter Harriet, they went back to New York "on the 3:48 from Modena." In 1909, Sol and family came home to Gardiner for Thanksgiving, arriving "via Modena." That year the Deyos in their old farmhouse had twenty-two at their Thanksgiving table. It was heaped with fowl, both turkey and duck, and pie, both pumpkin and apple.

The Deyos are not known to have ever taken a Bridge Route train to go as far as Boston, Washington, or Chicago. Occasionally, however, various Deyos took CNE trains to visit Sol Deyo and family in their summer home in Norfolk, CT. Moreover, once in 1907, the daughter Lizzie went east on the CNE, all the way over the bridge through Norfolk to Hartford, apparently visiting relatives there. "L[izzie] went to Hartford on noon train. Fare 2.55." The Deyos paid attention to train fares, as they did to other expenses. Since the distance from Modena to Hartford on the CNE was 129 miles, the fare of $2.55 would be approximately 2 cents per mile, which was about standard for the region's railroads at the time.

According to Deyo diaries, occasionally the Deyos took trains from Modena in the other direction, going "west" as the railroad called it, but really southwest. Once when Sol Deyo had brought his wife and daughter up from New York to visit the family in Gardiner, and Bertha had friends visiting as well, they made their return to their homes a social event—train travel was naturally a more social experience than automobile driving. They all went home by taking a CNE train from Modena to Campbell Hall, and changing there onto an Erie branch line for Goshen. According to Agnes' diary, "All went back in afternoon via Modena & Goshen, Sol's family too." This would likely mean that from Goshen, at least Sol and his family took the Erie main line to Jersey City, and from there ferried across the Hudson back home to Manhattan.

To go the six miles from their Gardiner house north to New Paltz, the Deyos usually either drove all the way, or drove only to the Gardiner station and from there took a Wallkill Valley train. However, once in 1911, when Aggie planned to attend a high school graduation in New Paltz, she experimented with a different, if roundabout, means of getting there. She took the CNE train from Modena northeast only as far as the Lloyd station, and from there took the trolley west to New Paltz—the Highland-New Paltz trolleys were still running. She wrote in her diary triumphantly, "Very good way. 30 cents."

ANDREW LE FEVER DEYO, farmer and Gardiner town supervisor (JACBS). His family used the Modena station both to travel and to ship out fruit. (His grandson, John K. Jacobs, is the author of a short story in this book).

CNE PASSENGER TIME TABLE, JULY 1, 1903 (Selected portions only)

QUALITY OF SERVICE ISSUES

Stations were numerous on the Bridge Route. In 1898, when the Philadelphia, Reading & New England (PR&NE) owned the Bridge Route, it had 65 stations. In 1916, when the Central New England (CNE) owned it, it had 135 stations.

Most Bridge Route stations in the CNE period were owned and operated by the CNE itself, but at junction points the CNE sometimes arranged with other railroads to share stations. For instance, in 1917, the CNE shared the passenger stations of the Ontario & Western at Campbell Hall, NY, of the New York Central at Boston Corners, Rhinecliff, and Beacon, NY; of the New Haven Railroad at Hartford, Simsbury, Winsted, Canaan, and Danbury, CT. Arrangements for sharing costs varied. In some cases the CNE paid for its use of stations according to its fluctuating share of the business handled, in other cases according to a fixed share of operating costs.[2]

In the evening, local residents might walk to a station to watch the trains come in. They might

- EXCURSION -
TO
Po'keepsie.

A ride over the New Railroad
AND ACROSS THE GREATEST
Railroad Bridge in
AMERICA.

LUCKEY, PLATT & CO.,

The great DRY GOODS and CARPET DEALERS of Poughkeepsie have enlarged and improved their store, and will have a

GRAND OPENING

**Thursday,
Friday and
Saturday,**
April 24th, 25th, 26th.

☞ During these three days Tickets can be purchased from LOYD to Poughkeepsie and return for 40 cents, or from NEW PALTZ, including a stage ride to and from Loyd for 90 cents.

☞ Ask for LUCKEY, PLATT & CO.'S Excursion Ticket which will entitle the holder to a return ticket to be furnished at their store in Poughkeepsie free of charge.

☞ Orders for the Stage may be left with JOSIAH D. ELTING.

☞ Stage will leave New Paltz at 9.30 A. M., and connect with train. Leaving Po'keepsie at 6 o'clock returning.

EVERYBODY IS INVITED.

**Luckey, Platt & Co.,
332, 334, 336**
Main-street,
POUGHKEEPSIE, N. Y.

---

want to hear the news, whether from travelers coming and going, or from Western Union. On the presidential election day in 1916, a large crowd at the Salisbury, CT, station, stayed late until telegraph dispatches satisfied them that Charles Evans Hughes had been elected president, only to learn the next day, after the results from California came in, that Woodrow Wilson had won after all. Bridge-related stations were not always orderly. At the ND&C station at Millbrook, NY, according to a railway official, when ladies and children who were "unattended" stepped off the trains they were sometimes disturbed by finding "hoodlums and loafers" crowding the platform. Also some stations were not well maintained. In 1892 a newspaper called the station at Tariffville, CT, the "shabbiest, dirtiest, and most disgraceful hog pen on the road between Hartford and Poughkeepsie." In 1910, a state railroad inspector, while reporting the CNE stations in general were in "fair to good condition," called the Poughkeepsie station—which was still, as it was when the bridge opened, a remodeled private house—"dirty and unkempt" and "entirely out of keeping with the policy of the road."[3]

Most CNE passengers took trains only short distances. In 1911, the average distance the CNE carried its passengers was only 16 miles, which was less than the average distance for several major railroads with which the CNE connected. For the New York Central it was 37 miles.

Workers rode CNE trains to work, as to the State Hospital in Poughkeepsie, using the Hospital Branch to get there, or to the rail yard at Maybrook, using special employees-only trains. In some locations, teachers rode the trains on their way to the ubiquitous rural one room schools. Children also rode the trains, carrying their lunch buckets, as they did, for example in Dutchess County, NY, to reach the Seymour Smith Academy in Pine Plains, and in Litchfield County, CT, when Norfolk children took a special CNE afternoon school train from the Gilbert School in Winsted to go back home. In Ulster County, NY, at such stations as Relyea, Modena, and Clintondale, chil-

Left, ADVERTISEMENT, *New Paltz Times*, April 23, 1890

THE BRIDGE ROUTE'S STARTLING PASSENGER STATION AT POUGHKEEPSIE, 1927 (McEN). Originally the Lown residence, it was remodelled as a station at the time of the Poughkeepsie Bridge opening, 1889. Despite complaints that such a station was inappropriate for Poughkeepsie, it was used sucessively by the CNE&W, PR&NE, CNE, and New Haven, hinting that the Bridge Route, whatever its follies, was little inclined to delusions of grandeur. The station faced Parker Avenue's streetcar tracks which show in the foreground. The station's rear faced the railroad tracks. As shown in the left background, a covered walkway led from the station to a covered railroad platform.

dren boarded the morning eastbound milk train to attend school in Highland. In 1911, however, when residents of Highland petitioned the CNE to run a train service across the bridge to Poughkeepsie and back for the convenience of children going to school and "mechanics" going to factories, the railroad responded that they would expect only thirty regular commuters on such a train, not enough to justify running it.[4]

The CNE encouraged its train crews to be courteous. In the early 1900s, train crews were sometimes said to be so courteous that they would stop not only as required when flagged at a flag station, but also when flagged in rural regions anywhere. One rainy day, according to an engaging story, a little old lady standing on a track between stations signaled a train to stop. The train screeched to a halt, and a CNE conductor stepped down from a car, intending to help the lady up.

But she only handed the trainman a letter, asking him to mail it. The conductor gamely accepted the letter.

In 1897, however, the PR&NE superintendent admitted that on the Bridge Route "we have received a great many complaints" that our coaches are "old" and the track "rough." Although a little later the CNE advertised that it had "new and luxurious coaches," much of the time passengers on the Bridge Route and related lines were likely to ride on old, rattling wooden coaches, sitting up stiffly in green plush seats which were not easy to keep clean, especially in summer when the windows were likely to be open, allowing smoke and cinders to blow in.[5]

On the Bridge Route and related railroads, patrons and crews gave the railroads nicknames which, if light-hearted, suggested ambivalence about them. Playing with the initials by which they were known, they sometimes called the CNE "Curves Never Ending," the Newburgh, Dutchess, & Connecticut (ND&C) "Nasty, Dirty, and Crooked," the Poughkeepsie & Eastern (P&E) "Perverse and Eccentric," the Ontario & Western (O&W) "Old and Weary," the Lehigh & Hudson River (L&H) "Late and Hungry."

At one time the Erie had such a reputation for its trains being late that it was said that when a passenger fell off a train, he thought as he rolled down an embankment that he might reach his destination sooner by falling off the train than by staying on it. The CNE, however, had a record for its passenger trains being fairly well on time. In the years 1910 to 1912, the average delay of its trains was only three to five minutes, which was worse than the New Haven but better than the O&W, New York Central, and Erie.

In the Bridge Route's early years, its passenger traffic grew significantly. For the CNE and its predecessor railways, the number of their passengers increased from 289,000 in 1890, to 568,000 in 1900, to 1,055,000 in 1917, part of this increase reflecting an increase in trackage. The CNE's annual revenue from passenger trains reached its peak with $461,000 in 1914. It remained substantial until about 1921.[6]

DECLINING PASSENGER SERVICE

As automobiles, trucks, and buses came into increasing use, governments often subsidized the improvement of highways for them. However, as J. W. Cuineen, the CNE superintendent, complained, governments did not correspondingly subsidize roadbeds for trains. Train traffic, especially passenger traffic, suffered.

To cope, in 1921 the New Haven Railroad, earlier than other railroads, was experimenting with running single-unit gasoline-powered passenger rail cars as a substitute for steam-powered passenger trains, and they soon became common on the Bridge Route. The gasoline-powered cars, often built by Mack or Brill, required only a motorman and conductor to man them, so they were less expensive to operate. Other bridge-related railroads tried similar gasoline-powered rail cars, including the L&NE from 1926, and the L&HR from 1928. However, the gas cars were uncouth, hard to heat, unreliable, and often broke down.[7] The gas cars did not stop the decline of the Bridge Route's passenger traffic.

By 1920, the CNE had stopped running any through passenger trains from Hartford via Canaan over the Poughkeepsie Bridge to Campbell Hall. Gradually, during the 1920s, the CNE abandoned certain small, unprofitable branches, until its total road was down from 304 miles in 1914 to 264 in 1926. It also reduced the number of cars on its passenger trains, often to only one car, a combination baggage and passenger car. In 1925, with both passenger and local freight traffic declining, the CNE asked permission of the New York State Public Service Commission, to discontinue its agents at eleven rural stations, including in Columbia County, Jackson Corners; in Ulster County, Lloyd and Clintondale; in Dutchess County, Billings, Brinkerhoff, Moore's Mills,

CENTRAL NEW ENGLAND RAILWAY (CNE) VITAL STATISTICS FOR 1917 Compiled from NY State, Pub. Serv. Com'n, 2nd Dist., *An'l Rep*, 1917, "An'l Rep.," 1917	
General Office	New Haven
Length of road (including trackage rights)	302 miles
Average no. of employees	1,889
Average no. of passenger trains run per day	23
No. of revenue passengers carried per year	1,054,526
Revenue per passenger mile	2.4 cents
Accidents: Persons injured Persons killed	261 3
Passenger cars	60

Freight cars:
  box
  flat
  coal
  caboose
  company service
	
	996
	52
	34
	29
	50

Steam locomotives
  including:
  Consolidations (2-8-0s)
  Americans (4-4-0s)
	78
	
	42
	11

Shekomeko, and Salt Point. Nevertheless, near the end of 1926, it was still keeping agents at these stations, even though some of them, including the Clintondale agent, were stretched to man several stations at once. By that year only one passenger train daily, each way, ran over the bridge, and it ran only between Poughkeepsie and Maybrook.[8]

In 1927, when the New Haven took over the operation of the CNE directly, and the CNE itself ceased to exist, the New Haven stopped passenger service on much of the Bridge Route's northern claw, from Millerton east to Hartford. Soon afterward, the L&HR had already stopped running any passenger service to Maybrook, while other lines that fed into Maybrook, such as the L&NE and the New York Central's Wallkill Valley line, were severely cutting back.

By early 1930, the only regularly scheduled passenger train running back and forth over the bridge was what had once been the Scoot, which had been an employees-only train, taking railroad employees to and from work in Maybrook. Over several years it had evolved into a regular passenger train, appearing in regular passenger time tables, and open to the public. A Poughkeepsie youth, not a railroad employee, recalls riding it once, shortly before it ceased to run. He rode it one morning for adventure, paying ten cents to ride from Poughkeepsie over the bridge to Highland. By August 1930, however, the Mid-Hudson highway bridge having opened, a bus was able to run from Poughkeepsie along almost the same route as the train all the way to Maybrook; the New Haven found itself without enough riders on the Scoot to make it pay, and discontinued it. However, when passengers protested, the Public Service Commission directed the New Haven to run the Scoot again, which it did in November and early December, 1930, but only as a gasoline-powered single car. Then the commission held a hearing on whether the New Haven should continue the service, a hearing at which the New Haven presented evidence that it was losing money on it. By December 19, with the commission's approval, the New Haven had discontinued the service again, this time permanently. Thus the inexorable increase in the use of highways finally brought an

TWO BRIDGE ROUTE CREWMEN. Top, WILLIAM J. REEL, BRAKEMAN, AGE 19. Photo by Ardron, Poughkeepsie, June, 1893 (FULLR). At the time, Reel was serving on PR&NE trains running between Winsted, CT, and Poughkeepsie. The insignia on his shoulders read "P&R," for Philadelphia & Reading, the major railroad of the Reading System of which the PR&NE was a part. Bottom, CONDUCTOR FRED W. SNOW (1874-1961) (BEAUJ). Photo by Deming, Winsted, in the 1890s, when Snow worked for the PR&NE, as the insignia on his shoulder directly indicate. He later worked for the CNE, often making runs over the Poughkeepsie Bridge to Maybrook where he is remembered for his love of wine and song.

CENTRAL NEW ENGLAND RAILWAY'S ROAD MILEAGE, BY STATE (Including mileage leased) From CT Railr's Com'rs *An'l Rep.*, 1900, 1910; CT Public Utilities Com'n, *An'l Rep.*, 1915, 1920, 1925 (Totals may be affected by rounding)				
	NY	CT	MA	Total
1900	114	67		181
1910 (after the CNE had opened the Springfield Branch, and absorbed both the Dutchess County and the Newburgh, Dutchess, & CT RRs)	194	77	8	279
1915 (after the CNE had leased track rights from Hopewell Jct., NY, to Danbury, CT)	213	83	8	304
1920	210	84	8	301
1925 (after the CNE had severed its Springfield Branch)	206	84	2	292

end to regular passenger service over the Poughkeepsie Bridge.

For several more years, the New Haven Railroad still operated a few passenger trains—some in the form of gasoline propelled cars—on various parts of what had traditionally been the Bridge Route. But by 1934, in the depth of the Great Depression, the New Haven Railroad, as it headed toward bankruptcy, had ended regular passenger service on all of the former CNE.[9]

AN UNGAINLY, SNOUTED "RAILBUS": A GASOLINE-OPERATED SINGLE-CAR, BUILT BY BRILL IN 1925 FOR PASSENGER SERVICE, SHOWN ON ITS LAST TRIP, SEPT. 9, 1933, at Pine Plains, Dutchess County, NY (STICK). By this time, the New Haven was close to ending all passenger service on what had been the CNE's Poughkeepsie Bridge Route. Among the railmen shown, probably all involved at different times in operating this railbus or its companion one, were, at the left, engineer Dennis Foley of Canaan, and, at the right, engineer Everett Sisson of Hartford. Both men figure elsewhere in this book.

FREE PASSES

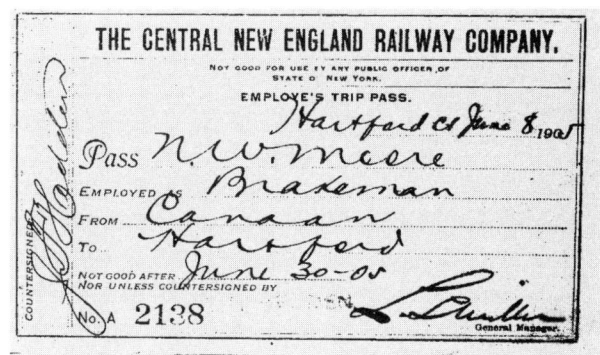

Following custom, bridge-related railroads issued free passes. They issued passes to officials of other railroads, as in the top left example (1896), hoping for favorable interchange of traffic. They issued passes to legislators, as in the top right example (1895), hoping for favorable legislation. They issued passes to their own employees, including station agents, telegraph operators, and train crew, as in the right example (1905) (all STICK). But when railroads discovered that granting passes was expensive, they found that curbing them could be awkward. In 1895 when the PR&NE refused a free pass to its former milk agent in the Canaan, CT, region, he retaliated by trying to persuade farmers not to ship their milk on the PR&NE. In 1898, the PR&NE signed an agreement with two other Bridge-related lines, the P&E and the ND&C, which, as the PR&NE's general agent interpreted it, prohibited "giving passes to everybody."(Quote: ND&C)

RAIL FAN PASSENGER EXCURSIONS OCCASIONALLY RAN OVER THE BRIDGE AFTER REGULAR PASSENGER SERVICE STOPPED like this one on May, 20, 1951, sponsored by the CT Valley Chapter, National Railway Historical Soc. The photo was taken by one of the passengers, Wm. D. Knauss of Poughkeepsie (KNAU).

TOP DAILY WAGES FOR CENTRAL NEW ENGLAND'S ROAD TRAIN CREWS From CNE, "An'l Rep.," 1917	
Engineers:	
On freight trains	$5.25
On passenger trains	4.25
Firemen:	
On freight trains	4.00
On passenger trains	4.00
Conductors:	
On through freigts	4.00
On local freights	4.50
On passenger trains	4.50
Brakemen and flagmen:	
On through freights	2.67
On local freights	3.00
On passenger trains	2.60

FRANKLIN D. ROOSEVELT SAILING HIS ICE BOAT "HAWK" ON THE HUDSON RIVER WHERE HE OFTEN COULD SEE THE POUGHKEEPSIE BRIDGE (FDRL). During the year of this photo, 1905, Franklin married Eleanor, studied indifferently at Columbia Law School, and continued his enthusiasm for the Hudson.

13. THE ROOSEVELTS AND THE BRIDGE

(1873-1945)[1]

He enjoyed saying that he had been brought up on the Hudson, "the most glorious river in the world," and that his family had lived near it for two hundred years.

OVER MANY YEARS, the Roosevelt family spun a web of connections with both the Hudson River and railroads. The Roosevelts were a New York City family of Dutch origin who became wealthy by refining sugar, including slave-grown sugar imported from the West Indies. Franklin Delano Roosevelt's great-grandfather, James Roosevelt, as an elderly man still living regularly in New York City, bought a farm on the northern edge of Poughkeepsie on a rise overlooking the Hudson and built a summer residence there. He called it Mount Hope. After he retired there in 1847, when the first railroad to run along the Hudson River was being planned to run on the river shore below his house, James Roosevelt, like many other landowners of his century, instead of being concerned that the railroad would degrade his landscape, encouraged building the railroad. He subscribed $5,000 to help build it.[2]

About twenty years later, before the drive to build the Poughkeepsie Railroad Bridge was seriously underway, the Mount Hope house burned. By that time it had been inherited by another James Roosevelt, the grandson of the earlier one. This James Roosevelt, who was to become the father of Franklin Delano Roosevelt, rather than rebuild the house, sold it, and it soon afterwards became the site of the Hudson River State Hospital.

To replace Mount Hope, James Roosevelt bought another farm, about two miles farther north in Hyde Park, but also overlooking the Hudson rail line and the Hudson River. He moved there in the late 1860s. From the piazza of his new house, which he named Springwood, James Roosevelt and his family could see in the distance, over their lawn as it sloped down to the river, the site of the proposed Poughkeepsie Railroad Bridge. Today from that lawn, visitors cannot see the bridge because trees have grown up to block the view, but from the second story of the house, they can still see it.

From the Hyde Park station, James Roosevelt took trains on his way to his law office in New York, trains which passed on the Hudson Line in front of his Hyde Park farm. When he was at home on his farm, he regularly rode horseback on its 500 acres, managing them, and was often in sight of the river. He hunted along the river, he fished in it, skated on it, kept a boathouse on it. He belonged to local yacht and ice boat clubs, and sometimes sailed downriver, passing the site of the bridge.

As a gentleman who modeled himself on British aristocrats, James Roosevelt served as an officer of his Episcopal church and on the board of the Hyde Park public school. He also served on the board of the Hudson River State Hospital. In 1875, when a group of Bostonians came to Poughkeepsie to investigate what a bridge there would do for them, he was among those showing them around the hospital.

In this great age of railroad building, James Roosevelt invested in railroads. In doing so, he became an associate of top Pennsylvania Railroad officers. In 1873, when the Pennsylvania Railroad was providing the major financial support to begin the construction of the Poughkeepsie Bridge, James Roosevelt joined Pennsylvania Railroad officers on a train which came up from New York to Poughkeepsie to attend the bridge cornerstone laying.

This being also a great age of coal mining, James Roosevelt also invested in coal. As an officer of a coal mining company, James Roosevelt grew to know his fellow officer Warren Delano II. After Roosevelt's first wife died in 1876, he courted Delano's daughter Sara, who was to become the mother of Franklin Delano Roosevelt.

The Delanos—a Huguenot family, their early name being De La Nouye—had been Massachusetts maritime merchants, and like other such merchants they had become wealthy in trade with China, including opium trade, of which the Delanos preferred to say little. After settling on the northern edge of Newburgh, in a house overlooking the Hudson River, the Delanos had become so active in investing in coal that a coal town in Pennsylvania's Lehigh Valley was named Delano after them.

When James Roosevelt was courting Sara Delano about 1880, he took her out boat riding on the Hudson River, by his home at Hyde Park. At that time if they could see any of the Poughkeepsie Bridge at all, it was likely to be only two stumps in the river, the pathetic remains of efforts, long abandoned, to build two bridge piers in the water. James and Sara married in Newburgh, and afterwards lived in the Roosevelt home in Hyde Park, where their son Franklin Delano Roosevelt was born in 1882.

When Franklin was about five years old, he and his parents could see the Poughkeepsie Bridge rising above the Hudson River from their lawn. About that time, a state court appointed James Roosevelt to be a member of a bridge-related commission. This commission was to settle any disputes about what the Bridge Company should pay landowners for the land it was seizing from them for the bridge and its connecting rail lines. The commission, including James Roosevelt, viewed many parcels of land on both shores of the river, took testimony about the value of such land, and made awards as small as $300 and as large as $10,000.[3]

As Franklin grew up, he had many opportunities to become aware of the bridge through both his family's experience and his own. He not only could see the bridge from his house, but he also swam in the Hudson below his house in sight of the bridge. While Franklin never became an accomplished athlete, his father taught him to row, sail, and ice-boat on the Hudson, and of course, once the railroad bridge was built, Franklin sometimes passed under it, on the water or on the ice. Later Franklin became an officer of the local ice-yachting club, sponsoring yachts whose runners shrieked and threw up sparkling arcs of ice spray as they slid under the bridge's long steel web.

Franklin's uncle John A. Roosevelt and his family, whose house overlooking the Hudson was a little closer to the railroad bridge than Franklin's house, were also conscious of the bridge. When the last pin was driven in the bridge in the summer of 1888, making it for the first time a continuous bridge across the river, a flyer was issued to commemorate the event; John A. Roosevelt's family preserved one of the flyers. John was a leader,

JAMES AND SARA DELANO ROOSEVELT, WITH THEIR ONLY CHILD, FRANKLIN, aged 17, in May, 1899 (FDRL). At the time Franklin was a student at Groton School, MA., which aimed to prepare the sons of gentlemen to become Christian stewards in public service.

along with the bridge promoter Aaron Innis, in improving Hudson River ice yachts until they were reputed to be the fastest such yachts in the world—many years later Franklin as president was still boasting about Uncle John's ice-yachting. As Franklin would likely be aware, during the annual intercollegiate rowing regatta, his uncle John made his boathouse available to the Columbia University oarsmen who rowed their shells down the river under the new railroad bridge, and his uncle's wife was a "patroness" of the annual regatta dance.[4]

Moreover, as Franklin grew older, he became conscious of railroads. The Hudson rail line passed on the river shore below his and his uncle's Hyde Park houses, and the West Shore line passed along the shore opposite them. Franklin's father was himself a long-term director of a sizable railroad, the Delaware & Hudson, which ran from Wilkes Barre, PA, through Binghamton and Albany to Montreal. When his father took Franklin long distances by train, as on summer visits to Campobello Island in New Brunswick, or to the 1893 World's Fair in Chicago, it was likely to be in the private railroad car which the Delaware & Hudson put at his disposal. The car had separate bed and sitting rooms, was trimmed in mahogany, and was regularly tended by the same black cook-porter.

When Franklin was courting his shy, insecure cousin Eleanor Roosevelt, he took her and other friends on a dinner cruise on the family's yacht on the Hudson, steering the yacht himself. As they made their way home in the evening, they were in sight of the lights of the bridge. The night after Franklin and Eleanor were married in New York, with Eleanor's uncle, President Theodore Roosevelt, giving her away, Franklin and Eleanor headed for Hyde Park by taking a train on the Hudson Line, a line which Franklin's great grandfather had helped to build. The train took them under the Poughkeepsie Bridge, a bridge which his father had helped to build.

In time, Franklin Roosevelt came to care like his father not only about the family farm and the Hudson River but the region. For several years he served as the Hyde Park town historian. He was a founding member of the Dutchess County Historical Society. He collected publications on local history, including an 1875 pamphlet advocating the building of the Poughkeepsie Bridge.[5]

In the 1920s, when automobile drivers were agitating for adding a highway onto the Poughkeepsie Railroad Bridge but it had became clear that doing so was not feasible, Franklin Roosevelt encouraged them to agitate instead for building a new, separate highway bridge beside the railroad bridge. At last in 1930, when Franklin Roosevelt was New York State's governor, he and his wife Eleanor opened a new highway bridge, the Mid-Hudson Bridge, just south of the Poughkeepsie Railroad Bridge, a noble companion bridge for it.

Eleanor Roosevelt cut the ribbons. Governor Roosevelt, speaking to the assembled people, reminded everyone that he himself had been brought up on the Hudson River, "the most glorious river in the world," and that his family had lived near it for two hundred years. Then he formally named the new bridge the Mid-Hudson Bridge.[6] Many years later in 1995, the name was expanded to include Roosevelt's name, making the new name the Franklin D. Roosevelt Mid-Hudson Bridge.

While he was president, Franklin Roosevelt, like other members of his family before him, took an interest in the annual Intercollegiate Rowing Regatta which sent rowers racing their shells on the river under the Poughkeepsie Railroad Bridge. He served on the Poughkeepsie committee which sponsored the regatta. In 1938, he planned to see the regatta, but was delayed, and missed it.[7]

As president, Roosevelt traveled on railroads more than any other president ever had. He found railroad travel relaxing. To help him enjoy it, and perhaps also to make it easier for him to move about the trains, disabled from polio as he was, he asked that the trains on which he rode run slowly—25 miles an hour was his favorite speed, 35 would make him angry, though he did not complain if the train made up time at night while he was sleeping.

Before World War II, President Roosevelt usually traveled from Washington back home to his beloved Hyde Park by either of two routes, neither of which included the Poughkeepsie Bridge Route. One route was by the Pennsylvania Railroad and its tunnel under the Hudson into Manhattan, on by tunnel into Queens, then across the Hell Gate Bridge into the Bronx, and finally up the New York Central's Hudson Line directly to his Hyde Park estate—which had a convenient private rail siding on the river below his house. The other route, the one the Secret Service preferred after the U.S. entered the war because it was less congested and easier to protect, was via the B&O to Jersey City, and then via the West Shore Railroad north along the Hudson to its Highland station, from which Roosevelt would then drive across the Mid-Hudson Bridge to Hyde Park.

Raymond Hegeman, who grew up in Highland on the slope going down to the Hudson River, recalls that he always knew a day ahead of time that Roosevelt was coming to the Highland station because Secret Service men would crowd into the neighborhood to monitor it for the president's safety. Perhaps two hundred of them would check out the nearby wharves, tracks, rocky hillsides, and the two Hudson River bridges.

One day, Hegeman recalls, when he was about eleven or twelve, as he was watching President Roosevelt being helped from a train into his limousine, the President's dog Fala broke away from the president and ran off. Raymond, an agile boy, quickly recovered the dog. When Secret Service men converged on Raymond, seeming to threaten him, Mrs. Roosevelt stepped forward, saying, "Let the boy bring the dog to me," which he did. She thanked him. He felt afterwards that she was agreeably "down to earth." Raymond continued to go to the station when the president was due, becoming well acquainted with Fala.[8]

In his early years as president, Roosevelt rode in a variety of private railroad cars, as other presidents had before him—the government leased the cars for him and paid the usual rates for cars to travel on railroad lines. After the U. S. entered World War II, the government remodeled for Roosevelt one of these private cars, called the Ferdinand Magellan, owned by the Pullman Company. As remodeled, the Magellan had four bedrooms, an observation room, a dining room seating eight, a kitchen, and servant quarters. To protect the president, the car, though painted green to look like other Pullman cars, was sheathed with bullet proof steel, except for the windows which were fitted with bulletproof glass. To assist the president to swing about the car, it was furnished with special railings. To allow him to enter and leave the car as conveniently as possible, the steps at one end of the car were removed and replaced by an electric elevator which would raise and lower him in his wheel chair. When the president traveled, he usually did so in his own Magellan car, which was usually placed last in his presidential train; altogether the train might have as many as 16 to 18 cars, most of them occupied by Secret Service guards, newsmen, communications men, and presidential staff. During the war, the president's travel plans were usually off the record, the number of newsmen allowed to travel with the president was limited, and their freedom to send out telegraph dispatches from the train was curtailed.

At many times in his life, Roosevelt might have crossed the Poughkeepsie Railroad Bridge, either in regular passenger trains, or in private cars, his father's or his own, but if he did cross it, a clear record of such a trip is not available. However, according to plausible claims, he came close to crossing it at least once in 1944 when he was traveling from Washington home to Hyde Park by way of Allamuchy, NJ.

According to the journalist Jim Bishop, writing in the early 1970s, Roosevelt traveled home twice in 1944 by significant portions of the Poughkeepsie Bridge Route, the first time being on July 13-14, just before the national Democratic convention, and the second time, after he had been nominated for president for a fourth term, on Labor Day weekend. According to Bishop, speaking of the July trip, when Roosevelt's train left Allamuchy, it departed "for Poughkeepsie," suggesting that it went on to cross the Poughkeepsie Bridge. Bishop also said, speaking of the July trip, that when Roosevelt's train arrived near his home at Hyde Park, Mrs. Roosevelt would ask why he was arriving on the "other side of the river." The usual side of the river on which Roosevelt arrived

WEST SHORE RAILROAD STATION, HIGHLAND, NY, WHERE PRESIDENT ROOSEVELT OFTEN ARRIVED BY TRAIN on his way from Washington home to Hyde Park (BONE). The Hudson River is on the right. Today the station is occupied by apartments but the track, even though it has been reduced to a freight only, single track, is busy.

during wartime was, as we have seen, the west side, and the usual place was the West Shore's Highland station. Then the "other side of the river" would seem to mean the Poughkeesie side, which would seem to mean that he crossed the Poughkeepsie Bridge and arrived at the New Haven's Poughkeesie station. However, corroborating evidence is not available to support Bishop's claim for the route of the July 13-14 trip. The official log of the trip specifies that Roosevelt's train went not by the Bridge Route but by what was then the usual way, the B & O to New Jersey's Hudson shore, then up the Hudson by the West Shore line. Moreover, two presidential aides, William D. Hassett and Jonathan Daniels, implied that Roosevelt traveled by Allamuchy and the Bridge Route only once, on Labor Day weekend, 1944.[9]

According to a newsman who interviewed Dewey Long, the presidential aide from Texas who normally planned Roosevelt's travel, Long seemed only "dimly" aware of the Poughkeepsie Bridge route as a way for the president to travel from Washington to Hyde Park.[10] According to Aide Hassett, who traveled with Roosevelt on the Labor Day weekend trip, it was not Long but Roosevelt himself who chose the Bridge Route for the Labor Day weekend. It may be that Roosevelt's familiarity with the Poughkeepsie Bridge and its connecting lines helped him choose the route. But it was for a special reason that he chose it, a reason not directly related to his interest in the Poughkeepsie Bridge, or to his deteriorating health, or to the war, or to his campaign to be elected for a fourth term. It was a reason related to Allamuchy, NJ.

On this Labor Day weekend, Roosevelt left Washington without Mrs. Roosevelt; she was waiting for him in Hyde Park. He left in his usual Magellan car, in his usual presidential train, but not as usual on the B&O. His train left on the Pennsylvania Railroad line and continued on it through Trenton and Phillipsburg, NJ, and then changed onto the Lehigh & Hudson River line, heading toward the Poughkeepsie Bridge.

On this trip, while Roosevelt's train was on the L&HR line in Warren County, in northwestern New Jersey, it stopped at a small rural station, Allamuchy. There, early in the morning, Roosevelt left his Magellan car, as usual with the help of its elevator, and drove off, accompanied by Secret Service men. For most of the day he stayed away while some of the passengers he left behind walked restlessly up and down beside the train, wondering where he had gone. When a reporter asked Presidential Aide Hassett where the president had gone, Hassett, who knew, nevertheless deliberately misled him, as the staff often did for the sake of the president's privacy and security, saying that Roosevelt was helping some relatives with their financial problems.

What Roosevelt did was to drive nearby to Lucy Rutherfurd's thousand-acre estate. Much earlier, Eleanor Roosevelt had employed Lucy, then called Lucy Page Mercer, as a secretary. Roosevelt, then a young man, had become enamored of Lucy. When Eleanor found out about it several years later, Franklin promised to break off his involvement with Lucy which he apparently did for many years. But in his old age, with his heavy war responsibilities pressing on him, and with Eleanor engrossed in world betterment, he was lonely. He allowed himself to see Lucy again, repeatedly, with the cooperation of his daughter Anna, the Secret Service, and some of his staff. By this time Lucy, in her fifties, was a widowed grandmother. She was dignified, warm, and suave. It was to see Lucy again that Franklin Roosevelt chose in 1944 to go home to Hyde Park by the Bridge Route, stopping at Allamuchy.

In mid-afternoon Roosevelt returned to his presidential train, which then resumed its journey. It rolled on to Maybrook, NY, where it switched onto New Haven tracks. It passed through the fruit growing region of Modena and Clintondale. It curved around Illinois Mountain. Shortly before reaching the Poughkeepsie Bridge, according to Aide Hassett, the train stopped in Highland. Hassett in his diary specified this meant the train stopped at the "upper station in Highland." That would certainly mean not the West Shore

PRESIDENT ROOSEVELT RECEIVING VISITORS ON THE REAR PLATFORM OF HIS PRESIDENTIAL CAR, THE MAGELLAN, in Kansas, April, 1943 (FDRL). The next year, 1944, Roosevelt rode the same car on the Poughkeepsie Bridge Route.

THE BRIDGE ROUTE'S STATION NEAR THE WESTERN END OF THE BRIDGE, AT HIGHLAND, NY, ON COMMERCIAL AVE. IT WAS HERE THAT PRESIDENT ROOSEVELT DETRAINED ON HIS WAY HOME FROM WASHINGTON, ON LABOR DAY WEEKEND, 1944 (LILL). Why didn't the president continue by train across the bridge to Poughkeepsie?

Highland station, low down on the Hudson shore, where the president usually detrained; the New Haven tracks did not have any connection there. It would mean instead the New Haven's Highland station, the Bridge Route station, up on the heights above the river in the village of Highland. It was there, Hassett said, that "we arrived and detrained," at a station "which I had never seen before." It was also there, Hassett said, that Mrs. Roosevelt came to meet her husband. He would have to explain to his wife and others why he arrived at this unusual station, by this unusual Poughkeepsie Bridge Route, which he presumably intended to do, as Bishop indicated he had done on return from his supposed earlier visit to Allamuchy, by saying that it was a variation of his usual travel route for the sake of security.

Why on this occasion did the president not ride on across the Poughkeepsie Railroad Bridge to Poughkeepsie, especially since detraining in Poughkeepsie would have been closer to Hyde Park? Perhaps a reason was that it would be difficult for the Secret Service to protect the Poughkeepsie Bridge for his ride over it. Still, if he got off his train in Highland, whether as usual at the West Shore Railroad station, or as he did this time at the New Haven Railroad station, to reach Hyde Park he had to ride by automobile across the Mid-Hudson Bridge, and wouldn't protecting that bridge be a concern too? Probably protecting the Poughkeepsie Railroad Bridge for Roosevelt to cross it, which secret servicemen were not accustomed to doing, would be more difficult than protecting the Mid-Hudson Bridge, which they were accustomed to doing.

According to Aide Hassett in his diary, the Secret Service was "satisfied" with the president's routing his train home by the unusual Poughkeepsie Bridge Route (Hassett called it "the Pennsylvania-Lehigh route, via Allamuchy"). The Secret Service encouraged the president, for wartime safety, to choose this alternate Poughkeepsie Bridge Route again, and Roosevelt said "he would try it again."[11]

Roosevelt didn't have much time to try it again. Seven months later, in April 1945, at the polio rehabilitation center in Warm Springs, GA, while Lucy Rutherfurd and others were visiting him, he collapsed with a cerebral hemorrhage. Within a few hours he died. His funeral train, carrying his body along with Eleanor Roosevelt and the new President, Harry S. Truman, passed through Washington and New York City, and up the Hudson Line. In Poughkeepsie, the train passed under the approaches to the two Roosevelt-related Hudson River bridges. In Hyde Park it stopped on the bank of the river Roosevelt loved, below the house he loved. His body was drawn up the slope toward the house by horses, accompanied by a West Point band, and was buried, as he had planned, next to the house, in the rose garden.

WEST POINT CADETS AT THE WEST POINT STATION, PREPARING TO ENTRAIN FOR A TRIP OVER THE POUGH-KEEPSIE BRIDGE TO NEW HAVEN, FOR THE ARMY-YALE GAME, NOV. 6, 1954 (Charles Gunn photo, in *Shoreliner*, 1991, by permission of NHRH&TA). Looking north; the Hudson River is behind the cars on the right. At the left, cadets carry band instruments. Though the station is the New York Central's, the train is the New Haven's, with two New Haven locomotives, DER-4s, numbers 791 and 798.

14. WEST POINT FOOTBALL SPECIALS (1921-1955)[1]

On the way from West Point to the Yale Bowl by the Bridge Route, cadets were ordered to wear "full dress grey," to keep on their coats, and be prepared to beat the "enemy."

THE UNITED STATES Military Academy had been accustomed to playing its major football games on its own campus. But the Academy's superintendent, General Douglass MacArthur, came to regard its stadium as too small and access to it as too limited. In 1921, he pushed instead for the Academy to play its major games away from West Point, at locations like the Yale campus where the stadium was larger and access easier. From this time, special trains often carried West Point cadets to Army-Yale football games in New Haven.

From 1921 through 1930, special West Point football trains often ran not directly from West Point, but from Beacon, on the east side of the Hudson River, across the river from West Point. Beacon was on a branch of the New Haven Railroad (at first literally a branch of the Central New England, which the New Haven controlled) and had adequate side tracks to handle special trains. However, beginning in 1931, such football trains, in the years that they ran, instead ran directly from West Point and crossed the Hudson on the Poughkeepsie Bridge. Whether the football specials left from Beacon or West Point, these trains were through trains, not requiring any change of cars, and they were New Haven Railroad trains, operated by New Haven crews, with New Haven equipment. Whether the trains crossed the Poughkeepsie Bridge or not, they ran much of the way on what can be considered to be part of the Poughkeepsie Bridge Route, or, as it was often called at the time, the Maybrook Line.

West Point's football players themselves usually went to New Haven a day early, to prepare for the game. But virtually all the rest of the West Point student body—the whole corps of cadets, about 1000 to 1300 men—were ordered, military style, to take the special football trains to see the games, whether they wanted to or not. The only cadets exempted from the order were those who were ill or being punished.

For the 1923 trip, on the Saturday of their game in New Haven, the West Point cadets, in their dress gray uniforms, with overcoats and gloves, marched down the bluffs from their stone Academy buildings at 6:05 AM to a dock on the Hudson river front. There, company by company, they entered a boat for the nine-mile trip upriver to Beacon. On the boat, the cadets were required to stay with their units. Arrived in Beacon, they debarked in the order required, and walked a few steps into their assigned train cars, to their assigned seats. The train was in two sections of eleven cars each. The band was assigned to the first section, accompanying officers and their families to the second section. On the train, the cadets' orders for the trip specified that they could unbutton their uniform coats but not remove them.[2]

From the Beacon station, close to the Hudson River, each train ran south a short distance, then curved east, crossing over the New York Central's Hudson Line tracks by a bridge. Then the train strained up a steep, twisting incline, along Fishkill Creek, passing close to brick factories, and continued on east. As the trains passed through Danbury, crowds cheered them on.

In the 1920s and early 1930s, after the cadets had arrived at the New Haven station, they usually formed into companies, marched to the Yale campus for lunch in the Yale dining hall, and then marched the rest of the way to the Yale Bowl, a total march of nearly three miles. At about 1:30 PM they entered the Bowl ceremonially, in what

103

became a high point of the Yale-Army football experience. With the band playing, the cadets spread out over the whole field, in military maneuvers, glowing in their grey, caped uniforms. While flags flew and the crowd of 45,000 to 80,000 cheered, they sometimes marched as long as half an hour. They marched, according to the *New York Times*, "with rhythm that was flawless." Under the gaze of the distant West Rock Mountain, their "white gloves flashed up in perfect unison."[3] Finally the cadets took their seats—the cadets were required to sit together—and joined the great crowd in singing, chanting, roaring.

In a variation in 1925, soon after the opening of the Bear Mountain highway bridge over the Hudson River just south of West Point, Academy officials sent the cadets from West Point to the Yale Bowl not by train but by bus. The buses took them across the graceful new Bear Mountain suspension bridge, then on, often by narrow twisting highways, through Peekskill and Danbury to New Haven. However, some of the buses had broken windows, chilling some cadets miserably. Also some of the buses broke down, and many cadets arrived at the Yale Bowl late. After the game, on the way home the seats shook so that many cadets could not sleep. Cadets complained loudly about the buses as they are not known to have complained about the trains. Academy officials tried buses again in 1932, 1933, and 1941, but far more often they sent the cadets to New Haven by train.[4]

In the 1930s and 1940s, the special football trains, in the many years when the Academy arranged for them, departed not from Beacon but directly from the West Point station. It was a New York Central station, located on the Hudson River shore below the bluffs on which the Miliary

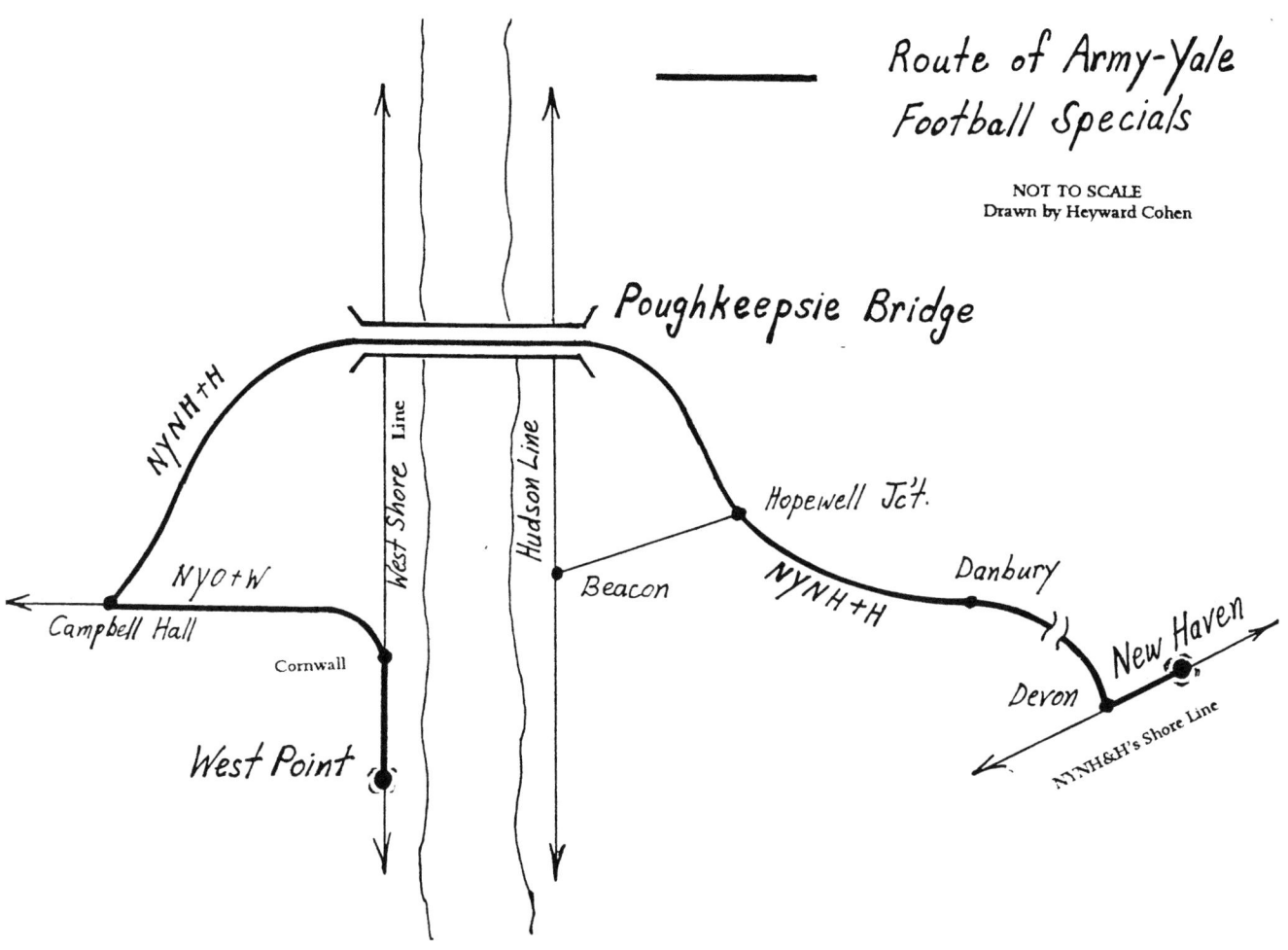

Academy buildings stood. Trains from there took the cadets over the Poughkeepsie Bridge directly to New Haven, without any change of cars.

For the 1935 trip, reveille was sounded for the cadets at 6:20 AM. The cadets were required to be in their dress uniforms, to have their rooms in order, and be ready to march to the West Point station at 8:00 AM. At the station they found the train to be in two sections. The first section was scheduled to depart at 8:20, the second one ten minutes later.

Before such special trains reached the Poughkeepsie Bridge, the New Haven's division engineer took care to arrange that if any repairs were being made on the bridge, they did not "interfere" with the passing trains. While the trains were running through Danbury, the cadets were issued a box lunch and ordered to eat it while still on their train. On arrival in New Haven—the first section of the train was scheduled to arrive at 12:30—the cadets left their trains, immediately formed into their units, and as had become customary by this time, marched only a short distance to waiting street cars. They were chartered street cars which took them directly to the Yale Bowl.

At the Bowl, by 1:15 PM the cadets were to form into units again, and then march in with their band, making their traditional colorful entrance. After the game, they returned by streetcar and train by the same route. It was a long day for the cadets. Often they endured cold, wet weather in the stadium, and loss of the game to Yale. Nevertheless, on their return trains, the cadets, in the view of railroad men, behaved well.[5]

As the crow flies, West Point and New Haven are only about fifty-five miles apart. But West Point is on the west side of the Hudson River, which meant that any through train from there to New Haven had to cross the Hudson somewhere. The nearest place to the south where such a train could cross the Hudson would be New York City, where it could cross from the New Jersey shore by a passenger train tunnel to Pennsylvania Station. If a train went that way, it would be a roundabout route, and it would pass through metropolitan congestion. The nearest place to the north where a train could cross the Hudson would be by the Poughkeepsie Bridge, which would also be a roundabout route, and the distance would be about the same, but the route would be less congested. Railroad officials chose to route these football specials over the Poughkeepsie Bridge for the same basic reason that they often chose to route other traffic over the bridge, to avoid New York City's congestion.

While West Point was on the New York Central's West Shore line, New Haven was on the New Haven Railroad line. These two lines did not interchange anywhere directly. The New Haven line which passed over the Poughkeepsie Bridge crossed over the West Shore line at the western end of the bridge, but as we have already noted, the two lines made no track connection there. At that site the difference in the height of the two lines, and the rocky terrain, would have made any connection difficult to build.

To run a train from the New York Central's West Shore line onto the New Haven's Poughkeepsie Bridge Route required running over a third railroad. West Point football specials ran from West Point on the West Shore line north along the banks of the Hudson seven miles to Cornwall, then west, away from the Hudson, on the Ontario & Western (O&W) line sixteen miles to Campbell Hall. Then they ran primarily east the rest of the way on the New Haven Railroad, crossing over the Hudson on the Poughkeepsie Bridge, and continuing on through Danbury and Devon to New Haven, for a total trip of 147 miles. Over the years, the scheduled time for the train trip direct from West Point over the Poughkeepsie Bridge to New Haven varied a little, but it was usually between four and five hours.

Running these special trains on the tracks of three different railroads required planning which could be expensive. Also when New Haven train crews were running their locomotives on unfamiliar "foreign" lines, locomotives were required to carry pilots from those foreign lines to guide the trains safely, and such pilots were costly.

While in the steam engine years of the 1920s and 1930s, the football specials regularly ran in two sections, in their last years, which were in the diesel engine years of the mid-1950s, the extra power of the diesels made it possible to run each train in only one, long section. At that time, most of the coaches were in the New Haven's traditional green, but they were new and well scrubbed for what the New Haven regarded as a prestigious excursion. The train included a stainless steel parlor car for the Academy officers, a baggage car for such accouterments as band instruments, and, at the end of the train, a deadhead coach for railroad personnel.

In 1954, on the Saturday of the Yale game, according to the West Point military orders, reveille for the cadets was at 4:45 AM, their breakfast at 5:20. Wearing "full dress gray with overcoats [and] carrying rain cap covers," the cadets, about one thousand of them, left West Point by a special New Haven train at 6:30 AM. It was one train of eighteen cars, and it passed over the Poughkeepsie Bridge. On the train, the cadets were served box lunches to eat at once. On arrival at New Haven station, which was scheduled as 11:10 AM, the cadets marched not to the streetcars of earlier years but to waiting buses. At 11:40 the convoy of twenty-six buses departed for the Bowl. At 12:30 the cadets made their ceremonial march into the Bowl, and by 1:00 were in their seats, where according to their orders they were to watch West Point beat the "enemy." After the game, the cadets took dinner at Yale, then bused to the railroad station. By just after midnight, they had returned to the West Point station, from which the weary cadets marched up the bluffs to their home barracks.[6]

In both 1954 and 1955, the final years of the football special trains, two diesel units pulled

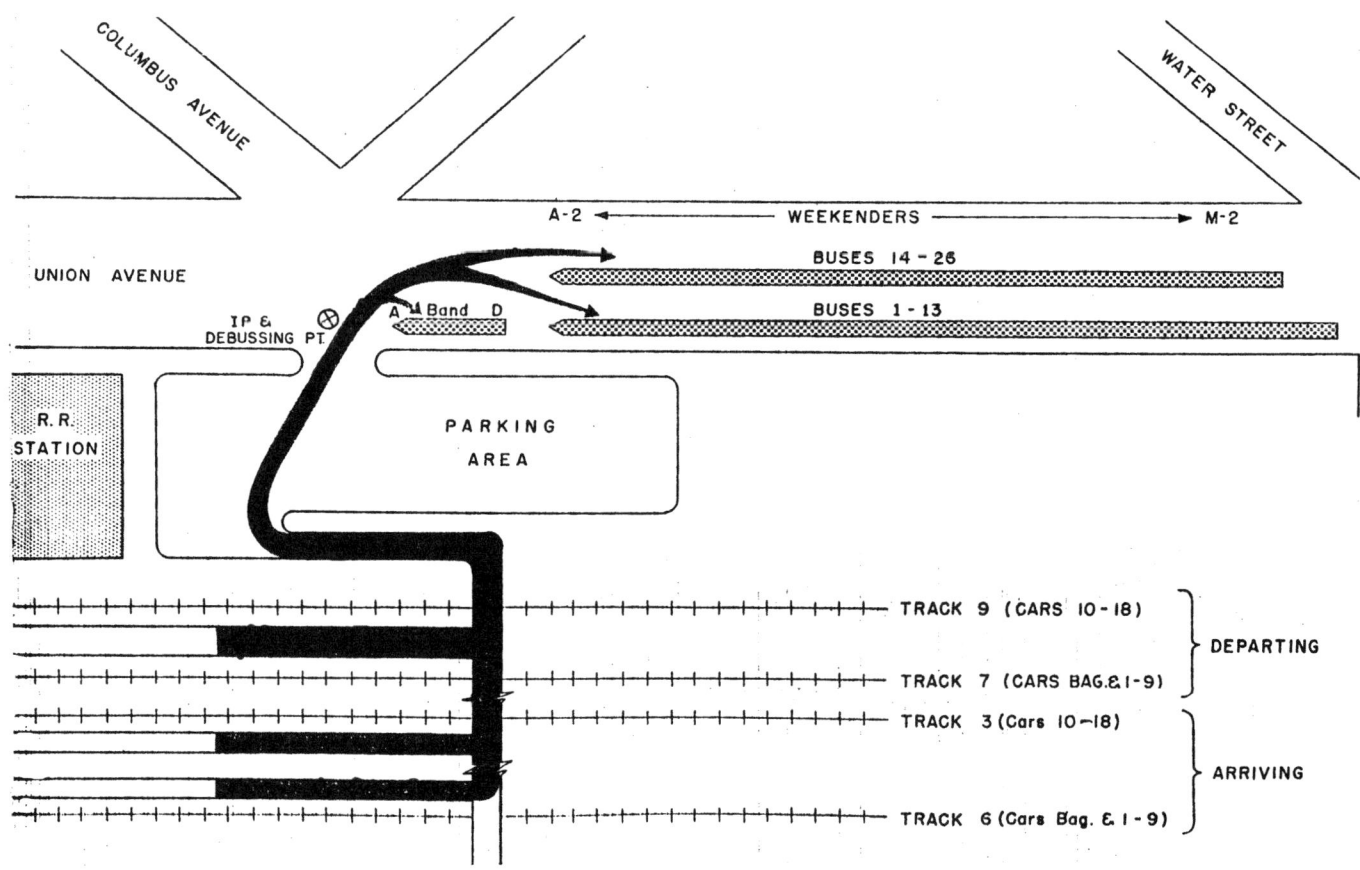

DIAGRAM OF THE NEW HAVEN RAILROAD STATION AT NEW HAVEN, CT, PREPARED BY THE US CORPS OF CADETS TO GUIDE CADETS AS THEY ARRIVED AND MARCHED TO ASSIGNED BUSES, ON THEIR WAY TO THE YALE BOWL, NOV. 6, 1954 (USMA). The heavy black lines indicate where the cadets were to march. The arriving and departing trains, each having 18 cars, were to occupy different tracks, and at least at the station, were to be divided into sections of 9 cars each.

THE YALE BOWL: When the West Point cadets marched in, their white gloves were flashing (YALE)

each football train. They were Fairbanks-Morse DER-4s, each with 2400 horsepower, the highest horsepower two-unit passenger locomotives the New Haven ever had. Two other such diesel units followed each train, for backup. They would provide extra power, if needed, on the steepest grades. They would also help speed up the process of reversing the direction of the train. When a train arrived from New Haven at West Point to pick up the cadets, it was headed south, and had to be reversed to head back toward New Haven. Also when the train arrived from West Point on the O&W tracks at Campbell Hall, it was headed west; it had to reverse direction again to head east on the New Haven tracks, toward the Poughkeepsie Bridge.

Soon after 1955, the O&W, having suffered years of financial agony, shut down, abandoning its tracks. The necessary track connection from West Point to Campbell Hall no longer being available, football specials over the Poughkeepsie Bridge stopped running. A memorable tradition came to an end.

Railroad Freight Use

15. MILK TRAINS (1889-1930s)[1]

Milk was precious. Millions of customers were waiting for it. Everywhere on the bridge-related rail lines there was pressure to get milk to its destination and to get it there safely and fresh.

BEFORE the Poughkeepsie Bridge was planned, when few rail lines had yet been built in the region, many farmers were unable to get their milk to metropolitan markets while it was still fresh. They processed much of it into butter or cheese. By the time the Poughkeepsie Bridge opened, however, with more rail lines available, and with many of them transporting milk in refrigerator cars, farmers had become better able to ship out their milk fresh, and had increased production.

In the early years of the Bridge Route, according to a boy who grew up in Litchfield County in the northwestern corner of Connecticut, farmers brought their milk to the Bridge Route's little Taconic station in the morning by team. They were likely to bring the milk in the standard forty-quart cans. The station agent would set the cans into a big vat, fetch ice from an ice house, and pack the ice around the cans. Then in the afternoon, before the milk-and-passenger combination train was to arrive, he would take the cans out of the vat and set them on the station platform.[2]

When the Poughkeepsie Bridge first opened, the Interstate Commerce Commission was regulating the rates railroads charged for shipping milk. For the New York City milkshed, the commission required railroads to charge the same rates for milk shipped from a distance to New York as for milk shipped from close to it. Those rates encouraged the shipping of milk to New York from hundred of miles away, and deprived the dairy farmers of the Mid-Hudson region of much of their natural advantage of being located close to New York. Nevertheless, by 1895, milk plants located along the Bridge Route were shipping out considerable milk, as in Connecticut, from Lakeville 53 cans daily; in New York State, on the east side of the Hudson, from Ancram 65 cans, Hibernia 65, Salt Point 25; and on the west side of the Hudson, from Modena 30.

In 1897, however, the Interstate Commerce Commission reversed itself, requiring railroads to charge rates which took distance into account, so that farmers located near New York, as in the Mid-Hudson region, again had an advantage.[3] By 1910, there were elaborate creameries along the Bridge Route, with accompanying rail sidings, icehouses, and coal-burning steam plants. Such creameries were often big enough to ship out milk by the carload, which was cheaper than shipping it by less than a carload.

Most of the Bridge Route was in the New York metropolitan milkshed. Since the Bridge Route

BORDEN'S CREAMERY, COPAKE, COLUMBIA COUNTY, NY, ON THE CNE (McEN). At one time, this creamery sent its milk to Hartford, CT, later to Port Chester, Westchester County, NY. The windowless portion of the building to the left is the ice house. The smoke stack was for the steam boiler that produced steam to sterilize the milk cans and the milk cooling pipes. Creamery workers often wore white uniforms to foster sanitation and wore boots because the steam tended to make the building wet.

and some of its related lines did not themselves reach directly into the metropolitan area, they often transferred their milk shipments to other railroads which could take them directly there. On the Poughkeepsie & Eastern line in 1906, shortly before it was absorbed into the Bridge Route proper (the CNE), a train on a regular milk run from Boston Corners, Columbia County, NY, to Poughkeepsie, picked up five loaded milk cars: one at Ancram Lead Mines (Borden's), two at Pine Plains (Borden's), one at Clinton Corners (Beakes), and one at Pleasant Valley. In Poughkeepsie, the loaded cars were switched onto New York Central's Hudson Line tracks, for delivery in New York City.

When these milk cars were New York Central cars, Conductor Nate Blodgett recalled, they "had platforms, like the old coaches, and a hand brake on each platform. The color was dark green." When these cars were Merchants Dispatch Transportation cars, however, they were white, and like many old box cars, they had the hand brakes on top of the cars. On their sides, they had "swinging" doors for loading the milk, and on their roofs, "trap" doors for dropping in ice. The ice was often harvested on local ponds or rivers, with the help of horses, hauled by railroad, stored in railroad-owned ice houses near rail stations, and dropped as needed through trap doors into bunkers at each end of refrigerator cars.[4]

On the CNE's Rhinecliff Branch, at one time there were normally two or three milk cars daily. At least one of them was en route to Rhinecliff, for switching there onto New York Central's Hudson line, to head south to New York. But at a different time on this branch, perhaps as many as ten milk cars were moving daily in the other direction, toward Hartford, some for the Bryant & Chapman milk plant there. At one period, Blodgett recalled, Bryant & Chapman kept two employees of their own on such milk cars to handle the milk just the way they wanted it handled.

The ND&C line, which the CNE absorbed by 1907, forwarded considered milk to metropolitan New York. This line ran milk cars from Millerton through Hopewell Junction to Dutchess Junction (near what was later called Beacon) where they were transferred onto the New York Central's Hudson Line, for delivery in New York. The line aggressively sought milk business. It erected creameries and then leased them out. It came to have creameries at such stations as Shekomeko (Lahey), Millbrook (McDermott & Bunger), Verbank (Locust Farm), and LaGrange (Beakes). It recruited farmers to bring milk to the creameries. It harvested ice and hauled it to them. It responded to the creameries' concerns about the time, cost, and safety of their shipments. In 1924, as many as fourteen loaded milk cars traveled south on this line daily.

Milk was precious. Millions of customers were waiting for it, including many children.

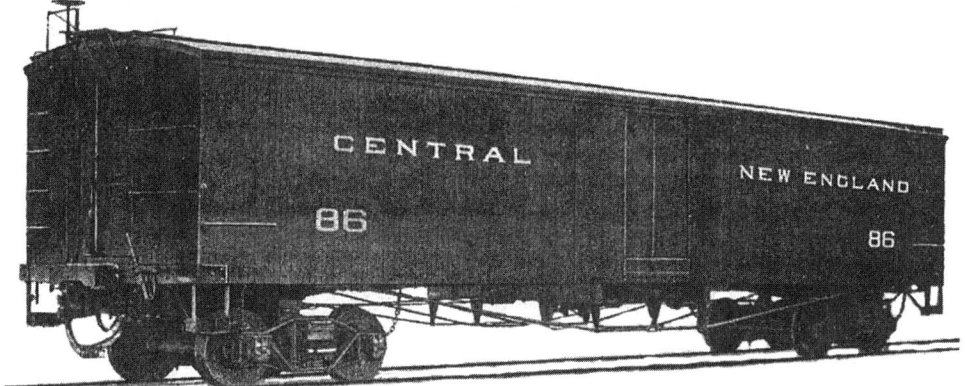

MILK CAR, ONE OF TWO BUILT FOR THE CNE BY WASON, OF SPRINGFIELD, MA., DELIVERED 1909 (Gardner, *CNE*). With a wood frame, tin roof, insulation inside. Length: 47 ft., 10 in.

BORDEN'S CREAMERY, HOPEWELL JCT., DUTCHESS COUNTY, NY, SOON AFTER IT OPENED IN 1901, on land leased from the ND&C (SHUKR). Over many years, this creamery sent its milk via the ND&C (later CNE) to the Fishkill Landing (Beacon) area, where it switched onto the NY Central's Hudson Line, for metropolitan New York. The track showing slightly at the midde left ran off left toward Fishkill Landing. The track showing more plainly at the lower right ran off right toward Danbury.

Everywhere on the Bridge-related rail lines, there was pressure to get milk to its destination, in New York or elsewhere, and to get it there safely and fresh. Under this pressure, railway officials could become edgy about how their milk shipments were handled. When the drivers who brought in milk to one station quarreled with each other, a railway official scolded the station agent for allowing this to happen. When milk was loaded into a rail car unevenly, weighting the car too heavily on one side, an official held the train crew responsible. When the conductor of a mixed milk-and-passenger train kept his train at the Hopewell Junction station for twelve minutes "to do the milk work," an official asked him why it took so long.[5]

Once when a shipper in Millbrook had only ten cases of bottled milk to ship daily, and he wanted them shipped to Yonkers, the ND&C tried a questionable means of shipping it. The Borden's plant at Hopewell Junction regularly shipped 500 cases of bottled milk daily to Yonkers, and for this purpose the ND&C assigned the plant a car. After this car was loaded daily in Hopewell Junction with Borden's large shipment, ND&C agents took the opportunity to add into the car the Millbrook shipper's ten cases, and then sealed the car. After the car had been transferred onto the New York Central's Hudson Line at Dutchess Junction and had reached Yonkers, the seal on the car was opened and the Millbrook shipper's cases were unloaded first. This was done, as Borden's complained, before the Borden's shipment had been unloaded, and without a Borden's representative being present. This meant that the car was "left unprotected, sometimes with the doors open," and Borden's milk disappeared. To stop this loss, ND&C officials reluctantly arranged to send the Millbrook shipper's milk separately from Borden's milk, by a different train.

While most milk moving on the Bridge Route went to the New York metropolitan region without crossing the Poughkeepsie Bridge, significant amounts of milk passed directly over the bridge. In 1891, the Bridge Route added a special milk car onto a passenger train which headed west from West Winsted, CT, daily just before noon. The train made connections with other milk trains as at Canaan on the Housatonic (Berkshire) line, and at Boston Corners on the Harlem line, and then crossed the Poughkeepsie Bridge. It reached Campbell Hall in the evening, according to a newspaper report, in order to attach its milk car "to the milk train on the NYO&W reaching Weehawken in time for delivery to the New York trade."

Similarly, according to CNE Superintendent Martin in 1901, a train which went west over the bridge—a train which carried a few passengers but was run, the superintendent said, "largely for milk"—left Norfolk, CT, at 2:50 PM, and had to hurry to reach Campbell Hall "in order to make the Ontario [&Western] connection for the milk to New York."[6] In the period 1916 to 1920, according to CNE time tables, a "1st class Milk and Passenger" train was running daily over the bridge from Poughkeepsie to Campbell Hall, arriving there about 6:00 PM.

In 1921, the CNE's annual revenue from milk reached a high of $192,000. About this time, according to New York State officials, big milk dealers had long been conspiring to reduce the prices they paid to dairy farmers. Many dairy farmers, including those living in the Mid-Hudson region, felt squeezed by the low prices they received. They also felt pressured to use more sanitary methods in handling their milk, but were not receiving higher prices to compensate them for doing it. Some farmers were quitting the dairy business, and those that stayed with dairying were turning away from shipping by railroad to shipping by truck. In 1923, the Bryant and Chapman creamery in Hartford was reducing its use of the CNE line because of poor service, they said, and was turning to trucks instead. About that time, several creameries on the CNE closed: for example, in Ulster County, the Modena creamery closed by 1921; in Dutchess County, the Locust Creamery in Verbank closed by 1924, and the Beakes Dairy at LaGrange by 1925. By 1926, when both passenger and milk traffic had drastically declined, CNE time tables stopped listing milk trains.

After the CNE was absorbed by the New Haven in 1927, Bridge Route lines were still picking up some milk. By the late 1930s, New York City was still receiving 40 percent of its milk by train rather than truck, but much of it came from far upstate. By about that time, milk cars on the Poughkeepsie Bridge Route were disappearing.[7]

PAGE FROM CNE TIME TABLE, FOR EMPLOYEES, SEPT. 24, 1916. As indicated, at this time trains which carried passengers might also carry milk.

16. BATTLE OVER THE SPRINGFIED BRANCH

(1871-1922)[1]

Trying "every ruse" and enlisting the "keenest legal minds"— for what?

IN 1901, the *New York Herald* declared that Montague Farm, "a little patch of unproductive soil," is the site of one of the "fiercest" battles in railroad history. It is "the Thermopylae of the American railway world today."

On one side of this battle, according to the *Herald*, the warriors are "the powerful combination of brains and money" who wish to lay their railroad tracks through this farm. On the other side, they are "an unseen adversary" who wish to prevent the tracks from being laid. The opposing forces have tried "every ruse," enlisted the "keenest legal minds," and are spending "fortunes" on their battle.

The circumstances which led to the battle reached back at least to 1871 when promoters of the Poughkeepsie Bridge were planning eastward rail connections for it. At that time, the promoters intended that their affiliate railroad, later called the Hartford and Connecticut Western (H&CW), would build a short branch line to Springfield, MA. At first it was planned to branch off from the H&CW's main line at Collinsville, CT. By 1875, it was planned to branch off instead at Tariffville, CT, which would be shorter. From Tariffville it would be only eighteen miles to Springfield.[2]

Having a friendly branch to Springfield, the bridge promoters believed, would provide the Poughkeepsie Bridge Route a significant outlet into New England in addition to the long-planned one at Hartford. In Springfield the Bridge Route would connect with two major railroads, the

UNION STATION, SPRINGFIELD, MA, SERVED AS THE TERMINAL FOR THE CENTRAL NEW ENGLAND'S SPRINGFIELD BRANCH. Postcard, postmarked 1906 (LILL). The station primarily served the New Haven, B&M, and B&A; the CNE entered it on B&A tracks. In use 1889-1925, the station was built of stone, as designed by the Boston firm, Shepley, Rutan, & Coolidge, successors to H. H. Richardson, in Richardson's rough-textured, asymmetrical Romanesque style.

Boston & Albany and the Boston & Maine. They would help the Bridge Route, its promoters came to believe, to escape the heavy hand of the New Haven Railroad which dominated Hartford and considered itself to be in rivalry with the Bridge Route.

For many years, financial troubles hampered the bridge forces from carrying out their plans to build their Springfield branch. By the late 1890s, however, although the Bridge Route owner at the time, the PR&NE, was entering bankruptcy, its officers had become aware that if it were selling out to some other railroad interest such as the New Haven, building the Springfield Branch could help it get a better price. The PR&NE president, John W. Brock, pushed forward to build the branch—it would surely be "of great benefit to this company," he said. By 1898, the PR&NE had already arranged to avoid the expense of building its own bridge across the Connecticut River to Springfield, by leasing the right to cross the Boston & Albany's bridge. Moreover, the PR&NE was

JOHN W. BROCK was a major figure in creating the Poughkeepsie Bridge and its Route, including the Springfield Branch. Top, BROCK'S LETTERHEAD WHEN HE WAS THE PRESIDENT OF THE PR&NE (Beaujon Collection, UCONN). Middle, BROCK HIMSELF (*Poks., The Bridge*, 1889). Below, IN 1899 WHEN THE PR&NE WAS REORGANIZED AS THE CNE, BROCK ALSO BECAME ITS PRESIDENT. While still CNE President, he invested in gold and silver mines in Nevada and by 1903 was president of a railroad serving those mines.

| CENTRAL NEW ENGLAND RAILWAY (CNE) SELECTED DIRECTORS, 1900 |||
From NY State Bd. of Railr'd Com'rs, An'l Rep., 1900		
Although at the time the CNE office was in Poughkeepsie, none of its directors was from that regiion.		
Brock, John W., pres.	Formerly PR&NE pres.	Phila.& Lebanon, PA
Sherwood, Jas. K. O., treas.	Formerly PR&NE receiver	NY City
Appleton, Julius H.	Paper manufacturer	Springfield, MA
Brock, Arthur	Iron manufacturer	Phila. & Lebanon, PA
Sinnott, Joseph F.	Distiller	Phila.

already said to have done the necessary surveying to build the branch from Tariffville to its connection with the Boston & Albany tracks, and to have acquired the necessary right of way, section by section.[3]

In fact, the PR&NE, operating through its affiliate H&CW, had not yet completed the purchase of one small section of the right of way. This section, near East Granby, CT, was an unimpressive strip of unused farmland, only 313 feet long.

The H&CW had already agreed with the owner of this farm on a price. The bridge forces were so confident that this transaction would soon be completed that they had already laid track across this section of the farm, when the owner suddenly told H&CW representatives that he had just sold his entire farm to someone else. He had sold it to Charles C. Montague.

When H&CW representatives looked for Montague, they found him in Amherst, MA. Montague said that he had never seen his new farm, but refused to sell any of it to the railroad at any price.

Why Montague had bought the farm and refused to sell any of it was a mystery to many observers until one day a gang of men appeared at the farm. They put up barbed wire fences at the farm's boundary lines, blocking the H&CW's new railroad tracks. Inside the fencing, they tore up the track already laid on the farm's land. They tore up the track skillfully, with standard railroad tools. They were not farm hands, but railroad men, and their tools bore a name.

Their tools bore the name of the New York, New Haven, and Hartford Railroad. At the time the New Haven Railroad dominated much of southern New England, was gearing up to control more of it, and perceived the Bridge Route's Springfield Branch as threatening its push.

The H&CW, trying to defend its Springfield Branch plans, began proceedings in the Connecticut courts to acquire the necessary strip of land by condemnation. In the proceedings, it was New Haven Railroad lawyers who defended Montague in refusing to sell his land.

Lower courts approved the H&CW's request to condemn the land, but on April 10, 1900, the Connecticut Supreme Court, with the encouragement of the New Haven's lawyers, dismissed the request after all. It had been too long, the court said, since 1889 when Connecticut authorities had approved the location of this branch. The PR&NE having recently been sold to the newly created Central New England Railway (CNE), it would be forced, if it wished to pursue the matter, to reapply, through its affiliated H&CW, for approval of the location of the branch.

Early in 1901, the CNE, with John Brock serving as its president (as he had of the PR&NE before), had virtually completed constructing the Springfield Branch except for the disputed farm section. The CNE decided to bypass the courts by asking the Connecticut General Assembly for permission to seize the farm property by condemnation. However, the New Haven Railroad lobbied the legislature against giving the permission. According to the *Waterbury American*, there was more lobbying on this issue than on any other issue before the legislature at the time. Moreover, the *Hartford Telegram* argued that the new branch would send trade to Springfield, reducing Hartford to "an insignificant station of the Poughkeepsie route."[4] The legislature refused to grant the CNE the right of condemnation.

Continuing to defy the New Haven Railroad, the CNE decided to construct a loop rail line around the farm, expensive though it would be. The CNE created a new railroad company for this purpose, and by December 1901, the new company had won Connecticut's permission to build a four-mile loop line. By September 1902, CNE, using the completed loop, was at last able to run regular trains on its new branch to Springfield.

The New Haven Railroad, having lost its struggle to prevent the completion of the branch, finally allowed the farm owner to sell his disputed farm land to the CNE. The CNE then rebuilt the branch line through the farm, and abandoned its loop line.

By this time, the CNE, partly because of its unexpected Springfield Branch expenses, was in sad financial shape, and the big New Haven Railroad bully—with the backing of the banker J. P. Morgan—was taking advantage of the CNE's weakness to push it to sell its most valuable asset, the Poughkeepsie Bridge. The New Haven wanted to buy the bridge, along with certain of its immediate connecting lines, enough, with other trackage rights which the New Haven could acquire, to allow the New Haven to create its own bypass around congested New York City.

But the CNE, looking out for its investors, refused to sell its Poughkeepsie Bridge or any other assets separately. It wanted to sell all its assets or none. By 1904, the New Haven, succumbing to the CNE's insistence, had bought control of the whole CNE, including what the New Haven did not want, its H&CW subsidiary and its Springfield Branch. The president of the New Haven, C. S. Mellen, then also became president of the CNE. The New Haven now operated the very Springfield Branch which it had fiercely tried to prevent being built.

The *New York Herald* had grandly predicted that when the branch opened, the Poughkeepsie Bridge would see "an endless procession of heavily laden cars journeying eastward over the fine structure," heading for Springfield, "bearing a large share of the enormous coal supply destined to consumers in New England."[5] Once the New Haven Railroad took control of the Bridge Route, however, much of the significance of the Springfield Branch for the Bridge Route was gone. Of course the New Haven Railroad was no longer clogging the Bridge Route's entry into New England.

Not surprisingly, the New Haven Railroad chose not to promote the use of the CNE's Springfield Branch. The New Haven even decided not to promote the use of the CNE's line from which the Springfield Branch ran, the line from the Poughkeepsie Bridge via Canaan to Hartford which had been the Bridge Route's main entry into New England. From the New Haven's point of view, this line was too mountainous, too rural, too little populated, and not strongly enough built. Instead the New Haven chose to push for the use of the more southerly route from the Poughkeepsie Bridge into New England, the route via Danbury. It was through less mountainous, more populated, more industrialized territory. It was the route more useful for a bypass around metropolitan New York. It was also headed more usefully toward connection with the New Haven's big artery, its Shore Line.

When the new Springfield Branch first opened, the Central New England sent a burst of five trains over it daily, each way, whether for passengers or freight. However, after the New Haven acquired the CNE, with the New Haven scarcely promoting the branch either for passengers or freight, the number of trains using it dropped. By 1906, there were only two regular trains, each way daily, on the branch, whether for passengers or freight; by 1917, only one. By January 1922, the New Haven had abandoned the tracks northeast of Feeding Hills, MA, cutting the branch. The Bridge Route no longer carried passengers, coal, or anything else over the branch to Springfield.[6]

In retrospect, the battle over the Springfield Branch may have been lucrative to certain railroad investors, but it was costly to railroad users. It was typical of how railroads and other ostensibly useful enterprises could struggle against each other deviously, fiercely, at great cost, to little public benefit.

CLASSIFICATION TRACKS, MAYBROOK YARD, Jan. 31, 1948, looking southwest (McEN).
At the left is the ice manufacturing plant; at the right, car shops; at the center left, the roundhouse smokestack.

17. MAYBROOK GATEWAY (1901-1940s)[1]

If the air was smoky, the work dangerous, and the men crusty, how could the yard be "paradise"?

THE MAYBROOK YARD, in Orange County, NY, became the gateway through which almost all traffic crossing the Poughkeepsie Bridge had to pass. It was the New Haven Railroad that led in turning Maybrook into a flourishing gateway. Beginning in 1904, when the New Haven Railroad took over the Central New England (CNE), it worked through the CNE to improve the Maybrook yard to make it, rather than the nearby Campbell Hall yard, into the CNE's major western terminus. In doing so, the New Haven also worked with the five other railroads which connected to the Maybrook yard and contributed to its operating costs. These five were the Erie, the Lehigh & Hudson River, the Lehigh & New England, the Wallkill Valley (controlled by the New York Central), and the Ontario & Western (controlled by the New Haven). Four of these railroads entered Maybrook by the same tracks from Campbell Hall.

A factor in the New Haven's choice of Maybrook as the site for a major yard was the abundant level land available there. Another factor was that the New Haven perceived the Lehigh & Hudson River (L&HR), the one railroad which joined the Bridge Route directly at Maybrook, as important for interchange with other railroads. By 1905, the New Haven had joined several other railroads which interchanged with the L&HR to purchase it. Eventually the railroads owning the L&HR came to include, beside the New Haven, the Pennsylvania, Lehigh Valley, Erie, Lackawanna, Central of New Jersey, and Reading. By 1947, there were thirty-eight regularly scheduled trains coming in and out of Maybrook daily, the New Haven Railroad providing the largest number, and the L&HR the second largest. By that time, the New Haven and its associated railroads had developed the Maybrook yard to reach three miles long and include seventy-one miles of track, and it was considered one of the largest and most efficient rail yards east of the Mississippi.

To help the Poughkeepsie Bridge do its job of providing a bypass around New York City, the New Haven not only improved the Maybrook yard but also improved the rail line from Maybrook eastward to the Poughkeepsie Bridge. From there it also improved an entry line into New England—not the bridge's original northern entry route into New England, through Canaan to Hartford, but the bridge's southern entry route, through Hopewell Junction and Danbury. At first much of this route was only single tracked, but by 1914 it was at least double tracked all the way from Maybrook across the bridge through Danbury to New Haven.[2] While once this Maybrook-to-New Haven line had been regarded as part of the Poughkeepsie Bridge Route, gradu-

ENGINES LINED UP BY THE CNE TURNTABLE, AT MAYBROOK YARD, postcard, about 1906 (SIMMS). The turntable displays handles for turning it by hand. The edge of the roundhouse shows at the right.

ally Maybrook became so dominant as the entry point for it that it became popularly known as the Maybrook Line.

FACILITIES

The Maybrook yard was organized, in accordance with the usual railroad practice, as if all the trains running through it ran either east or west, although actually they ran in a variety of directions. At both its east and west ends, the yard had receiving and departure yards. In the middle it had classification yards, in which engines pushed cars over a hump from which, with the help of a brakeman riding each car, they coasted onto appropriate tracks. Scattered about were a shop for repairing cars, a roundhouse with twenty-seven stalls for repairing locomotives, a turntable, a stockyard, coaling and watering facilities, icing platforms, and freight-transfer platforms.

The cars carrying freight of less-than-carload lots, as they arrived in Maybrook, were often sent to the transfer platforms. There the cars would be unloaded, and their cargo sorted and redistributed into other cars, so that, as much as possible, all the cargo in any one car was to go to only one destination.

Refrigerator cars—which railmen called "reefers"—rumbled into the Maybrook yard in large numbers, bringing meat from Chicago, fruit from California, vegetables from the South. By 1910, after an icing plant had been built at Maybrook, Swift, the meat-packing company, kept an expediter in Maybrook to see that its refrigerator cars were promptly iced and sent on. In 1940, as one trainman recalls, the Million Dollar Train, the OB2 which took meat and fruit daily from the Midwest to Boston, came to Maybrook via the Erie line, and was iced fast. This train usually had about ninety cars. A locomotive might push thirty of them onto the icing platform at once. There, conveyor belts would bring in ice cakes from the adjoining artificial ice-making plant. For some refrigerator cars, men would use spiked poles to nudge the cakes into the cars through their side doors. For other cars, they would drop the cakes through roof openings into ice bunkers at the end of each car. In accordance with the specifications in each car's waybill, the men might break up the ice somewhat and add scoops of rock salt for colder temperatures.

If live animals arrived at Maybrook in stock cars, railmen might slide movable ramps into the car doorways and guide the animals out, prodding them if necessary. As Brakeman Sam Christian recalled, railmen would guide the animals into pens to feed, water, and exercise them.

CNE DEPOT, ON MAIN ST, MAYBROOK, postcard, postmarked 1907 (SHUKR). The engine is CNE switcher #100, 0-6-0, built in Schenectady. After about 1930 when the depot ceased to be used as a depot, it was moved nearby to Jewell Ave. where it became a residence.

EARLY CNE YARD OFFICE BUILDING, Maybrook, in the yard on Main St., conveniently near the depot. Postcard, ca.1906 (SHUKR)

Hay and feed were kept available at the yard, as appropriate for different animals, whether they werehogs, horses, cows, or sheep.[3]

YARD WORK

Maybrook yard work, like railroad work generally, was inherently dangerous. An Italian immigrant was working with a mechanical worm lifting coal from a coal pile when his coat sleeve caught in the worm. He could not pull it out. The worm drew in his hand and mangled it. It had to be amputated. When Jack Harris of Campbell Hall, aged thirty-two, was on top of a car handling ice, he lost his balance. He fell off the car, hit his head on a rail, and fractured his skull. Rushed to Goshen Hospital, he soon died, leaving three small children. When a yardman was sitting on a track near a string of coal cars, unexpectedly a switcher shoved those cars over him, killing him. Once in the car shop, a crane was moving a pile of heavy steel sheets, lifting them by chain. To adjust the chain, Phil Favaro was reaching his hand between some of the steel sheets when they happened to close on his fingers, crushing them. Favaro screamed. A worker nearby, "Big John" Grisinger, grabbed a bar. In a feat of strength he was never afterwards able to repeat, he pried up the steel sheets and jerked Favaro's fingers out.

A function of the Maybrook yardmen was to check rail cars passing through. Once George B. Hess of New Paltz was checking a string of coal cars, peering as usual underneath them, when his clothing became caught in a journal box. When the cars began to move, he could not get his clothing free. He was dragged to his death. A Maybrook-based conductor used to tell his crew, "Eat your pie first, boys. You are liable to get killed any minute."[4]

Another function of Maybrook yardmen was guiding cars up over the hump, and then as they coasted down the other side, switching them onto appropriate tracks. To brake them as needed, each day perhaps fifteen to thirty men rode on top of cars as they coasted down the hump. Switchmen set the switches for the cars, to put them into "station order," so that they would be placed in various trains conveniently for dropping off at the appropriate stations. They would place the cars to be dropped off first, closest to the engine, then next those to be dropped off second, and so on.

Albert Alexander recalls that in his early years working as a road brakeman on trains running out of Maybrook, he occasionally volunteered for a day in the Maybrook yard, helping to switch cars.

The work he was asked to do in the yard not being familiar to him, he recalls, he needed advice as to how to do it, but he found the regular yard crew scarcely helped him. In fact, it seemed to him that especially because he was not a regular yardman, the yardmen tried to "embarrass" him.

Once Alexander was assigned to be a brakeman on top of an old car packed with condensed milk cans. It was being pushed over the hump, to coast down into the eastbound classification yard, onto a designated track. Before Alexander's car was cut loose to coast down the hump, Alexander sat on the roof of the car with the brake wheel in front of him, testing the brake, as was customary, by winding the wheel. When a yardman called up to him, "Do you have a brake?" he answered too quickly, "Yes, I have a brake." The yardman then uncoupled the car, letting it roll down the slope and switch into its designated track, track 3. When Alexander twirled his wheel to brake the car, as he recalled afterwards, he gradually realized the brake was not taking hold as it should. Later he understood that the brake chain had a kink in it which was caught, preventing the brake from working.

Meanwhile, the heavy car was picking up speed, probably going at least thirty miles an hour. Though Alexander considered himself young and supple, he believed the car was going too fast for him to jump off safely. He could see that farther down the slope it was going to crash into the first of the string of cars already occupying track 3—the cars onto which his car was supposed to couple. He crawled to the middle of the car's roof. Timing himself carefully, just as his car was about to crash into the string of cars, he jumped up in the air. By the time he came back down, the crash was over. He rolled on top of the car, but stayed aboard and was not hurt.[5]

Much of the Bridge Route's freight tonnage was coal, and when coal arrived in Maybrook, the New Haven forwarded most of it over the Poughkeepsie Bridge, for distribution eastward. However, the New Haven would keep some of the coal at the Maybrook yard, some to fuel locomotives, some to fuel the yard's powerhouse, and some to be stored for emergencies such as coal strikes. To store the coal, an engine would push coal hopper cars up onto a high coal trestle, which would hold forty cars at a time, and they would dump their loads onto piles. The stored coal was usually soft coal. Despite the greater smoke which soft coal spewed into the air, railroads normally burned soft coal in their locomotives. It was cheaper than hard coal, and stoker locomotives, which loaded coal into fireboxes by the use of mechanical worms, would only work with soft coal.

The yard kept records of the cars and cargoes which passed through. In the yard's early decades,

LEHIGH & HUDSON RIVER RAILROAD CHRONOLOGY	
1882	As organized, it reached northeast from Belvidere, NJ, on the Delaware Rover to Greycourt, NY, at a junction with the Erie Railroad.
1889-1890	To connect with the new Poughkeepsie Bridge Route, it extended itself north from Greycourt to Maybrook. At its southern end, to reach into the Lehigh Valley coal region, it secured rights over the Pennsylvania Railroad line from Belvidere, NJ, south along the Delaware River, to Phillipsburg, NJ, and helped open a new bridge from there over the Delaware river into Pennsylvania.
1905	Several railroads took control over the L&HR to ensure their access by it to Pennsylvania coal (at various times these railroads included the New Haven, Lehigh Valley, Pennsylvania, Central of NJ, Reading, and Erie)
1941-1945	Because of a wartime surge in its traffic, the L&HR borrowed locomotives from other railroads

if a shipper in Virginia shipped out a carload of motors to Boston, the consignee in an assembly plant in Boston waiting for those motors would not know for several days where the motors were—the only record of their progress along the rail lines would be a paper "waybill" record, copies of the waybills being attached to the cars they described, and being passed from conductor to conductor by hand along the route. At the end of World War II, the New Haven installed in its Maybrook yard, as well as in its yards in New Haven and Boston, markedly improved long-distance car tracking facilities. It installed IBM punch card devices by which car numbers, cargoes, and destinations were recorded, and a teletype system, by which senders in the form of typewriters transmitted this information to teleprinters which wrote out the messages on a continuous line of paper. With the new system, once the carload of motors arrived in Maybrook, the consignee would be notified of its arrival thus far, and from there on he would be kept informed of its progress along the line so that he would know within a couple of hours when it was likely to be delivered in Boston.[6]

SMOKE, DRINK, AND FAMILIES

In the Maybrook yard's early years, its accompanying village was small and it remained so. Even in the yard's glory years through the 1940s, the village still had only about 1,400 residents.

In the yard's earliest years, many of its employees came to work by foot, bicycle, or horse and wagon. By the 1910s, however, if Maybrook employees lived at some distance in the direction of Poughkeepsie, they could ride to work free, on a special CNE train for employees, popularly called the "Scoot," which ran from Poughkeepsie over the bridge to Maybrook and Campbell Hall. It and other trains stopped in East Walden, where they might pick up additional passengers who had reached there by an electric trolley line from Newburgh. If Maybrook employees lived in the direction of Middletown, by 1921, the CNE had contracted with the O&W for special cars to shuttle them free to work in Maybrook. By the late 1920s, however, as railroad passenger service

WEST YARDS, MAYBROOK, looking northeast, postcard, postmarked 1913 (SHUKR). The Maybrook station is slightly visible behind trees, to the left of the bend in the tracks. Main Street crosses the yard in the center of the picture, from near the station eastward. Guards protected Main St. traffic from passing trains. The smoke showing in the center appears to be from an L&HR train coming in from the right, from New Jersey.

everywhere declined, special passenger services for the convenience of Maybrook employees were disappearing (the Middletown shuttle may have stopped running by 1926, and the Scoot, as we have already seen, stopped running in 1930). By about that time, many Maybrook employees had learned to drive automobiles to work instead. That's about when the yard storekeeper Ed Burns, who lived in Middletown, bought a car to drive to work—it was a1916 Studebaker touring car which in winter, he recalled, even when he put up its side-curtains, was less than cozy.

Up to the 1910s, if road crewmen who were based elsewhere had schedules which required them to stay overnight in Maybrook, they often stayed in boardinghouses. Conductor Nathan Blodgett and some of his crew, when they were working as "extras," stayed at the boardinghouse of Jim McCormick who was an L&HR section foreman. McCormick's had space for twenty-five boarders, at five dollars a week for room and meals, usually several boarders to a room. The boarders, apparently all railroad men, had "a big room to hang out in, "Blodgett remembered, while they waited for "the call boy" to call them to work. In that room they played Pinochle: "No money involved, just a pastime." They also boxed.

One of the boarders, a conductor, "bought a set of boxing gloves one day when he was in Poughkeepsie. We all chipped in and repaid him," Blodgett recalled, "and used them a lot in our recreation room. They were a big help and no one ever lost their temper."[7]

During World War I, the Scoot train, as it took workers to Maybrook, became so crowded that when the 1918 flu epidemic swept the region, it was blamed for spreading the infection, increasing the number of deaths. At that time, the Maybrook yard itself had become so busy that according to Conductor Blodgett, one eastbound train was held up for a month before officials could find a crew to take it to Danbury.

At about the same time, crewmen who ended their work runs in Maybrook found it so difficult to find a place to stay overnight there, that the railroads concerned built a YMCA in Maybrook especially for their benefit. (In the 1870s, YMCAs were beginning to be adapted for railroaders, partly with the hope of providing them recreation that would keep them away from drinking, and by the 1910s there were already more than 200 American railroad YMCAs.) When crews arrived from New Haven with more than twelve hours to lay over before returning to New Haven, the railroad would give them a meager allowance for staying somewhere overnight, in the YMCA if they wished. The YMCA was located conveniently next to the rail yard, and offered a restaurant, pool tables, reading room, and "the best gym around." During World War II, the YMCA served meals to men standing at the counter "four deep," it was said, and when the rooms were all taken, railroaders slept in the lobby and corridors. By 1950, the YMCA had seventy-eight rooms.[8]

The residents of Maybrook became accustomed to living with the snorting of engines, the hissing of air brakes, the crash of cars coupling, and smoke—they dubbed Maybrook "Smoky Center." They also got used to living with crusty railroad men, men who worked at all hours of the day and night, men who could be both fiercely protective of one another and fiercely competitive. They were also men who, living constantly on the edge of danger, were sometimes called to be heroic. Railroading, said President Theodore Roosevelt, demands "heroic virtues."

Almost every family in or near Maybrook

TURNTABLE AT MAYBROOK YARD, WITH AN ENGINE ABOUT TO GO ONTO IT, AND, AT CENTER RIGHT, CONSTRUCTION OF NEW ENGINE HOUSE UNDERWAY, perhaps about 1912 (construction was authorized May 19, 1911, according to CNE, Bd. of Dirs., Minutes). Postcard, postmarked 1921 (SHUKR).

came to have someone working for the railroad. Sons often followed their fathers into railroading. Frank Doolittle, the engineer, had a son of the same name, who before he was sixteen, would sometimes shovel snow from yard switches and ride with his father in his engine cab over the Poughkeepsie Bridge "just for the ride." By sixteen he was a yard messenger, and later, though he was not a railroad enthusiast, he became a brakeman on trains running out of Maybrook over the bridge. Bert Griffin, another CNE engineer, had five boys all of whom went into railroading. One of those boys, Ken, quit school in the 1910s to become a "call boy" at the yard. His job was to go from house to house calling the "extra" men, those not already scheduled, to come to work when needed. He worked his way up to being a yard conductor, and when he retired from the New Haven Railroad in 1966, he was a Maybrook yardmaster.[9]

Maybrook's first labor union, affiliated with the Brotherhood of Railroad Trainmen, was formed in 1904. Blodgett, in his recollections of his twenty years of service on the Bridge Route, recalled only one strike at Maybrook, by the yardmen. It was a painful strike. Two of Blodgett's fellow road conductors, in sympathy with the striking yardmen, refused to work in the yard and thereby lost their jobs. Blodgett recalled one of them as saying, "My name is John Myers. They call me Dingbat. But nobody is going to call me a scab."

Brakeman Christian, like other railroad men, recalls the unions as improving wages, hours, and vacations for Bridge Route workers, including Maybrook yardmen. Partly because of the activity of unions, during Maybrook's heyday its railroaders were well paid, as railroaders generally were. They were often proud of their jobs. They delighted in taking their children to visit the Maybrook yard when they could. Edward F. Burgess, for instance, a brakeman who lived nearby in Campbell Hall, took his little daughter to the yard in the 1940s, and she still remembers that it was exciting to ride an engine as it pushed cars over a hump.

Though alcohol and railroading were a vol-

THE SCOOT, THE NEW HAVEN'S EARLY MORNING PASSENGER TRAIN TO MAYBROOK, traditionally for employees only. Shown in 1929, near the end of its life, parked in the Cottage Street yard, Poughkeepsie (McEN). The long, low, dark building to the left was the Pouvail-Smith plant which made steering wheels for automobiles.

atile mix, Maybrook village came to have what one old-timer recalled as "a saloon on every corner." Railroaders, redolent with "grime and glory," could frequent the Red Onion, Green Turtle, Blue Goose, or House of Blazes, and find prostitutes. A brakeman recalls that when he was young he and his railroad buddies would spend long hours at local dives, where they knew nearly everyone, joining them in drinking beer, playing guitars, and singing country music. A fireman recalls that in the 1960s when the Maybrook yard was declining, if you went in the evening to Whitey's Bar, you could still find plenty of railroaders to share your troubles.[10]

WORLD WAR II

During World War II, the trains passing through the Maybrook yard grew both longer and more frequent. According to Albert Alexander's recollection, freight trains of seventy to one hundred cars, fully loaded, kept heading east out of Maybrook toward the Poughkeepsie Bridge at the rate of one per hour. The number of employees at the yard itself rose to about 1,500 and continued nearly as high into the 1950s.

During the war, Sam Christian recalls that war supplies streaming through the Maybrook yard on

L&NE ENGINE 101, CLASS I-1, WHEELS 0-8-0, 174,300 LBS., NEWLY BUILT BY BALDWIN, Jan., 1913 (McEN). This engine was a camelback. That is, its cab was placed in the middle of the engine, in front of the boiler, rather than as usual behind it. This was intended to help the engineer in the cab see better. However, it meant that the fireman, instead of stoking the boiler's fire as usual from the cab, stoked it from between the boiler and the tender, leaving the engineer to work alone in the cab which in an emergency might be dangerous. Nevertheless, such L&NE engines often brought Pennsylvania coal to Maybrook until from 1947 they were replaced by diesels.

their way to cross the bridge were often headed to New England ports such as New London, Providence, or Boston, to be shipped overseas. They included carloads of jeeps and carloads of food like chocolate bars intended for military rations. They also included solid trains of gasoline tank cars. These, often previously discarded cars brought back into service, were rusty, and if they were found to be leaking on arrival at Maybrook, might have to be repaired on the spot. A smaller portion of war supplies moved through the yard in the other direction, some heading for the Pacific Coast, for shipment overseas from there.

Troop trains also came through, usually heading for eastern ports, taking fresh young men to God only knew what horror. Also, in smaller numbers, prisoner-of-war trains came through, carrying Italians and Germans, many of them captured in Africa. While Christian himself was born in Maybrook, his parents were born in Italy (his family name was originally Christiano). He could speak a little Italian, and he felt an impulse to talk to the Italian prisoners. But the cars were guarded and their windows shut. However, Frank Doolittle Jr., then a yard clerk, remembers German prisoners giving him letters to mail—he mailed letters for some of them, watching out that he didn't get caught doing it.

During the war and later, there was some pilfering at the yard, sometimes by railroad men themselves. They might snitch from the cargo of a car a rationed item such as chocolate or soap. They might hide it nearby, and return at night to pick it up.

LEHIGH & NEW ENGLAND RAILROAD (L&NE) CHRONOLOGY	
1882	The South Mountain Railroad reorganized as the Pennsylvania, Slatington, & New England (PS&NE). It soon graded most of its route from Slatington, PA, to Pine Island, NY, intending to connect with the proposed Poughkeepsie Railroad Bridge
1885-1887	PS&NE went into receivership and was reorganized as the Pennsylvania, Poughkeepsie, and Boston (PP&B), with the Poughkeepsie Bridge promoter W. W. Gibbs as president
1890	PP&B inaugurated service from Slatington 131 miles through to Campbell Hall, NY, connecting there with the Poughkeepsie Bridge Route
1894-1895	The PP&B reorganized as the Lehigh and New England (L&NE). Once the Maybrook Yard was developed, it often carried coal to Maybrook, for forwarding over the Poughkeepsie Bridge
1961	Largely because of the decline in coal traffic, the L&NE ceased operating

TWO BOYS

In the late 1940s, two New Jersey boys, teenagers, kept hearing about Maybrook as an exciting yard where several railroads met. As rail fans acquainted only with the Erie line on which they lived, they were eager to know other rail lines, and especially their engines; the boys even knew particular engines they would like to see. One morning they set out for Maybrook, carrying brown-bag lunches, railroad time tables, and cameras. They rode the Erie Railroad as far as Goshen, and from there they thumbed auto rides.

When they arrived at the sprawling Maybrook yards, they sought the main office. They went in cautiously, uneasy as to how they would be received.

To their surprise everyone seemed friendly. "Especially helpful" was one official who handed them "all the old employee timetables he could find."

The boys went into the engine shops. There they saw one of the steam locomotives they were especially looking for—an L&HR Baldwin 4-8-2, No. 12. They were in awe of it. It was more modern than any of the steam engines on the Erie which they knew. Built in 1944, it had "73-inch drivers and a tractive force of 67,000 lbs," and was called a "Mountain." They also saw a New Haven 2-10-2, No. 3230, "feed water heater and all," which was "impressive." Behind the engine house, apparently on its way to take a train out of the yard, they saw L&HR No. 82 with a "dog house," that is, a brakeman's shanty on its tender.

Farther on, they noticed the large railroad YMCA, which they thought was for railroaders only, and so did not go in. It was only on later visits that they found they were welcome to go in to take meals there, sit next to railroad "heroes," and listen to their stories.

Out in the yards, they watched a New Haven switcher do its "endless switching chores." They also watched an arriving L&HR freight "cut its well-maintained wooden caboose off on the fly, always a sight to see." By the time the boys headed home, they were enthusiastic about Maybrook, more than some Maybrookers were. Maybrook, they decided, was "a rail paradise."[11]

MAYBROOK YARD CHRONOLOGY	
1904-1913	CNE improves the yard, including tracks, signals, ashpit, water storage, transfer platforms, repair shops, icing and coaling facilities
1914	CNE completes doubletracking the "Maybrook Line" from Maybrook over the bridge through Danbury to New Haven
1914-1918	During World War I, the yard booms. The CNE extends the yard's tracks, builds new transfer platform
1930	The New Haven ends all passenger service over the bridge to Maybrook, including the Poughkeepsie-Maybrook train for yard employees
1939-1945	During World War II, yard traffic booms again, as in munitions, food, fuel, and troops
1960-1968	Pressured by competition from autos, trucks, and airplains, the New Haven shrinks the line from Maybrook over the bridge to Danbury from double to single track and closes sections of the yard

BRAKEMAN SAM CHRISTIAN, RIDING ON THE BACK OF A CABOOSE, about 1973 (CHRIS). During World War II when he saw Italian prisoners of war passing through Maybrook, he wanted to talk to them.

FREIGHT TRAIN HEADING OFF THE BRIDGE, POUGHKEEPSIE, 1892: CNE&W LOCOMOTIVE #14, 4-6-0 (STICK). While the engine, built by Baldwin in 1891, at first served the CNE&W, during most of the 1890s it served the PR&NE, and from 1899, the CNE. Standing was engineer Edward Smith (known as "Hungry Ed" because of his craving for bananas); seated on the cowcatcher was his fireman. The barrels to the left of the train (one near the tender, one farther back) were water barrels placed at intervals along the bridge, ready for fire fighting.

18. FREIGHT TRAINS (1890s-1950s)[1]

Apples for Boston, oil for Fall River, planes for the West Coast.

IN 1905, some 400 freight cars a day were crossing the Poughkeepsie Bridge. By 1921, 2,000 cars were. By 1943, when World War II increased railroad use, nearly 3,500 were.

In weight, the freight carried by the principal Poughkeepsie Bridge Route railroads also grew dramatically over the years, though not as much as the bridge promoters had hoped. While separate records for the freight which crossed the bridge are not regularly available, in the years 1891 to 1903 the Bridge Route railroads' annual freight ranged from about 300 to 700 thousand tons. After the New Haven Railroad took control of the Central New England, however, the CNE's freight increased by 1917 to 7 million tons, and *Engineering News* called the bridge traffic "heavy." By 1952, though railroads were being squeezed by competition from trucks, freight crossing the bridge alone had gone up to nearly 17 million tons, and many of the trains were so long that they stretched all the way across the river from shore to shore.[2]

Long through freights, carrying the bulk of the tonnage, sometimes ran at night, while shorter local freights, carrying comparatively little tonnage, often ran by day. Some freights were likely to be slow, including those devoted to coal or oil. Other freights were likely to be fast, including those carrying meat or fruit.

In 1900, the CNE itself owned 383 freight cars, most of them likely to be wooden. By 1913, it owned 1195 cars, many of them likely to be steel. Among shippers who sent their own private cars over the CNE in 1910, meat-packer Cudahy paid

FREIGHT TRAIN CROSSES THE BRIDGE TOWARD POUGHKEEPSIE, AS A POUGHKEEPSIE-HIGHLAND FERRY PASSES UNDERNEATH. Postcard, undated (SHUKR). The ferry stopped running in 1941.

TERMINAL FOR A HUDSON RIVER FERRY WHICH COMPETED WITH THE POUGHKEEPSIE BRIDGE: THE NY&NE'S RAIL-CAR FERRY TERMINAL AT FISHKILL LANDING (NOW BEACON), 1892 (SHUKR). Both the NY&NE and the ND&C Railroads brought rail cars here to the ferry. In the center of the river is the ferry "Hart" which moved such rail cars across the river to the Erie Railroad's terminal in Newburgh. The tracks in the foreground are the NY Central's Hudson Line, with which the NY&NE and the ND&C connected; the Hudson Line's passenger station (at the same site as the present Beacon rail station) was slightly to the north, off the picture to the right.

the largest amount for doing so, meat-packer Armour the second largest, the cars of both being refrigerator cars. Berwind White Coal Mining paid the third largest amount, its cars being coal cars.

As on many east-west rail routes, more freight crossed the Poughkeepsie Bridge eastward than westward, many cars going west being empty. In the early decades of the 1900s, the traffic moving into New England was likely to be raw materials, especially coal, but also oil, grain, lumber, beef, and farm produce, some from as far away as Texas and California. The traffic moving out of New England was likely to be lighter, consisting especially of manufactured goods, but in season including Atlantic fish, Maine potatoes, and Cape Cod cranberries.

Overwhelmingly, most freight tonnage on the Bridge Route originated not on the Bridge Route itself but on connecting lines. Because most freight was through freight, freight trains often rumbled through Bridge Route stations without stopping. Nevertheless, local freight, either freight originating on the Bridge Route or destined for it, while small in tonnage, was often valuable, and was important to the local communities concerned.

In the 1890s and early 1900s, the Newburgh, Dutchess, and Connecticut (ND&C) interchanged considerable traffic with the Bridge Route, as in Dutchess County, NY, at Hopewell Junction, Millerton, and Pine Plains. The ND&C also competed fiercely with the Bridge Route. It tried to bring in Pennsylvania coal by crossing the Hudson not by the Poughkeepsie Bridge but by the Newburgh-Fishkill Landing rail-car ferry, and in cooperation with the P&E, to distribute the coal in Dutchess County. The ND&C also tried to use the Newburgh ferry rather than the Poughkeepsie Bridge to ship out the products of Dutchess County factories, such as those located along the ND&C tracks in Matteawan. In the course of their competition, the Bridge Route's General Freight Agent W. J. Martin accused the ND&C of deliberately intending to "jab" the Bridge Route, while the ND&C's General Freight Agent William Underhill accused the Bridge Route of not adhering to its published rates, but instead doing "vicious" secret rate cutting to favor certain customers.[3]

By 1904, the natural advantage of the Poughkeepsie Bridge had drawn so much freight away from the Newburgh rail-car ferry that the New Haven Railroad, which by this time controlled the ferry, closed it. At the same time, the Bridge Route, by being taken over by the New Haven Railroad, became part of an extensive and efficient rail system. Thereafter, the New Haven, focusing on the Bridge Route's potential for freight, not passengers, strengthened the Poughkeepsie Bridge, developed the Maybrook yard, by 1907 acquired both the ND&C and the P&E and handed them over to the CNE to operate, and by 1912, having long before gobbled up the NY&NE, also handed over to the CNE what once had been the NY&NE's Hopewell Junction-Danbury line.

FRUIT

Since in its early years the Bridge Route directly served a largely rural region, of course much of its local freight was farm-related. In the early 1900s, along the east end of the CNE's northern claw in Connecticut, in the flatland near Hartford, the farm towns shipped out tobacco— Bloomfield shipped out some, Griffin shipped out more. In New York State in Columbia County in 1900, farmers in the Ancram and Copake region shipped out 34,000 pounds of Thanksgiving turkeys.[4] However, fruit-growing regions in New York State both east and west of the bridge provided bigger business for Bridge Route trains.

East of the bridge, such Dutchess County stations as Clinton Corners and Red Hook shipped out apples by the Bridge Route, Red Hook as many as ten or eleven carloads of them a day. These apples were likely to go south to New York City without crossing the Poughkeepsie Bridge. Just west of the bridge in Ulster County's fruit-growing region, where fruit was the major industry, in the Bridge Route's early years, much of the fruit shipped out was grapes, later more often apples. Catering to the fruit growers, the railroad posted market quotations daily in appropriate stations, keeping farmers aware of where they could obtain the best prices for their fruit, whether by shipping across the bridge or not.

In 1892, during the fruit season, two trains daily picked up shipments of fruit along the Bridge Route between the bridge and Maybrook. One train went east across the bridge, taking fruit for delivery in New England, as in Waterbury,

HOPEWELL JUNCTION YARD, postmarked 1909, when it served the CNE & the New Haven (postcard, SHUKR). The CNE tracks at the left, formerly ND&C, ran northeast (upward) toward Millerton and southeast (downward) toward the Fishkill Landing (Beacon) area. In the background center left, are the white buildings of Borden's Creamery (cf. Ch. 15)

LOADING GRAPES FROM FARM WAGONS INTO A TRAIN at the Bridge Route station, Clintondale, Ulster County, NY. Daniel Donaldson's vineyard is in the background (WEBER). Donaldson family tradition dates the photo 1891; it was published in CNE, *Summer Homes*, 1899. The cars at the left are marked Lehigh Valley, to the right Central Railroad of NJ.

Hartford, Providence, and Boston. The other went southwest to Maybrook, taking fruit for delivery especially in New York City and Philadelphia. Both trains made their pickups in the afternoon, for delivery the next morning.[5]

In 1908, the CNE, struggling to attract fruit business in competition with Hudson River shipping, advertised that it shipped apples to New York City. From anywhere west of the Poughkeepsie Bridge, the CNE offered a route over the bridge, then by New York Central's Hudson Line to New York City. Instead of switching such loads onto the Hudson Line immediately in Poughkeepsie, the CNE preferred to keep them longer on its own lines, often switching them onto the Hudson Line only after sending them south through Hopewell Junction to the Fishkill Landing area. During the fruit seasons of the 1920s, when the fruit-producing region just west of the bridge was said to be the second most productive such region in the nation, second only to southern California, as many as ten refrigerator cars of fruit per night went out from the Clintondale station.[6]

A farm family situated west of the bridge, the Andrew L. F. Deyo family of Gardiner, often shipped out their fruit by train. Living about equidistant from the Wallkill Valley Railroad station at Gardiner and the Poughkeepsie Bridge Route station at Modena, the Deyos shipped out by both rail lines. In the fall of 1898, they picked five hundred barrels of apples. One day in October, according to a Deyo family diary, "Andrew is sending off apples. He has taken two car loads to Gardiner and next will take a car load to Modena." The apples sent via the Modena station were more likely to go to New England; those sent by the Gardiner station were more likely to go to New York City.

During the apple season of 1917, the Deyos employed as many as forty-eight farm workers, local men, among them a few African Americans. These men not only picked apples but also sorted and packed them, all right on the farm. They packed them occasionally in baskets, more often in barrels. They packed more than 4,000 barrels that season. One November day, the Deyos' hired man Charles drove a wagon six times to Modena, just over a mile, hauling twenty-four barrels at a time.

By 1920, the Deyos had abruptly changed how they shipped out their apples. A fruit growers'

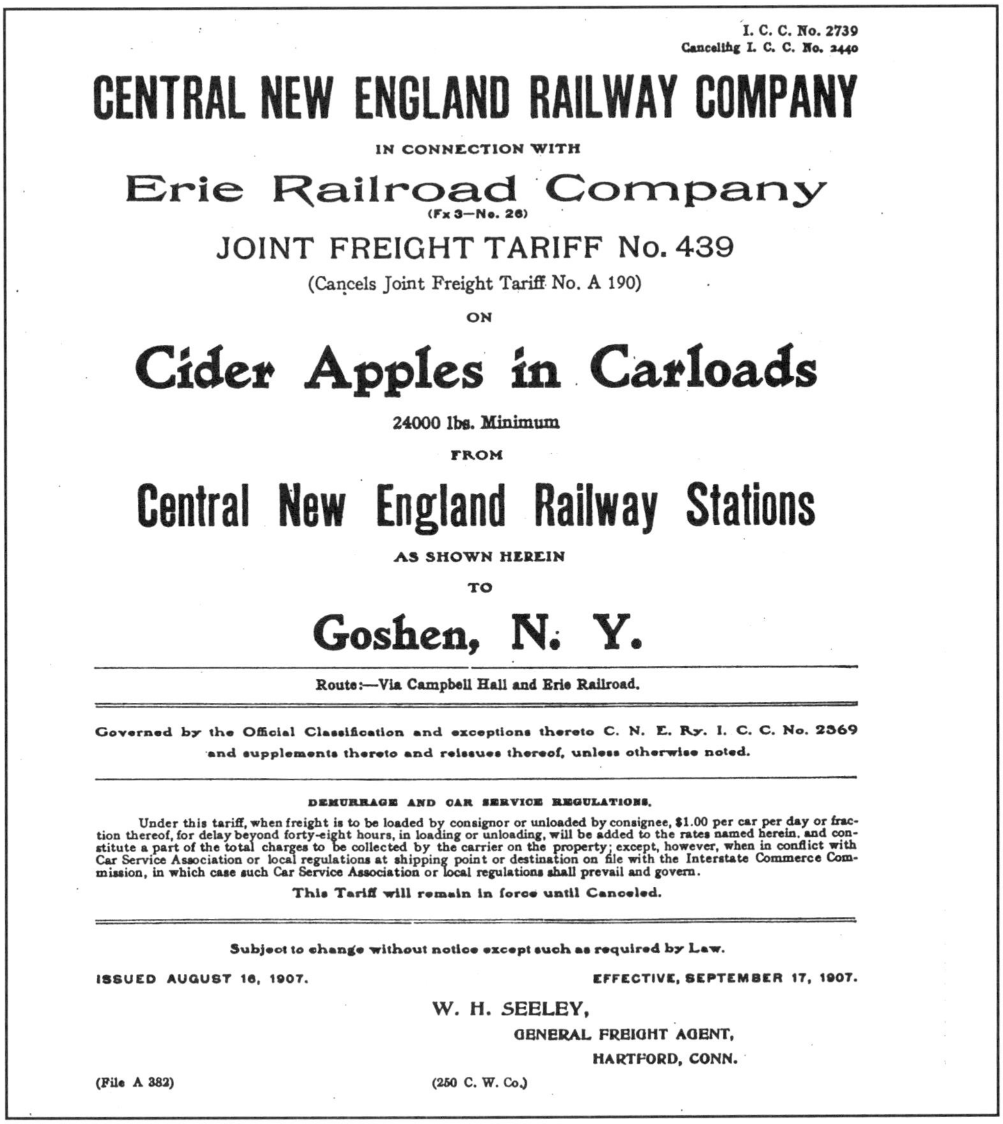

I. C. C. No. 2739
Cancelling I. C. C. No. 2440

CENTRAL NEW ENGLAND RAILWAY COMPANY

IN CONNECTION WITH

Erie Railroad Company
(Fx 3—No. 26)

JOINT FREIGHT TARIFF No. 439

(Cancels Joint Freight Tariff No. A 190)

ON

Cider Apples in Carloads

24000 lbs. Minimum

FROM

Central New England Railway Stations

AS SHOWN HEREIN

TO

Goshen, N. Y.

Route:—Via Campbell Hall and Erie Railroad.

Governed by the Official Classification and exceptions thereto C. N. E. Ry. I. C. C. No. 2569 and supplements thereto and reissues thereof, unless otherwise noted.

DEMURRAGE AND CAR SERVICE REGULATIONS.

Under this tariff, when freight is to be loaded by consignor or unloaded by consignee, $1.00 per car per day or fraction thereof, for delay beyond forty-eight hours, in loading or unloading, will be added to the rates named herein, and constitute a part of the total charges to be collected by the carrier on the property; except, however, when in conflict with Car Service Association or local regulations at shipping point or destination on file with the Interstate Commerce Commission, in which case such Car Service Association or local regulations shall prevail and govern.

This Tariff will remain in force until Canceled.

Subject to change without notice except such as required by Law.

ISSUED AUGUST 16, 1907.　　　　　　　　EFFECTIVE, SEPTEMBER 17, 1907.

W. H. SEELEY,
GENERAL FREIGHT AGENT,
HARTFORD, CONN.

(File A 382)　　　　(250 C. W. Co.)

JOINT CNE AND ERIE RATES FOR SHIPPING CIDER APPLES TO GOSHEN, Orange County, NY, 1907, published in accord with Interstate Commerce Commission rules (SHUKR). The ICC, created in 1887, originally had little power. But by 1907, public awareness of railroad abuses had led to giving it more power to regulate railroad rates.

cooperative had built a cold storage plant—a mechanically cooled plant—beside Clintondale's Bridge Route station, which was about five miles from the Deyo farm. The Deyos took part in building the plant, running it, and often took their fruit there. They and their neighbors for miles around could conveniently store their fruit there and ship it out by the Poughkeepsie Bridge Route whenever they could find the best prices, which might be in the winter or spring. The co-op's office was at first located in the station, enabling W. J. Margraf to double as station agent and co-op manager.

Through 1924, the co-op shipped out all its fruit by the Bridge Route. In the fruit season of 1925, when it was still shipping out five train cars of grapes a day, it also began to ship out fruit by truck. By 1937, when despite the depressed economy, the Clintondale co-op was doing more business than ever, it was still receiving carload lots of farm supplies by train, including chemical fertilizer. But it was no longer shipping out any fruit by train, only by truck.[7]

WEST END LOCAL

A train which was too slow to be efficient in carrying fresh fruit was the CNE's West End Local Freight. Carrying miscellaneous cargo, it left Canaan, CT, at 7:35 AM, heading west for the Poughkeepsie Bridge. It made frequent stops at small stations along the line, doing whatever switching was necessary at each station. At Salisbury, CT, as Nate Blodgett, who was its brakeman from about 1911 to 1913, recalled, its crew sorted out the less-than-carload-lot local freight on a platform which was open to the weather. They would sort it into a different cars, as into one to be set off at Silvernails, NY, for the Rhinecliff branch, another to be set off at Stissing, NY, for the Beacon (Fishkill Landing) branch (formerly the ND&C), another to be set off at Pough-

L&HR ENGINE 91 PULLING FREIGHT ACROSS THE O&W LINE AT BURNSIDE, Orange County, NY, north toward Maybrook, for forwarding over the Poughkeepsie Bridge (McEN).

keepsie, and another to be set off at Highland but intended for Maybrook.

After the local crossed the Poughkeepsie Bridge and reached its final destination, Highland, at 5:00 PM, Blodgett and his crew would drop any remaining freight cars, and then run their engine and caboose back across the bridge to Poughkeepsie. There they would tie up for the night, in a yard with better facilities to service their engine than were available in the little yard in Highland. Blodgett and another of his crew would usually plan to sleep overnight in their caboose, and in the meantime, walk out to take their supper in a restaurant at the corner of Main and Washington Streets. They would sit at the counter. Blodgett would ask for steak and potatoes. After he had consumed them, Blodgett, still hungry from a day's hearty work, would ask for a second order of the same.

In the morning in the Poughkeepsie yard, the yardmaster would tell the crew what freight he had ready for the return east-bound trip, so the crew could allow for that in making up their return train at Highland. By 8:10 AM the crew would have taken their engine and caboose back over the bridge to Highland, and have made their train ready to take off eastward, across the bridge once again, back toward Canaan.[8]

IRON AND LIME

From the 1890s into the 1920s, the Bridge Route fostered the mining of both iron and limestone in the northwest corner of Connecticut and over the border into nearby New York State. Bridge Route trains brought in coal to fire the region's lime kilns. They brought in wood for coopers to make the barrels into which the kilns packed their lime. They also carried out the packed barrels. At busy times, the Connecticut Lime Company alone shipped out more than one hundred barrels of lime a day.

Bridge Route trains moved iron ore from this region's mines to its blast furnaces. They also brought charcoal over the bridge to fire those furnaces—in 1902, there were often twenty to forty cars loaded with charcoal waiting at Maybrook to be shipped to the Barnum-Richardson furnaces in East Canaan. When those furnaces were ready to ship out their pig iron, a company weigh master would direct the loading of it into rail cars—often the same rail cars which had brought in charcoal. To get the correct weight into each car, the weigh master would sing out, in a ritual, either "throw off some," or "throw on some." When finally he was satisfied, he would sing, "seal her up." The Bridge Route sent out many such cars of pig iron to Chicago to be manufactured into railroad-car wheels, until the local iron ore became depleted and Mesabi Range ore flooded the markets.[9]

SYRUP, PAPER, BOOKS

Once the New Haven Railroad had come in control of the Bridge Route and was actively improving it, new factories sometimes sprang up

---

Barnum=Richardson Co.

SOLE MANUFACTURERS OF

SALISBURY CHARCOAL

PIG IRON,

ESPECIALLY ADAPTED BY REASON OF ITS TENSILE STRENGTH AND CHILLING PROPERTIES, FOR CAR WHEELS, ORDNANCE MACHINERY AND OTHER SIMILAR HIGH-CLASS WORK.

✤ ✤ ✤ ✤

ALSO MANUFACTURERS OF

CAST CHILLED

CAR WHEELS,

MADE FROM THE SALISBURY IRON IN THE
BARR CONTRACTING CHILL,
WITH TREADS GROUND TO AN ACCURATE CIRCLE.

✤ ✤ ✤ ✤

GREY IRON CASTINGS

OF ALL DESCRIPTIONS.

✤ ✤ ✤ ✤

GENERAL OFFICE,

LIME ROCK, :-: CONN.

MILO B. RICHARDSON, President.
CHARLES W. BARNUM, Vice-President.
PORTER S. BURRALL, Sec'y and Treas.

120

ADVERTISEMENT in CNE, *Summer Homes*, 1900

along it. In 1911, the *Poughkeepsie Eagle* reported that "nearly all our new factories are located along the CNE tracks, and have been induced to come here because of the additional transportation facilities the great Poughkeepsie Bridge affords." In this period, transportation lines, including railroads, boats, and trolleys, helped to concentrate industry, commerce, and population in cities, and as a result, for several decades Poughkeepsie's downtown, like that of many other Northeastern cities, flourished.

In Poughkeepsie one of the plants located along the Bridge Route was the Smith Brothers Cough Drop plant, on North Hamilton Street. In the 1940s and 1950s, the Smith Brothers often brought in over the Poughkeepsie Bridge tank cars of corn syrup from Decatur, IL, to make their cough drops, and wax paper from Kalamazoo, MI, in which to wrap them. They also shipped out over the bridge carloads of their cough drops to distribution warehouses, as in Chicago and Seattle. It could take two or three days for delivery in Chicago, six for Seattle.

In the early 1950s, when Arthur B. McComb was the Smith Brothers' traffic manager, the New Haven Railroad president, Frederick C. Dumaine Jr., invited him and other traffic managers to lunch in the dining car of his presidential train. In a dramatic gesture, Dumaine arranged to have his train move out to the middle of the bridge and let it sit there while lunch was served, high over the Hudson.[10]

THE SMITH BROTHERS, OF POUGHKEEPSIE: THEIR ADVERTISING MADE THEIR BEARDS KNOWN WORLDWIDE. They used the Bridge Route to bring in corn syrup and ship out cough drops (McCO).

About the same time, Brakeman Sam Christian recalls, he crewed on trains which were forwarding boxcars of paper from Maybrook across the bridge to Poughkeepsie for Western Printing. His crew "spotted" cars for Western Printing right inside their factory building, which was on the Bridge Route's Hospital Branch; its doors would open for them to shove cars right in. Western Printing shipped out books by both the New York Central and the Bridge Route—among the books it shipped out on a large scale were the famous Little Golden Books for children.

As early as the 1910s, Poughkeepsie meatpackers, the Knauss Brothers, brought in live hogs by train. In the 1930s to 1950s, they brought them in especially from Ohio and Indiana, usually weekly, in two or three double-decker cars. According to the recollection of one of the Knauss family, they usually brought them in via the Erie Railroad to Maybrook and then by the New Haven Railroad over the bridge. The Bridge Route offered Knauss slightly better prices for bringing in hogs than NY Central did, especially because Knauss was located right on the Bridge Route's Hospital Branch. Knauss also brought in coal from Pennsylvania by the Bridge Route to fuel their plant, one car every three or four months. In addition, once or twice a year they shipped out over the Bridge Route tank cars of "grease," a by-product of their hog slaughtering. They shipped it to a factory in New Jersey which used it to make soap.[11]

Over the long term, however, Poughkeepsie expected more of the Bridge Route than it delivered. Bridge Engineer Gustav Lindenthal, speaking in Poughkeepsie in 1921, told Poughkeepsie Bridge promoters that they had been "gullibies" not to be skeptical about the benefits the bridge would bring them, "with trains passing over your head 200 feet high," not stopping, but "passing on." In 1925, the Poughkeepsie Chamber of Commerce, recalling that Poughkeepsie had led in fighting "hard battles" to build the bridge, doubted that the bridge had brought the "substantial benefits" which Poughkeepsie had expected. Poughkeepsie's population did not double in the five years after the bridge opened, as historian

Lossing had predicted. In fact, more than sixty years later in 1950, it had still not quite doubled. Poughkeepsie and many other Northeastern localities gradually learned that their manufacturing was declining, and that rail lines passing through them did not assure them prosperity as much as they assured them noise, smoke, and danger. While in 1951 the *Poughkeepsie New Yorker* said that Poughkeepsians regarded their railroad bridge with "affection" and it was still "serving nobly," by about that time the automobile, with its dispersal effect, was already bleeding many city centers, including Poughkeepsie's, foreshadowing blight to come.[12]

COAL AND OIL

It would be difficult to overstate the importance of coal for the Bridge Route's early years. The proportion of coal in the route's traffic tonnage went up from 42 percent in 1890 to 61 percent by 1918, and 62 percent by 1926.

During the depression of the 1930s, according to the recollection of the Mylod family who lived in Poughkeepsie next to the bridge, there was a concentration of people in the neighborhood who were unemployed and sometimes unable to heat their dwellings. Some trainmen identified with them. During the harshest periods of winter, passing trains seemed "accidentally" to spill coal down through the bridge to the ground. It quickly disappeared.

During World War II, when the threat of German submarines cut into the usual shipping of coal and oil by water along the Northeast coast, a fierce demand for fuel developed in New England. To meet the demand, increased coal and oil shipments passed over the bridge. Under the continued pressure of this demand, in 1943, a new oil pipe line—known as the Big Inch—was built from Longview, Texas, north to Norris City, in southern Illinois. There, a flood of oil from the pipeline was pumped into tank cars, often rusty, grimy old cars, resurrected from storage. From

Selected Commodities as
Percent of Total CNE Freight Tonnage
From NY State, Bd. of Railr'd Commrs.,
An'l Rep., 1901; CNE, "An'l Rep.," 1918, 1926

	1901	1918	1926
Coal, Coke	56.4%	61.0%	61.8%
Metal Products (pig iron, machinery	7.3	4.1	4.8
Cement, Brick, Lime		3.9	7.6
Grain, Flour	4.5	3.5	2.7
Lumber	3.8	1.8	2.1
Meat	1.6	1.1	1.9

CENTRAL NEW ENGLAND RAILWAY (CNE) CHRONOLOGY

1899	CNE succeeds PR&NE as owner of the Poughkeepsie Bridge and connecting lines
1902	CNE opens branch to Springfield, MA
1904	New Haven Railroad takes control of CNE but allows it to continue its separate existence
1904-1913	CNE develops Maybrook yard as major gateway from the west and south over the Poughkeepsie Bridge into New England
1906-1907	CNE strengthens Poughkeepsie Bridge
1907	CNE acquires the Poughkeepsie & Eastern, the Newburgh, Dutchess, & Connecticut, and the Dutchess County railroads
1912	CNE extends its southern approach toward New England by leasing the New Haven's line from Hopewell Jct. to Danbury, CT
1917-1918	During World War I, CNE, with its traffic booming, further strengthens the Poughkeepsie Bridge
1927	The CNE, beginning to shrink, is absorbed by the New Haven, and ceases to exist

ADVERTISEMENT in CNE, *Summer Homes*, 1900

there many of the cars moved in long trains over New York Central tracks as far as Marion, Ohio, then over Erie tracks to Maybrook, NY, and then over the New Haven tracks across the Poughkeepsie Bridge, and on to such coastal ports as Providence and Fall River. These oil trains, like coal trains, were slow, their round trip from Illinois to the Atlantic coast and back taking ten days.[13]

SYMBOLS AND "HOT SHOTS"

In the early 1900s, the Central New England claimed that its Poughkeepsie Bridge Route was "in a position to move all freight East and West of the Hudson River with perfect despatch. "In doing so, it explained, the Bridge Route connected with "fast freights" which reached vast areas of the continent. From Campbell Hall, for example, the CNE said, it connected through the Erie Railroad with the "Erie Despatch," and through New York Central's Wallkill Valley branch with the "Grand Trunk Despatch." From Maybrook, it

FREIGHT PASSING RELYEAS STATION, NEW HURLEY RD., ULSTER COUNTY, NY, HEADING FOR MODENA AND THE POUGHKEEPSIE BRIDGE, 1936 (SHUKR). In the early 1920s, this station, a very rural one, had a White Cross Creamery beside it, but by this time it was gone.

STATIONS		AGENTS OR REPORTING STATIONS
p‡Hartford, Jt.,N.Y.,N.H.&H.	Conn.	W. S. Kelly, Ticket. R. T. Lambert, Bag. Mas. C. A. Frost, Freight.
336N	"	
Cottage Grove	"	Bloomfield.
Bloomfield	"	E. W. Miller.
Griffins	"	H. V. Perry.
Barnards	"	Griffins.
Tariffville	"	Thomas McDonough.
Springfield Branch		
East Granby	"	W. H. McNamara.
Sheldon Street	"	
West Suffield	"	W. J. McLaughlin.
Freemans	"	
Feeding Hills	Mass.	C. J. Mulvihill.
Agawam Junction	"	
Hoskins	Conn.	Simsbury.
Simsbury, Jt.,N.Y.,N.H.&H.	"	H. J. Osborne.
Strattons	"	Simsbury.
Canton	"	Collinsville.
High Street Junction	"	
High Street	"	
†Collinsville	"	W. F. Mack.
Cherry	"	Collinsville.
Pine Meadow, Jt., N.Y.,N.H.&H.	"	R. E. Coderre.
New Hartford, Jt., N.Y.N.H.&H.	"	John Keefe.
Barkhamsted	"	New Hartford.
East Winsted	"	
p‡Winsted,Jt.,N.Y.,N.H.&H.	"	W. J. Walker.
West Winsted	Conn.	
Lawrence	"	
Grant's	"	
Summit	"	
x†Norfolk	"	C. W. Van Buskirk.
Haystack	"	
Whiting River	"	East Canaan.
East Canaan	"	J. L. Merritt.
p‡Canaan, Jt.,N.Y.,N.H.&H.	"	E. C. Finney.
Washining	"	
Twin Lakes	"	Taconic.
Taconic	"	L. E. Stark.
Salisbury	"	E. Ashman.
xLakeville	"	T. Martin.
Ore Hill	"	Lakeville.
State Line	N.Y.	
Mount Rigs	"	
Boston Corners, Jt., N.Y.C.R.R.	"	Boston Corners. R. H. Lind.
Tanner's	"	
Halstead's	"	
Ancram Lead Mines	"	C. J. Hoysradt.
N. D. & C., Junction	"	
Briarcliff Farms	"	Pine Plains.
Attlebury	"	Stissing Jct.
Stissing Junction	"	F. H. Wilsey.
McIntyre	"	Stanfordville.
Stanfordville	"	B. Wheeler.
Willows	"	Stanfordville.
Upton Lake	"	Clinton Corners.
Clinton Corners	"	O. M. Berrington.
Webb Farms	"	Clinton Corners.
Salt Point	"	R. Champlin.
Pleasant Valley	"	G. S. House.
Van Wagner's	"	Pleasant Valley.
Poughkeepsie Junction	"	
Hopewell Branch		
Manchester Bridge	"	Poughkeepsie.
Briggs	"	Poughkeepsie.
Didell	"	Hopewell Jct.
Fishkill Plains	"	Hopewell Jct.
Hopewell Jct.	"	S. W. Lesher.
Stormville	"	W. J. Mahar.
Green Haven	"	
Poughquag	"	Poughquag.
West Pawling	"	D. C. Wooden.
Whaley Lake	"	Poughquag.
Holmes	"	C. H. Cole.
West Patterson	"	Holmes.
Towners	"	J. J. Lynch.
Dykemans	"	
Brewster	"	S. Gallagher, Ticket. W. C. Hogan, Freight.
Mill Plain	"	T. J. Murphy.
Fair Ground	Conn.	Danbury.
p‡Danbury Jt.,N.Y.,N.H.&H.	"	F. W. Pierce, Ticket. J. D. Cunningham, Frt.
x†Poughkeepsie	N.Y.	A. Fraleigh.
Highland	"	J. R. Melius.
Loyd	"	W. J. Margraf.
Clintondale	"	W. J. Margraf.
Elting	"	Clintondale
Mowbray	"	Modena.
Modena	"	Rufus Jenkins.
Relyeas	"	Modena.

connected via the Lehigh & Hudson River with both the "Southern States Despatch" and the "Central States Despatch."

During World War I, two "extra" freight trains moved regularly out of Maybrook over the bridge, and then onto the CNE's northern claw through Canaan, CT. One headed for East Hartford, CT. The other, heading for Northampton, MA, was set

STATIONS		AGENTS OR REPORTING STATIONS
St. Elmo	N.Y.	East Walden.
East Walden	"	H. A. Beatty.
Berea	"	Maybrook.
†Maybrook,"Campbell Hall, Trans." Jt.,N.Y.,N.H.&H. R.R.	"	J. F. Shields.
" Jt., Erie R. R.	"	
" "Lehigh&H.R.Ry.	"	
Campbell Hall, Jt. N.Y., O. & W. Ry.	"	C. F. Clifford, Ticket.
" " Jt.,West Shore, R. R.	"	B. Chase, Freight.
†Port Morris Tfr. Jt. D.L.& W.R.R. Jt. N.Y., N.H. & H.	N.J.	F. S. Myers.

RHINECLIFF DIVISION

STATIONS		AGENTS OR REPORTING STATIONS
Copake	N.Y.	E. T. Hotchkin.
Cooks	"	Copake.
Ancram	"	Miss M. S. Hoyt.
Gallatin	"	Silvernails.
Silvernails	"	M. J. Wheeler.
Ice Pond	"	Pine Plains.
P. & E. Junction	"	
Mount Ross	"	Jackson Corners.
Jackson Corners	"	C. A. Henry.
Elizaville	"	G. F. Fisher.
Cokertown	"	
Fraleighs	"	Red Hook.
Red Hook	"	C. H. Beatty.
Rhinebeck	"	J. J. Creed.
Slate Dock	"	
Rhinecliff, Jt., N.Y.C.R.R.	"	C. E. Secor, Freight.
Rhinecliff, Jt., N.Y.C.R.R.	"	Morris Lewis, Ticket.
State Line	"	
xMillerton	"	E. S. Webber.
Winchells	"	Shekomeko
Husted	"	
Shekomeko	"	H. Flinn.
Bethel	"	Shekomeko.
‡Pine Plains	"	P. Clifford.
N. D. & C. Junction	"	
Stissing Junction	"	F. H. Wilsey.
Bangall	"	H. M. Kniffen.
Ansons Crossing	"	
Shunpike	"	Millbrook.
‡Millbrook	"	H. A. Elting.
Oak Summit	"	Millbrook.
Verbank	"	M. M. Joyce.
Verbank Village	"	Verbank.
Moores Mills	"	G. S. Wells.
Billings	"	G. S. Wells.
Lagrange	"	J. T. Angell.
Arthursburg	"	Lagrange.
Clove Branch	"	
Hopewell Junction	"	S. W. Lesher.
Brinckerhoff	"	Fishkill.
Fishkill	"	H. G. Bowman.
Glenham	"	J. T. Masterson.
Groveville	"	
Matteawan	"	H. J. Cooper.
Wicopee Junction	"	
Beacon	"	F. L. Lawson, Freight.
Jt., N.Y.C.R.R.	"	G. I. Goodwin, Ticket.

CNE STATIONS AND STATION AGENTS. (From NYNH&H, *Official List of Officers, Stations, Agents*, Nov. 1, 1926). If an agent served the CNE jointly with another railroad, his station's name is marked "Jt." and the other railroad's name is given, abbreviated.

off by CNE crews at Simsbury where New Haven crews picked it up to take it the rest of the way north to Northampton. However, as early as 1911, when the CNE was already emphasizing its southern rather than its northern claw, the CNE and the New Haven Railroad together ran twelve through freights daily from Maybrook east over the bridge through Hopewell Junction. One such freight went to New York City via Danbury and South Norwalk. The rest were headed to New England destinations. Of these, three went via Waterbury to Hartford. Another went via Plainville and Westfield to Northampton, another went via New Haven to Providence, and still another went all the way to Boston. All these trains had prestigious "symbol" designations of their own. Trains to Boston were designated OB-2, to Providence OP-2, to Hartford OA-2, to Northampton OX-2. Crews based in Maybrook might take these trains from Maybrook right through the yards at Hopewell Junction and Danbury, all the way to New Haven or Waterbury.

One day in August 1931, F. P. Sweeney was the conductor of an Erie train which he brought from Port Jervis, NY, by Erie tracks to Maybrook. Sweeney recorded, as conductors were expected to do, what was in each car of his train and what the ultimate destination of each car was. Each of the cars was fully loaded with one product only, and headed for one destination only. In the Maybrook yard, these cars were classified into appropriate trains, in accordance with their destinations. All of the cars were headed via New Haven tracks across the Poughkeepsie Bridge except for one car of feed which was headed only to Highland. A few of the others were also destined for stations in New York State, such as Pine Plains (cattle), and Poughkeepsie, Hopewell Junction, and Matteawan (coal). By far most of them were destined for New England, and most of these were refrigerator cars, with waybills indicating whether they required to be iced along the way. These New England-destined cars included cars for New Haven (eggs), Bridgeport (meat, lumber), Willimantic (feed), Hartford (lettuce), Wallingford (coal), Springfield (meat), and Boston (meat, lettuce, melons, grapes).[14]

Once in 1934, Arthur Bixby was crewing out of Maybrook on a Boston-bound New Haven train, an OB-2, which Bixby considered to be a "hot shot." In Maybrook, as he recalled it afterwards, its steam engine, class L-1a, wheels 2-10-2, collected a train of 67 cars. These cars had come into Maybrook especially from the Erie, Central of New Jersey, and Lehigh & Hudson River lines. Many of them were refrigerator cars, containing perishables. Though the OB-2 was scheduled to leave Maybrook at 8:30 AM, it left late. Since the New Haven Railroad guaranteed its arrival in Boston by 11:00 PM, ready to unload for the next day's markets, its crew was under pressure to make up time.

The OB-2 crossed the Poughkeepsie Bridge, passed through Danbury and continued on toward New Haven. At Devon, as it prepared to leave a double-track line to enter the New Haven's heavily-used, four-track, electrified Shore Line, its

COMMON ON THE BRIDGE ROUTE WERE LUMBER AND COAL YARDS LIKE THIS ONE AT MODENA, ULSTER COUNTY, NY: THE JOSEPH E. HASBROUCK YARD, built about 1895 near the center of Modena. Photo about 1950 (PLATT). The tracks, visible across the center background among the trees, lead southwesterly (upper right) toward the Modena station, almost half a mile away, and on to Maybrook. A boxcar shows on a siding between two yard buildings. Fifty years later, most of the buildings are still on the site, the one in the middle housing several businesses, the one on the center right housing a medical office. In the left center, the road showing an auto on it is highway Rt.44-55 heading left toward Highland.

SELECTEED BRIDGE ROUTE (MAYBROOK LINE) STATION AGENTS, 1941	
Arrranged by location from East to West. All are New Haven Railroad agents except as noted. Compared to the 1926 list above, the railroad route related to the bridge has changed, bringing out a considerably different list of stations, and for the stations which remain the same, few of the same station agents appear. From NYH&H, *Official List of Officers, Stations, Agents,* May 1, 1941	
New Haven, CT	G. E. Bagre J. J. Powers T. E. Bradley O. A. Weber
Derby	W. C. Hogan
Botsford	O. H. Tarbox
Hawleyville	E. E. Gray
Danbury	F. W. Pierce C. V. Garrity
Brewster, NY	M. J. Lamere
Towners, West Patterson, and Holmes	C. H. Cole
West Pawling, Poughquag, and Green Haven	D. C. Wooden
Stormville, Hopewell Jct., and Fishkill Plains	H. J. Cooper
Brinkerhoff and Fishkill	W. J. Mahar
Glenham, Groveville, and Matteawan	Florence M. Leion
Beacon (jointly with NY Central)	A. M. Ross
Manchester Bridge, Poughkeepsie, and Hudson River State Hospital	A. Fraleigh
Highland	F. H. Wilsey
Clindondale	W. J. Kelly
Modena	D. J. Murphy
St. Elmo, East Walden, Berea, and Maybrook (jointly with Erie and L&HR)	S. L. Scott
Campbell Hall (jointly with O&W)	C. F. Clifford

several crossovers to a track heading east. As their steam engine chugged triumphantly on, the crew noticed passenger trains stalled along the tracks, and at the passenger station at New Haven, crowds of passengers waiting. They learned only later that a fire near the tracks in Bridgeport had cut off the power for electrified trains. Steam power had advantages.

On the day after the Pearl Harbor attack, Dec. 8, 1941, a dispatcher told the Erie agent John P. Sweeney at Campbell Hall to stay on duty until a special train arrived from Connecticut over the Poughkeepsie Bridge. Agent Sweeney was to see that there was no delay in hand-switching the train onto the Erie line heading west. When the train came, according to Sweeney, it "was a solid consist of flat cars, each carrying a naval fighter plane with folded wings," each car tended by an armed guard. Sweeney called up to a guard, "Where are you going, to Tokyo?" But the guard wouldn't answer. The train's destination was meant to be secret. The agent believed this shipment was one of the first post-Pearl Harbor rail shipments of war materials on its way from New England to the Pacific Coast.[15]

DERAILMENT OF FREIGHT ON-4, ON ITS WAY FROM MAYBROOK AND OVER THE BRIDGE TO NEW HAVEN, SEPT. 13, 1953 (McEN). At Towners, Putnam County, NY. In the foreground, the metal grating is the catwalk on top of a box car. At center left, the car labelled PX is a leased tanker. Far back to the left is a car labelled "New York Central System," and a car to the right of it is labelled "CB&Q" (Chicago, Burlington, & Quincy).

crew knew that several Shore Line passenger trains were due to pass by at about this time, so they expected their OB-2 would have to wait. But surprisingly the signals for the OB-2 were clear. The engineer at the controls, aware of the need to hurry, widened his throttle and proceeded over

19. CENTRAL STATE DISPATCH (1952)[1]
by Wallace W. Abbey

Rolling from Maryland over the bridge toward Boston, they "slither" around the curves, listen to the incessant "chant" of the wheels, and, surprisingly, stay "ahead of schedule."

THE CENTRAL STATES DISPATCH, a special long-distance freight train, drew freight from such mid-continent railroad centers as Chicago and St. Louis. It collected the freight together in western Maryland and usually carried it over the Poughkeepsie Bridge into New England.

The Dispatch began running over the bridge in 1892. Over the years, it occasionally went by alternate routes which did not cross the bridge. If the Dispatch's freight was especially destined for portions of Connecticut close to New York City, such as Bridgeport, then the Dispatch might be routed through Jersey City and New York congestion. If its freight was especially destined for northern New England, it might be routed through Scranton and Schenectady. But ordinarily its freight was destined more for New Haven, Hartford, Providence, and Boston, and then it was likely to follow its traditional route, via Allentown, the Lehigh & Hudson River Railroad, Maybrook, and over the Poughkeepsie Bridge. It was especially the advantage of the Poughkeepsie Bridge in avoiding the congestion of New York City which nearly always bound the Central States Dispatch to the Bridge Route. It kept it there even into the 1970s.[2]

When journalist Wallace W. Abbey rode the Central States Dispatch in 1952, he kept a record of his trip which helps to explains why the Dispatch survived so long:

"This is Cumberland, sir." The porter takes your suitcase to the vestibule, and you look out of the window on a city virtually imprisoned in the breathtaking Maryland Alleghenies.

You are arriving in Cumberland, Maryland, from Pittsburgh on the Baltimore & Ohio. From here, you are going to ride a regular, daily freight train to Boston, a train which is ordinarily not available for passengers. As any B&O freight agent will be glad to tell you, this train is highly regarded by Midwest shippers for moving freight into and out of New England. Called the Central States Dispatch, it is operated jointly by seven railroads, and it crosses the Poughkeepsie Bridge. It will leave Cumberland at 8:30 tonight.

Service over the Central States Dispatch route dates back to September 1, 1892, when eleven railroads combined to give shippers a service that extended beyond their individual limits, and to

THE CENTRAL STATES DISPATCH PASSED THIS STRIKING RURAL STATION AT GREAT MEADOWS, NJ (McEN). On the tracks of the Lehigh and Hudson River Railway, ten miles east of the Delaware River. The absence of platforms suggests the station was little used.

put themselves collectively into a better competitive position with larger, more direct railroads. Through the years the C. S. D. route has come to comprise parts of the B&O, the Western Maryland, the Reading, the Jersey Central, the Pennsylvania, the Lehigh & Hudson River, and the New Haven Railroads.

At first, the C. S. D. route had its own general manager, freight solicitors and audit bureau. Each road that formed part of the route supplied a share of the rolling stock and painted it with a distinctive C. S. D. emblem. C. S. D. solicitors and auditors were discontinued years ago, and now the B&O and other roads handle the tub-thumping in their regular freight agencies. The special cars are gone, too, and today a C. S. D. freight enjoys little visual distinction from any other, except that it might have a "foreign" road's engine on it.

Part of the success of the C. S. D. service, as you will see, lies in the fact that in certain instances train crews don't climb off their charges at the end of their particular company's property, but deliver them to a more logical interchange point. In a couple of cases that point is considerably beyond their bailiwick. It all adds up to this: The Central States Dispatch will put you into Boston on the second morning after the B&O starts you out of Cumberland.

You might say that the Central States Dispatch route begins in the yard at Cumberland. But in a larger sense, Cumberland is merely the point where blocks of cars collected from anywhere on or beyond the fingers of the B&O—fingers that stretch out across Ohio, Indiana and Illinois—are classified over the hump into solid trains.

You eat dinner—your last hearty meal for quite a while—in an obscure but eminently satisfactory home-cookery tucked away on a side street, and you go back to the yard and watch through the blue-green glass windows of the control tower as cars of eastbound coal roll off the hump into 16 tracks.

Now you and the B&O trainmaster, and a like official of the Western Maryland, stand on the back porch of your caboose, waiting for the highball that will start you on a 33-hour, 632-mile train ride. The day is waning, and although the yard is quiet under the floodlights, you can hear the gentle urging of a switcher as it works someplace over beyond the hump.

You are riding what is, officially, First CSD-96 out of Cumberland, behind steam engine 7620, a mammoth EM-1. Your crew was called for 7:30 p.m.; your leaving time is 8:30. But there is one thing you're going to have to get used to on this trip, and that's barreling across country ahead of schedule. In most cases, symbol freight trains run extra on the roads you'll be on, and there's no reason why, if they can pick up some time here and there, they shouldn't run ahead of the schedule. So now, quietly and with no slack action, you roll past the tower and out onto the main line at 8:19 p.m. —already 11 minutes ahead of time.

You hang onto a couple of handrails. Your engineer soon takes advantage of the 50-mile-an-hour fast freight speed limit, whenever the meanderings of the right of way, which follows the Potomac River, will allow it. The flagman comes out on the platform and fastens a pair of red lanterns to the floor.

The B&O share of the C. S. D. route is the 64 miles between Cumberland and Cherry Run, W. Va. In that 64 miles are some of the quaintest names for stations and cuts and curves to be found in American railroading, names which generations of mountain folk have chosen. They include Okonoko, Paw Paw, Magnolia, Doe Gully, Great Cacapon, Sleepy Creek, Turkey Foot Curve, Grasshopper Hollow Curve. They make railroading sound poetic.

You stay on the platform until the acrid smoke inside a tunnel forces you into the caboose, which is a veteran and shows it. You sit at the scarred desk, and the conductor turns up the kerosene lamp and slips the string off the package of waybills. This, in a hip-pocket-size package, is the story of First CSD-96.

The bills show that your train contains principally high-grade merchandise and perishables. You see that most of the cars are going to be with you for the major portion of your trip. You have five cars destined for the Jersey Central at

Allentown, Pa., and 65 for Maybrook, N. Y., where the New Haven takes over and distributes them throughout New England.

Now you're near Cherry Run. You consult your watch—9:34 p. m. Not bad! You're not due by Cherry Run until 10 o'clock! You roll across the Potomac for the last time and onto Western Maryland tracks. Here is the first of those instances where a crew takes its train "out of bounds." The B&O crew, engine and caboose head right on into Hagerstown, Md., 19 miles up the Western Maryland.

The run into Hagerstown is short, and, considering Williamsport Hill, rapid. It's only 10:14 when the flagman cuts off your caboose on the fly as you head into the yard. Even though you're not due to leave for more than an hour, the yard diesel, a big Baldwin road-switcher, attacks your train as if it were considerably behind time. The yard crew couples on a spotless Western Maryland caboose, and you climb aboard to find coffee brewing on the stove. It's 10:50 p.m. when you begin to roll. You're now running as Western Maryland No. 96, behind engine 1402, a 4-8-4, headed for Lurgan, Pa., and Rutherford, the Reading's terminal on the outskirts of Harrisburg. You're 35 minutes ahead of time.

The conductor motions you up onto the cushions across the cupola from him. You slide back the windows and let the clean night air blow on your face. You're speeding northeast across Pennsylvania now, through country that lies comparatively flat between distant mountain ridges. It's double track most of the way.

Then you're in Chambersburg, crawling the length of a quiet city street. The rolling cadence of the wheels is multiplied and thrown back at you by the dark, quiet buildings. You think of the people who have long since learned to sleep through that racket.

At Lurgan, the transfer of your train from the Western Maryland to the Reading is merely a matter of streaking past a telegraph office. You see the operator seated at his interlocker, and you can imagine that he'll call the dispatcher now and report you going by at—let's see—11:57 p. m.

The lullaby of the rolling caboose is slowly putting you to sleep. The cupola sways gently, the wheels chant incessantly. Without realizing it, you doze off.

Something—perhaps a change of pace of your train—registers on your subconscious, and you awake and look out of the window on the yard at Rutherford, just beyond Harrisburg. What happened to Harrisburg? Must have slept through one of the biggest cities on the run!

For the second time, your Central States Dispatch goes through the ritual of changing engines and cabooses. This time you get a diesel, and you leave steam power behind for the rest of your trip. You leave Rutherford as Reading train HO-6.

This, the third leg of your junket, is a fairly long one. It's 84 miles from Rutherford to Allentown, the next terminal. As you doze now and then, you scoot through such Pennsylvania communities as Hershey, Lebanon, and Reading itself, the nerve center of your present host. In the purest sense of the term, yours is a through train—no stops to Allentown.

The next thing you know, your Reading train HO-6 is standing still at Allentown, stretched across the Lehigh River, waiting for a track in the Jersey Central yard at East Allentown. A train man turns you over to Superintendent Paul W. Early of the Lehigh & Hudson River while a yard engine gives the train its first reshuffling. Early, undoubtedly the veteran of countless all-night freight train rides, thoughtfully produces a sweet roll and a cup of coffee for you from a paper bag.

The L&HR's caboose is decidedly the homiest of the hacks that you've been in yet. But you accept Early's invitation to see his railroad from the engine, so you stow your suitcase in the caboose and the two of you walk up through the yard. At the head end you find two road switchers—two elevenths of the L&HR's motive power, Early tells you. He adds that his line is dieselized with Alco-GE road-switchers, and that two more are on order.

The bray of an air horn signals that your train is ready to go. You swing aboard as the train begins to move, and you settle yourself in the cab of the second diesel.

Your guide outlines the way that the Central States Dispatch—now L&HR train HO-6—is routed out of Allentown. It is to travel over the Jersey Central's Central Railroad of Pennsylvania to Easton, Pa.; cross the river to Philipsburg, N. J.; and follow the Delaware River on the Belvidere Branch of the Pennsylvania to Belvidere. There it is to start over the L&HR itself. As you begin this section of the trip, snaking out of the East Allentown yard onto the three-track main line at 5:55 a.m., you're still ahead of time—40 minutes ahead.

For the first and only time, you get a look at some of industrial Pennsylvania in the daytime. The track closely follows the Lehigh River to Easton, and you get a good view of the immense Bethlehem Steel Works across the river at Bethlehem. But the Central States Dispatch disdains to handle much of the freight born of this industry. You've still got that train of Maybrook cars, plus enough coal picked up at East Allentown, to give you 79 cars.

On the Pennsy, the scenery changes abruptly. You're out in the country again, and the Delaware River banks are wild. Presently Early speaks close to your ear so that he can be heard over the sound of the diesels—you're at Belvidere and entering onto L&HR tracks. Right away you see one difference—the L&HR has color-light automatic block signals, whereas the Pennsy branch line had manual block operation. You detect the fact that your ride is smoother, too.

It doesn't take you long to decide that the L&HR is a handsome little railroad. It's well ballasted and well maintained. You climb steadily for the 10 miles to Great Meadows, and then the grade tapers off and you're ambling through rural vistas and farming communities that give no indication of their proximity to teeming centers of civilization. You pick up new running orders at Andover; as the train passes, a trainman on the ground holds the orders up with a wire frame for the passing crew to grab. You wave at everybody as you pass the L&HR general office at Warwick, just over the line into New York State. The grades, mostly descending now, are steeper, and the hills on the horizon a little higher. The 72 miles of the Lehigh & Hudson River are slipping rapidly behind you. Now you're coming around the last curve into Maybrook Yard.

Maybrook, N.Y., is obviously much more of a center of railroad activity than a center of population. From your cab window you see all track and no town. You pull into the eastbound receiving yard next to an Erie train that you tentatively identify as New England 98, a hotshot perishable job. Your train is too long for the yard track, so you run up on the hump and double your head cars over onto another track. L&HR Superintendent Early leaves you in the care of the secretary of the Railroad YMCA, who gives you soap, a towel and a knowing smile. It dawns on you, presently, that the begrimed, unshaven mug staring at you out of the washroom mirror is yours. You look as if you'd been riding freight trains all night. It feels good to be clean again, and it feels good, too, to lean on a lunchroom counter and dig into a true railroad man's meal.

The New Haven Railroad gives the Central States Dispatch a thorough reclassification here at Maybrook, for it must pigeonhole its cars according to their destinations throughout New England. It must also work into trains the cars brought to Maybrook on the other lines—the Erie, the New York, Ontario & Western, and the Lehigh & New England. Consequently, when the conductor of train OB-4, which is your C. S. D.'s new name on the New Haven line, shows you his wheel report, it lists: 5 for Hartford, 1 for Springfield, 6 for Worcester, 54 for Providence, 16 for Boston. Thus, with a total of 82 cars of "high class" freight and pulled by three Alco-GE diesel units, you leave Maybrook at 5:15 p. m.

On this "Maybrook line," a certain feature sticks in your mind long after the hills and stone fences and operating details become part of your back-of-the-mind experience. It is the Poughkeepsie Bridge.

As we approached the bridge, our caboose came up opposite the picturesque depot at Highland and slid past, leaving it in the shadows of the waning day. All was quiet except for the metallic song of our wheels.

"You know, people come for a long way to see

this bridge, but I cross it twice a day and never think a thing about it." The conductor had joined me on the caboose platform and was gazing with me back along the track as it curved gently and passed beneath some arched viaducts. He tried to make you feel that the Poughkeepsie Bridge means little to him. But you notice he's out there pointing up and down the river, showing you all there is to see. It's almost as if you were flying over the river—in slow motion, of course. "It is quite a bridge, though," the conductor adds. "Two hundred and twelve feet down to the river!"

Two hundred and twelve feet is a long way for a heavy freight train to be separated from a river by a spindly looking trestlework, I thought. But the Hudson is a mighty river and its palisades tower like tremendous castle walls above a moat. To throw a bridge across it at Poughkeepsie today would be no small job, and this bridge was completed in 1889!

The solid earth below the track dropped away, and I looked down at the wooden planking of the bridge floor. I knew that up ahead our 82 cars were stretched out across the Hudson, and that our train must have looked like a toy to anyone on the water far below.

As I watched the twin rails unroll from beneath our caboose, the two tracks converged and overlapped in a gantlet. I asked the conductor how come. "Bridge used to be double track, but

DIESEL LOCOMOTIVES FIRST CAME INTO USE ON THE SEVERAL RAILROADS WHICH COMBINED TO RUN THE CENTRAL STATES DISPATCH AT DIFFERENT TIMES. As Abbey indicates, on his trip in 1952 the B&O and Western Maryland engines which pulled him were still steam engines, while the Reading, L&HR, and New Haven engines were diesels. In 1967 when this picture was taken, the Central States Dispatch was being pulled over the Poughkeepsie Bridge by New Haven diesel units like these three shown coming onto the bridge from Maybrook, units 2518, 2505, 2558. Photo by J. W. Swanberg (SWANB)

when we got the big engines the tracks were gantleted to keep the weight in the center." The big engines would be the 3200-series 2-10-2's which did the Maybrook-New Haven work until the diesels came.

We were out over the river now, and I found myself thinking that it didn't seem quite right for a railroad train to be up there in the middle of space—for that's the way it seemed—where you kind of forget about the engine and cars ahead of you and the steel and concrete below you and fancy yourself floating blimplike from one bank to the other. But our train was up there, as were some 4531 trains in 1952, hauling a total of 16,721,523 tons of freight over a route which has been vital to New England commerce ever since the Poughkeepsie Bridge Company built its long-dreamed-of span.

"That's the West Shore Railroad down there," announced the conductor, pointing to the tracks which paralleled the river and clung close to the rocky cliff. He stood nonchalantly very close to the edge of the caboose platform, I thought. I stood exactly in the center. "Up that way a few miles is Hyde Park—President Roosevelt's place. He's buried there."

I looked up the river through the hazy twilight, past where a tiny river craft was making a minor disturbance in the broad quiet surface. I wished that it didn't take a lot of special arrangements and a long ride on a freight train to get to cross Poughkeepsie Bridge. I knew that anyone who loved scenery would want to do what I was doing.

"That's Poughkeepsie and the New York Central main line." I saw the station and the yard and the buildings on the east bank mantled in a deep dusk. Poughkeepsie looked like a toy village.

"Well, that's it!" The conductor opened the door and went inside. I took one last look and then I climbed up into the cupola. Twice a day he crosses it, I thought, and he doesn't think a thing about it.

The measured cadence of the wheel clicks dissolved into the lilting music of a fast train's theme. We were back among the trees and rock cuts now, and suddenly it was dark.

We climb through West Pawling Mountain, slithering around hairpin curves. We pass Danbury, Conn., and from the cupola you watch the left side of your train as it practically falls down into the Housatonic River bottom toward Derby. Then you come out onto the electrified main New Haven line at Devon, and you recall the warning plate that was bolted to the cab wall of a diesel that you saw at Maybrook—keep away from the wires, it said in effect, unless you think you'd look better broiled.

You slip unobtrusively through New Haven's passenger station, and then you are engulfed in Cedar Hill Yard. OB-4 dives directly into the middle of the layout and halts in the Shore Line departure yard. You find the yard office among the floodlight towers and scurrying diesel switchers, and introduce yourself to the man at the desk. He phones someone and then tells you to wait, that there'll be a switch engine along shortly to take you down to the caboose for the run from Cedar Hill to Boston.

So you wait. The yard engine shows up, and you hoist yourself onto the deck. You find that OB-4 has been moved, and that its engine is on what had been the rear end when you pulled in. This time you have a pair of the 2000-horsepower Alco-GE dual-service units that started the New Haven toward dieselization before World War II. Your chauffeuring switcher clumps through the yard adjacent to OB-4 and stops alongside its caboose.

"Well, are you going to stay awake or sleep?" the conductor asks. You cogitate—last night you made an attempt to stay awake and see what was going on. But that was last night, and today has been a long day. Tonight, you decide, you're just too tired to argue with the sandman.

The conductor gives you a quick summary of your new consist—44 cars for Providence and 46 for Boston — and bunches up his coat for a pillow on one of the padded leather cushions. You watch until OB-4, otherwise known as the Central States Dispatch, is out of the yard and onto the New Haven's Shore Line headed for dawn and Boston. You check the time—12:40 a.m. Then you climb aboard that caboose-cushion cloud and float off to the lullaby of singing wheels.

While you sleep, OB-4 rolls along the aptly named Shore Line. If it were daylight, and if you were awake, you would see Long Island Sound lapping at the Connecticut shore almost continuously from Cedar Hill to Westerly. Then you turn inland and cut across Rhode Island to Providence, the state capital and the hub of many New Haven operations.

You awake momentarily at Providence, after your train has set out its cars for that city and is getting under way again. But the interruption is brief, and then, almost as if a Pullman porter had joggled your berth curtain, something tells you it's time to get up.

The feeble light of dawn enables you to spot a milepost, and you check it against the thick book that contains operating schedules for the entire New Haven. You're practically in Boston Town! The double-track main line splits, and you take the right fork, the freight line which ends in a yard down close to the Commonwealth piers. The conductor comes down out of the loft and offers, when OB-4 is tied up in the yard, to call a taxi for you. And you, being a stranger to Boston, readily accept.

And then you're at the end of the line. The conductor admonishes you to stay aboard the caboose until it has stopped, in view of the fact that you don't make a living by swinging off of moving trains, and then he drops off at the yard office. By the time you've lugged your suitcase back the dozen carlengths to the office, a yard diesel is busily cutting OB-4 into more easily manageable chunks.

You crawl into the taxi. You find its ride is the wildest of your long journey. As recollections of your Central States Dispatch journey pass in review, you decide that the C. S. D. is not necessarily the best way for a passenger to get from Cumberland to Boston, but that it's mighty fine for through freight.

	NEW HAVEN RAILROAD CHRONOLOGY
1872	Created as the New York, New Haven and Hartford Railroad
1893-1898	Reaching for monopoly in southern New England, it acquired the NY&NE which controlled the Bridge Route's southern entrance into New England through Danbury
1898-1902	Tried unsuccessfully to prevent the Bridge Route from building its Sprinfield Branch
1904	Bought control of the CNE, including its Poughkeepsie Bridge
1904	Bought control of the O&W which fed Pennsylvania coal to the Bridge Route through Campbell Hall
1904-1906	Doubletracked the Bridge Route's entrance into New England from Hopewell Jct. through Campbell Hall
1917	With the Pennsylvania Railroad, completed the Hell Gate Bridge over the East River, reducing the need for the Poughkeepsie Bridge
1927	Absorbed the CNE, so that the CNE ceased to exist
1935	Hurt by the depression and competition from trucks and autos, it went bankrupt
1932-1938	Abandoned most of the Bridge Route's northern approach into New England, through Canaan to Hartford
1961	Hurt by the improvement of highways and the long decline of heavy industry in the Northeast, it went bankrupt again
1968	Absorbed by Penn Central, the New Haven ceased to exist

20. CIRCUS TRAINS (1889-1960s)[1]

Chariots, roustabouts, "monster" horses, "sleek" tigers, "genuine" Indians, all traveled the Bridge Route.

ACCORDING to the recollection of the CNE conductor Nathan Blodgett, while he was growing up near Waterbury, CT, several members of his family joined the Adam Forepaugh Circus. For the circus, one of his brothers drove six ponies. Another brother drove eight horses. Nathan marveled at how efficiently their circus crews operated, day by day. They would arrive in a town by train in the morning, he recalled, and on the same day they would "unload all of their equipment, haul it to the circus grounds, set up their tents, give a street parade at 10 a.m., [and give] an afternoon and evening performance." That same evening they would immediately "dismantle everything, haul it to the railroad yard, load it on the train," and travel on the train that night. The next day in another town, they would "do it all over again."

In 1890, only a year after the Poughkeepsie Bridge had opened, the Forepaugh Circus came to Poughkeepsie by the Bridge Route. It came in three train sections. As each section arrived, even though it was early in the morning, "hundreds of people" were at the Bridge Route depot on Parker Avenue "to witness the unloading of the wagons, chariots, elephants, horses." Later in the morning, on Main Street, while perhaps 10,000 people eagerly watched, the circus paraded, showing off "genuine Indians" in "war paint," Cleopatra in a chariot, "sleek" tigers, fourteen elephants, and countless horses, In the afternoon and evening, the circus performed in a big tent on Mansion Street, a feature being Adam Forepaugh Jr., driving forty horses at once. While the performances were going on, the circus train cars were being switched

Po'keepsie Bridge

TRAINS TO AND FROM

BARNUM'S CIRCUS !

FOR THE ACCOMMODATION OF THOSE WHO WISH TO VISIT

Barnum's Great Show,

AND ALL OTHERS WHO DESIRE TO

Cross the Bridge

AND ENJOY THE MAGNIFICENT VIEW FROM IT,

THE PO'KEEPSIE BRIDGE COMP'NY WILL

Run Hourly Trains each Way

To-Day, May 27th.

Leaving Po'keepsie at 8 A. M. and every hour thereafter till 7 P. M.

Leaving Highland at 8.30 A. M. and every hour thereafter till 7.30 P. M.

The regular train will leave Po'keepsie at 11 o'clock P. M. for Campbell Hall, stopping at all Way Points.   N. B. TURNER,
General Passenger Agent.
S. B. OPDYKE, Jr., General Superintendent.
1t

ADVERTISEMENT,
Poughkeepsie *Daily Eagle*, May 27, 1889

from the Bridge Route onto the Hudson Line. When the circus was over, its crew loaded all its animals and equipment back onto its train cars, all the circus personnel boarded, and then during the night the circus train moved south for a performance the next day in Yonkers.[2]

In July 1903, the Barnum and Bailey Circus, even though it was larger than the Forepaugh Circus, was following the same one-day-in-each-city routine. One day it put on a performance in Danbury, CT, with difficulty because the grounds were wet, but that night its loaded rail cars moved out of Danbury on the Poughkeepsie Bridge Route, reaching Poughkeepsie early enough the next morning to be almost on time for its performances there. The circus was so huge, including as it did over 1000 employees, along with animals, tents, and other heavy equipment, that it traveled, the circus claimed, in a train of ninety-two cars, divided into five sections.

In the early morning, the circus crews put up their tents on upper Church Street at a baseball park, and on additional nearby land which they also hired—the circus needed altogether nearly twelve acres. First they put up cook tents to feed their own employees. Then they put up two mammoth tents to house their five hundred horses, mostly work horses which drew the circus wagons from the train to the performance site. Then they erected the principal performance tents: the menagerie tent to display both wild animals and "human freaks," and the main tent, said to be "the largest tent ever erected," with 14,000 seats.

By late morning, the circus paraded on Main Street showing off a steam calliope, a "herd" of elephants, "carved golden cages of rare beasts," and floats, among them floats illustrating nursery rhymes. One float was so heavy that it was drawn by forty "monster" horses. Others were so high that telephone wires had to be lifted to let them pass.

Visitors came to Poughkeepsie to see this circus by special trains, including Bridge Route trains, one bringing 400 passengers from the west over the bridge, another bringing 500 from the northeast through Stissing and Clinton Corners, and another 500 from the southeast through Hopewell Junction. After the circus folded its tents that evening, crews loaded them and all their animals in their usual rail cars. When the circus train, in its various sections, pulled out of Poughkeepsie, it was still on the Bridge Route. The train moved northeast to Canaan, CT, where it switched onto the New Haven's Berkshire line, on its way to a performance the next day in Pittsfield, MA.[3]

In the summer of 1907, the CNE offered to transport the small circus of Frank A. Robbins in sixteen cars from Canaan, CT, to Lakeville, only 9.5 miles away, for $120, and from there 74.5 more miles over the Poughkeepsie Bridge to Maybrook, NY, for an additional $250. Similarly, in the summer of 1908, the CNE offered to move the small circus of the Victor Amusement Company, in ten cars, from Maybrook across the bridge to Hartford, 139 miles, for $355. In both cases, if the circus delayed in loading or unloading more than forty-eight hours beyond the agreed upon time, the CNE would make an extra charge for each car.[4]

According to Thomas A. Flynn who grew up in Poughkeepsie in the 1910s and 1920s, circus trains arriving in Poughkeepsie often came from the west over the bridge, and stopped to unload their equipment at the CNE's Smith Street yard (originally the P&E yard), near Cottage Street. When they left town, they usually went on southeast, toward New Haven.

Austin McEntee, who grew up in Poughkeepsie at about the same time, recalls that when he watched circus trains arrive at the Smith Street yard, circus crews usually used horses to unload them. Horses pulled the circus wagons off the train's flatcars, then hauled the wagons through the streets to the performance site. While the circus was being unloaded, McEntee recalls, he and other boys carried pails of water for the circus animals to drink. The circus paid him and other boys for doing such chores by giving them tickets, not tickets to the circus proper, but to little side shows.[5]

Still another circus enthusiast growing up in Poughkeepsie at about the same, Frank V. Mylod, recalls that when he and other boys helped to set up circuses, they were sometimes rewarded by sodas, or for more work, free tickets to enter the main circus. Of course, once they had their tickets in hand, they wanted to see the circus right away. On one occasion when Mylod and his friends rushed to see a performance, they found that they were not prepared. They had taken no food with them, had no money to buy any, and so could not stay as long as they wished.

After Mylod had become a lawyer, he was fascinated not so much by circus performances, as by how circuses traveled on trains, unloaded from trains, and set up for performances. He would go to the Smith Street yards to watch circuses unload.

At about this time, while circuses still moved primarily in wagons, the wagons themselves traveled from city to city on railroad flatcars. To unload a circus train for a performance, the roustabouts knocked out the chocks from under the wheels of the wagons to free them to move, laid planks or steel plates across the gaps from

Central New England Railway Company

LOCAL FREIGHT TARIFF No. 635

ON

CIRCUS OUTFIT

COMPRISING 10 CARS

(of Victor Amusement Co.)

FROM

MAYBROOK, N. Y.

TO

HARTFORD, CONN.

RATE:—$355.00 FOR COMPLETE MOVEMENT.

Governed by the Official Classification I. C. C. O. C. No. 32 and exceptions thereto C. N. E. Ry. I. C. C. No. 2369, and supplements thereto and reissues thereof, unless otherwise noted.

DEMURRAGE AND CAR SERVICE REGULATIONS.

Under this tariff, when freight is to be loaded by consignor or unloaded by consignee, $1.00 per car per day or fraction thereof, for delay beyond forty-eight hours, in loading or unloading, will be added to the rates named herein, and constitute a part of the total charges to be collected by the carrier on the property, except when otherwise provided by regulations at shipping point or destination on file with the Interstate Commerce Commission in which case such regulations shall prevail and govern.

ISSUED JUNE 24, 1908   EFFECTIVE JUNE 28, 1908

Expires June 30, 1908, unless sooner canceled, changed or extended.

W. H. SEELEY,
GENERAL FREIGHT AGENT,
HARTFORD, CONN.

THE CNE OFFERS TO TRANSPORT A CIRCUS
in conformity with ICC regulations, 1908 (SHUKR).

CIRCUS WAGONS RIDING ON FLATCARS, heading toward the Poughkeepsie Bridge. They are on the New Haven's line near West Pawling, Dutchess County, NY, May, 1940. The cars are marked: "Ringling Bros. and Barnum & Bailey Combined Shows." (Photo by John P. Ahrens, SWANB)

flatcar to flatcar, and then, with the use of rigging, employed horses—or more rarely elephants—to pull the wagons off the train. The horses might pull the wagons off one by one, with ropes, each wagon rolling in a continuous line from one flatcar to another, until from the last flatcar the last wagon rolled down a ramp onto the ground.

Mylod would follow the circus wagons as they moved to the vacant land where the circus was to perform, and continue to watch as the circus crew set up their tents for a performance. Mylod might bring along his little son John, and both might continue to watch in awe for hours.

By the 1950s, as John Mylod was growing up, he recalls a combined Ringling Brothers, Barnum & Bailey Circus coming to Poughkeepsie every year or so. It would come by various train routes, but often by crossing the Poughkeepsie Bridge, perhaps coming from Philadelphia. From Poughkeepsie, he believes, it usually went on to Danbury or New Haven.

At the Smith Street yard, the circus train crew worked quickly, it seemed to John Mylod, to pull one wagon after another off the flat cars. At this time, it was no longer horses which provided the power to pull the wagons off. It was little Mack trucks which themselves had arrived on the flatcars of the circus train. After the Macks had pulled the wagons off the train, they hauled the wagons about a mile to the performance site, near St. Peter's Cemetery.

John Mylod recalls that he and his friends helped handle the ropes to put up circus tents. Sometimes they did it in cooperation with an elephant which walked under the canvas, lifting it and at the same time pushing tent poles into appropriate positions. Mylod also recalls that he and his friends helped to put up seats. The seats came in by train, on trailers riding on flatcars. Trucks pulled the seat-loaded trailers to the circus site where they were arranged inside the big top. Then while the seats remained on the trailers, the boys helped open them out into bleacher seats which faced toward the rings in which the principal acts were to be performed.[6]

Circus flatcars were often double the normal length of rail cars. That is, they were about seventy-two feet instead of the usual thirty-six feet (since railroads charged for transporting circuses by the "car miles," it was to the circuses' interest to lengthen their cars), so each flatcar might carry four wagons. Many wagons would carry tents or seats. One wagon would be for a calliope, and another, called the "pie car," would serve both as a dining car and a recreation center for the crew. While practice varied over the years, other wagons were likely to carry cages of small menagerie animals; in rough weather, these wagons might be covered with canvas for protection. Larger animals were likely to be carried directly in stock cars. Elephants or horses were likely to be loaded into stock cars sideways, and expected to remain standing as long as they were in the car.

Circus trains usually carried not only flat and stock cars but also passenger cars for the circus personnel. For the performers (acrobats, stunt riders, jugglers, clowns, musicians) the cars might be substantial, with comfortable bunks for sleeping. For the roustabout crew, in the early years the cars might be infested with lice and lack any sanitary or sleeping facilities. Roustabouts sometimes preferred to sleep on the flatcars in the open air,

CIRCUS CAR AT MAYBROOK YARD: ELEPHANT LET OUT TO BE WATERED, FED, AND EXERCISED, 1956 (Photo by Ralph Aiello, MRM). The full name on the car was "Barnum & Bailey Combined Shows." Aiello recalls elephants as being let out for only an hour or two. The circus crew hold prods in their hands, to use in controlling the elephant. There is hay on the ground for feed.

underneath the wagons. In later years, the roustabouts' cars also had bunks for sleeping, if poor ones.

In 1911, there were thirty circuses moving about on American trains, one of the largest being Barnum & Bailey and another being Ringling Brothers, both of which had about eighty train cars each. By the 1940s, although there were only six American circuses traveling by train, they were larger, often traveling in trains of over one hundred cars, but divided into sections. By the 1950s, circuses were fewer, smaller, and some of them were traveling by truck.

In the Northeast, circuses performed only in spring, summer, and fall. In the winter, circuses went into winter quarters. While there, circuses rested their animals, practiced their acts, repaired and painted their train cars—they owned their own train cars. In the early 1890s, some Barnum circus animals wintered in northern Dutchess County, NY, in the Millerton-Pine Plains area. As winter ended, in March 1892, seven Barnum stock cars filled with horses moved out of Pine Plains on the Bridge Route, heading via Winsted for Bridgeport. In October 1892, the ND&C superintendent proposed to move Barnum stock cars from Millerton to Shekomeko and Bethel at six dollars per car; he made his proposal on the assumption that the stock would winter in the area again, and that the ND&C would "rail the same stock next spring on its return, at the same price."[7] At the time, however, P. T. Barnum wintered the main part of his circus at his headquarters in Bridgeport, close to the New Haven Railroad line on which it depended. Barnum's circus continued to winter there until in 1927, after it combined with the Ringling Circus, it moved its winter quarters to Sarasota, Florida.

Over many years, the combined Ringling Brothers, Barnum & Bailey Circus brought its circus by train north from Sarasota each spring to near Albany, crossing the Hudson River there to make a circular tour of New England. On the way back, while as the years passed it no longer stopped in Poughkeepsie, since "The Greatest Show on Earth" now performed only in large cities, it customarily crossed the Poughkeepsie Bridge, and made a rest stop in the Maybrook rail yard.

When circus trains stopped in Maybrook as they often did, they created a stir. From the 1940s into the 1960s, the elementary schools in Maybrook, and sometimes in nearby Montgomery and Walden as well, might be let out for the occasion, if only for half a day. Throngs watched as trainers led trumpeting elephants out of their rail cars, but the trainers would not let out tigers—that would be too dangerous. Throngs watched as the trainers fed their animals, watered them, exercised them, and cleaned their quarters. They watched as the local fire department pumped water to help the trainers hose down the animals, while children gawked, and parents photographed their children gawking. Throngs also followed the circus crews about town. The crews were used to Maybrook and knew where to go. Some headed for Watt's Pharmacy for ice cream or soda. Others headed for bars which became wonderfully crowded.

In 1968, a circus train came as usual from Boston over the bridge to Maybrook. According to a New Haven Railroad telegram which helped arrange for moving this train, it was a train of the combined Ringling Brothers, Barnum & Bailey Circus. The train was somewhat short, as circus trains were likely to be by this time, one reason being that since circuses performed in arenas, they no longer carried tents. The train included four stock cars, six baggage cars, and fourteen passenger cars. In Maybrook, the train crew replenished their supplies, including fresh water for their animals. When such trains pulled out, trainmen recall, they usually left by the Lehigh & Hudson River line, en route to a performance in Philadelphia.[8]

Other Use

21. JUMPING OFF THE BRIDGE (1888-1895)

"Here goes the King."

A YOUNG SEAMAN, Sam Patch, helped to establish the American tradition of daredevil jumping from high places. At first he jumped from the masts of ships. Later he jumped from halfway up Niagara Falls. In Rochester in 1829, at about the age of twenty-two, he was jumping from the Genesee Falls when he died.

The Brooklyn Bridge, after it opened in 1883, attracted jumpers. A swimming instructor from Washington, DC, Robert Odlum, jumped from it in 1885. He wore a bright red shirt. As he jumped he raised one arm straight over his head. He died.

By the next year, Steve Brodie, an unemployed New York Irishman in his early twenties, was talking about jumping from the Brooklyn Bridge. In July 1886, it was announced that he had jumped from it and lived. Several of Brodie's friends said they had seen him do it, but skeptics claimed that a dummy had been thrown off the bridge while Brodie merely swam out from shore to be picked up by a passing boat. Anyway, Brodie took advantage of the publicity about his claim by demonstrating his jump on the stage, and by opening a bar on the Bowery where sightseers flocked to see him. When he was asked why he did not jump from the Brooklyn Bridge again, in front of reputable witnesses, to remove any question about whether he really had done it, he customarily replied, "I done it oncet."[1]

Meanwhile in the spring of 1887, Brodie claimed that he swam the Hudson River all the way from Albany to New York, passing, of course, under the then-being-constructed Poughkeepsie Bridge. By the next year, when the Poughkeepsie Bridge was nearly completed, Brodie said he wanted to jump from it. After all, it was higher than the Brooklyn Bridge, and no one, Brodie understood, had ever jumped off such a high bridge. The *National Police Gazette*, a sensational sports newspaper, while advising him not to jump off the bridge, countered its own advice by offering him a gold medal if he did.

Talk that Brodie might try to jump off the Poughkeepsie Bridge alerted its chief engineers,

STEVE BRODIE PREPARES TO JUMP FROM THE POUGHKEEPSIE BRIDGE (This and the two following drawings about Brodie are from the *National Police Gazette*, Nov. 24, 1888)

Dickinson and O'Rourke. They instructed the bridge guards to watch for Brodie and prevent him from going out on the bridge. Brodie made several trips to the bridge site to figure out how he could elude the guards.

Finally on the evening of November 8, 1888, Brodie, a few friends, and a *Police Gazette* reporter took a West Shore Railroad train north along the Hudson River. According to the *Gazette* and other newspapers, they got off their train at about 11:30 PM in the pouring rain at the Highland station, on the waterfront near the west end of the bridge. Guided by someone with a lantern, they walked behind the station to Dean's Hotel, where they stayed the night.

Early the next morning, still at the hotel, Brodie's friends prepared him for jumping. They stripped him. They rubbed him, first with whiskey and then with oil. They bandaged his ankles. They placed protective layers of cotton batting on his chest, back, and loins. Then Brodie dressed himself inconspicuously. He wore a dark coat, dark trousers, and a workman's customary Derby hat.

While it was still dark enough to help him avoid attention, Brodie, alone, walked up the steep slope near the west end of the bridge, and then, according to newspaper accounts, chose a section of the bridge approach trestle and climbed to its top. Nearby bridge watchman Charles Doxie was so busy trying to start an engine that Brodie slipped past him unnoticed and walked out on the bridge. Some of his friends rowed a boat out on the water near the bridge and waited. With the help of the "tremendous glare" cast by an iron foundry across the river in Poughkeepsie, several of his friends said they thought that they could glimpse Brodie on the bridge.

According to reporters, once Brodie was well out on top of the bridge, near the first pier which stands in the water, he took off most of his clothes, and put on a rubber suit he had brought with him. He blew up rubber life preservers, and tied them on. He kept on his shoes which were lead-soled to help him fall feet first into the water. After checking that his friends were waiting for him in their row boat below, reporters said, he held his hands down, close to his side, and jumped.

BRODIE JUMPS. According to newspaper accounts, Brodie wore protective layers of cotton batting, a rubber suit, and life preservers. A West Shore train shows below.

His friends below saw his splash. They rowed over to the spot. When his inflated life preservers brought him up to the surface of the water, they pulled him into their boat. He was "gesticulating wildly." Blood was trickling from his ears and nostrils, they afterwards reported, and he was moaning. But they rowed him to the shore by the Highland station, and carried him into the station waiting room. He was exhausted and shivering. He yelled with pain. He fainted twice. But at 6:37 AM his friends carried him onto a train for New York.

The *Police Gazette* announced that Brodie, having jumped from over 200 feet above the water, was "the champion jumper of the world." But the bridge's Chief Engineer O'Rourke doubted that

Brodie had really jumped from where he said he had jumped. It was a question if any reliable witnesses saw him jump. He could have jumped from lower down. According to the *Poughkeepsie Eagle*, Brodie believed he had broken his ribs, and expected to die, but no local doctor examined him, and later a doctor who examined him at his home in New York reported that he had no broken ribs. At any rate, he recovered.

Soon afterwards, O'Rourke and some of the bridge workmen made sport of Brodie's supposed jump. According to one of the workmen, they made a "stuffed" Steve Brodie. O'Rourke "pushed" him off the bridge.[2]

Over the next few years, Brodie became a popular actor, doing daredevil jumping stunts on the stage. He grew rich. In 1898, he hoaxed New York with false stories of his death out West. When he really died at the age of thirty-six, it was from diabetes, not from jumping.

Another jumper who fixed his attention on the Poughkeepsie Bridge was a former British army captain, Montague Martin of the Bengal Lancers. He claimed he had jumped from bridges in England and India, and intended to surpass the most daring deeds of American bridge jumpers, including Steve Brodie.

On Friday night, Oct. 13, 1894, Martin was arrested in Poughkeepsie for being drunk. On Saturday evening about 6:00 PM, he left a Poughkeepsie saloon in a cab, accompanied by his manager and friends. Newsmen followed him in another cab, and Martin seemed to welcome them. Both cabs drove to the dock at the foot of Main Street where all of the passengers — Martin, his friends, and the newsmen — stepped into two waiting rowboats. They rowed north along the shore and then out to the nearest bridge pier in the water. As those with him told it afterward, Martin crawled out of his boat and onto the stone pier and began to climb the ladder attached to the pier. They watched him climb above the stone of the pier into the metal work of the pier's tower. Then they lost sight of him. According to the *New York Times*, "It was very dark and the iron work was very slippery from the rain which had been falling all day," but Martin said he climbed up 130 feet altogether, to a truss span, and then walked out fifty feet on a girder which was "not over two feet wide." Martin said it was from that girder that he jumped. He jumped, he explained, protecting his feet with rubbers, standing erect and holding his hands above his head. He struck the water, he said, with his feet first.

When those in the boats pulled him from the water, he vomited; that much the observers in the boats could clearly see. Boatmen rowed Martin back to the Main Street dock, and soon afterward

STEVE BRODIE,

The "Police Gazette" Champion Aerial Jumper of the World.

Martin, to recover, retired to his room at the Nelson House hotel. Later in the evening, although he said his chest was sore and his knee joints were swollen, he walked around town.

According to the *Poughkeepsie Eagle* reporter who was present at the jump, "It is the general opinion of the people that the jump was made from the top of the [stone] pier," that is, about 30 feet above the water rather than the 130 feet which Martin claimed. He could "easily" have done this "in the darkness without detection."

Early the next morning at about 4:00 AM, when Martin and his manager went to the station to take a Hudson Line train for New York, they were arrested for not having paid their Nelson House bill.[3]

The next year, 1895, still another jumper showed up at the bridge. He was the Irish-born, New York City bartender Patrick Callahan, aged twenty-seven, who called himself "the King." He claimed that several years before he had jumped from the Black Friar's Bridge in London. He also claimed that just three months ago he had jumped from the Brooklyn Bridge, on a wager. Though injured and hospitalized, he had recovered.

On Sunday, October 27, at 6:08 in the morning, Callahan arrived from New York by the West Shore Railroad at its Highland station. He arrived with his trainer, a number of friends, and several New York reporters, and was met at the station by local reporters as well. His friends told reporters that he had made no wager for this jump, but that he hoped that the fame he would win from jumping would allow him to make money by going on the stage.

Callahan wasted no time in preparing for his jump. At a barn behind the station, he removed his coat, but otherwise kept on his usual clothing, including a derby hat. He strapped on a cork life preserver. Then while two of his friends rented a boat, and rowed out into the river to wait, Callahan himself, with three friends and a reporter, climbed from the station on a "rough path through the hills" to a rise where the western end of the bridge came close to the ground. From there, according to the *New York Times*, they climbed onto the surface of the bridge approach, and walked east along the rail tracks toward a watch tower. Anticipating a challenge from a guard at the tower, the four men "grouped themselves" about Callahan "to hide his equipments," and walked forward boldly. According to the *Poughkeepsie Eagle*, one of them called out, "We are painters."

"Where are your passes?" asked the guard.

"Our boss is coming behind us, and he has got our passes," replied one of the men, and though the guard shouted for them to stop, they pushed on by him. By 6:45 AM they had walked well out on the bridge over the water, and Callahan had chosen a place from which to jump. According to newsmen who were watching from below, it was about fifty feet west of the most westerly pier in the water.

Callahan shook hands with the men around him, stepped over an iron guardrail, brushed through four lines of telegraph wires, and walked out on one of the girders that jutted from under the board walk. Then shouting, "Here goes the King," he let himself fall.

According to the newsmen who timed his fall, it took three seconds. There was no doubt that Callahan had jumped from the top of the bridge.

He struck the water with a boom, like the boom of a bass drum. It could be heard plainly on the shore, 500 feet away.

Friends waiting in their boat quickly pulled Callahan out of the water. Once he was inside the boat, they stripped off his clothing and rubbed him with alcohol "to prevent a collapse." He was "pale and shivering." But when a friend on the bridge above called out, "Are you alive, King?" he stood up and replied, "You bet I am." As his friends cheered, he waved his arms above his head.

When his boat reached shore, he was able to get out without assistance, but as he walked along the West Shore railroad tracks toward the Highland depot and the nearby Dean's Hotel, he needed help. He was bleeding from a laceration near his groin. At the hotel, they laid him on the barroom floor, rubbed him with alcohol again, and dosed him with brandy. By now he was bleeding profusely.

DEAN'S HOTEL, ON THE HUDSON RIVER SHORE, at Highland Landing, behind the West Shore's Highland depot (LLOYD). It was at this hotel that Brodie stayed overnight before jumping from the bridge. It was here that Callahan was taken after he jumped from the bridge, and here that Callahan died. It was also here while the bridge was being built, that James Roosevelt (FDR's father) met with other court-appointed commissioners to decide what should be paid landowners for land seized for building the bridge.

Then despite Callahan's protest, his friends put him to bed. They summoned a doctor from Highland and later two additional doctors from Poughkeepsie. The doctors stitched the wound near his groin. They discovered he was paralyzed from the waist down. They believed he was suffering from shock and loss of blood.

By 5:15 PM, all of Callahan's friends who had come up from New York with him had taken trains back to New York, leaving him behind because he was not well enough to travel. Hotelkeeper William Dean felt obliged to hire someone to stay with Callahan. As the evening went on Callahan seemed to grow stronger. He was able after all to move his legs, and insisted that he would jump from the bridge again next week. However, at 11:25 that night he died.[4]

22. WALKERS ON THE BRIDGE (1889-1920)[1]

To provide the poor "an airing place." To provide poets "moonlit" strolling for "conjuring up sonnets." To provide everyone "perhaps the most restful, placid, pleasing river view in the world."

WHEN TRAINS began running on the bridge, the Poughkeepsie Bridge Company's Chief Engineer Dickinson, who was in charge of bridge operation, required walkers to have passes. But he issued only a few.

Some people resented this limitation. They were aware that the bridge company's original charter empowered the company, though it did not clearly require it, to open the bridge not only for trains and carriages but also for "foot passengers." They felt they had a right to walk on the bridge.

When the bridge was still under construction and the bridge company needed an extension of time to complete it, the company's president, Watson Van Benthuysen, recommended amending its charter to provide for pedestrians to walk across the bridge for free. He believed that it would "be a popular move" which "might disarm opposition" and help the legislature grant the extension of time. However, such a move proved unnecessary. The company succeeded in getting its extension of time with an amendment omitting any provision for carriage or pedestrian traffic on the bridge. The company then completed the bridge without providing for either carriage or pedestrian traffic.

However, the company did build maintenance walkways on the top of the bridge, on each side of it. Being aware of these walkways, substantial numbers of people felt that the company should give the public permission to walk on them, even if it was not required to do so. Responding to public pressure, as early as March 1889, the *Poughkeepsie Eagle*, which was the bridge company director John I. Platt's newspaper, said that for pedestrians to be permitted on the bridge with the payment of a toll "would of course be entirely satisfactory to all."[2]

Since the bridge company did not act on the matter, in 1890 public feeling led to a proposal in the state legislature to amend the company's charter to require it to permit pedestrians on the bridge. While the original charter had allowed the company to charge up to twenty-five cents per pedestrian, this bill, as if reprimanding the company for failing to permit pedestrian traffic, drastically reduced the limit it could charge to seven

ENTRANCE TO THE BRIDGE OVER WASHINGTON ST., POUGHKEEPSIE, MARKED BY TWO GUARD HUTS, looking from the bridge eastward (CNE, *Poughkeepsie Bridge Route*, 1903). The planked walkway which ran along the southern edge of the bridge appears to begin at the entrance. When Joseph Seymour tried to walk across the bridge through this entrance, a guard stopped him. Believing the public had a right to walk on the bridge, Seymour then sued.

cents, about the same as the charge for passengers on the Poughkeepsie-Highland ferry. Nevertheless, the bridge company seemed open to permitting pedestrians on the bridge. Its board of directors, on the motion of *Eagle* Editor Platt, authorized its executive committee "to open the bridge to foot passengers if they see fit." However, the state legislature refused to adopt the bill requiring the bridge be open to pedestrians. The legislature refused, according to the *New Paltz Times*, because Poughkeepsie-Highland ferry interests feared opening the bridge to pedestrians would reduce ferry traffic.

The next year, 1891, with considerable public support, again an amendment requiring the bridge be open to public walking was introduced, and again the bridge company seemed to support it. The company merely added a provision, through the efforts of the company's lawyer, Milton A. Fowler, that the company could prohibit smoking and loitering on the bridge, and prohibit children under fourteen walking on it unless accompanied by an adult. This time, in April 1891, the legislature passed the amendment opening the bridge to walking, and the *Poughkeepsie Eagle* welcomed it as "good news to the public." However, when immediately after the amendment was adopted, "a large number" of Poughkeepsie people went to the bridge to walk across it, a bridge watchman refused them admission.[3]

A month later, a Poughkeepsie constable, Joseph Seymour, had occasion to serve papers on someone across the river in Highland. He found a law clerk to go with him, a hint that he knew it might be wise to have a credible witness with him for what he was about to do. The two men went to Poughkeepsie's Bridge Route depot, on Parker Avenue. As Seymour recounted it afterwards, they walked up to Ticket Agent Theodore Neide and asked him for two tickets to walk across the bridge.

Agent Neide explained that the bridge was not yet open to pedestrians, and, pointing to a watchman's guard house along the tracks at the entrance to the bridge, said Seymour could check it with the watchman. Neide added, 'If you attempt to pass, he'll club the life out of you." Neide afterwards claimed that he was talking to Seymour "in a joking manner."

As Constable Seymour reported it, he then walked to the watchman's guard house, and told the watchman that he wanted to walk across the bridge. He offered him seven cents for himself and seven for his friend.

The watchman refused the money, saying "You can't go; we are not selling any tickets." Seymour laid the money on the guard house window sill, and began to walk across the bridge anyway. However, as Seymour recalled it later, the watchman caught up with him, and turned him around, saying, "Now don't attempt to go; you shall not, for I won't let you." Seymour and his associate then walked away. By the next morning, a lawyer, acting on Constable Seymour's behalf, had formally commenced a suit against the Poughkeepsie Bridge Company, claiming the law required the bridge be open to walking.

By June, the Bridge Company, through its lawyer Fowler, answered the suit. The company said that once the legislature enacted a law requiring the company to allow the public to walk across the bridge, the company "had proceeded with reasonable diligence to construct the entrances for foot passengers, but at the time complained of in this action no such entrance had been erected." In the meantime, "watchmen were employed to keep persons from entering upon the bridge" to prevent them from "being run over by the railroad trains." The company asked that the complaint be dismissed.

In October 1891, six months after the law opening the bridge to public walking had been enacted, the judge handling the case issued a decision. By this time, the Bridge Company was openly balking at carrying out the requirement of the law, and the judge was sympathetic to the company's doing so. The judge explained that the footpaths on each side of the bridge were built for maintenance and were not safe for public walking. To make them safe, the paths would have to be widened and a guard rail built to separate them from passing trains. It was a question whether the legislature had a right to "impose" the "great expense" of such improvements on the Bridge

Left, WALKWAY ALONG THE BRIDGE'S SOUTHERN EDGE, looking toward Poughkeepsie. Showing indistinctly, upper left, are two women pedestrians on the bridge's northern walkway (PR&NE. *Summer Homes*, 1895). Right, ENLARGED DETAIL OF THE TWO WOMEN, one with a parasol. Water barrels, for fire fighting, show between the two sets of railroad tracks.

Company. Besides, "the company has not had time to comply with the law and it is not certain that compliance is possible." Thereafter the CNE&W railroad, which controlled the bridge at the time, made it more difficult for anyone to secure a pass to walk on the bridge, requiring applicants to apply in Hartford.[4]

But pressure to open the bridge for public walking continued. Two years later, in April 1893, the legislature again amended the Bridge Company charter to require the company to open the bridge to walkers. This time the legislature, as if it was increasingly exasperated with the company, required the bridge be open by a specific date, October 1, and reduced the amount the company could charge walkers still further, to five cents per person.

The *Poughkeepsie Eagle* reported that people were pleased the amendment had become law, and predicted that while the walk across the bridge was "most too far" for business purposes, "for pleasure walking it will be a great success." The *Poughkeepsie News-Telegraph* speculated that while a police patrol would be necessary on the footway to prevent its being abused, the footway would provide an "airing place for the poor," "moonlit" strolling for poets "conjuring up sonnets," and an opportunity to recall Longfellow's words:

> I stood on the bridge at midnight,
> As the clocks were striking the hour,
> And the moon rose o'er the city,
> Behind the dark church-tower.

In the long run, the *Telegraph* predicted, the "foot-passage" over the bridge, with its "bracing air" and "perhaps the most restful, placid, and pleasing river view in the world," was "destined to be an important factor in the advancement of our city."

By this time, the bridge was owned by the Philadelphia, Reading, & New England Railroad. Editor Platt was no longer one of the bridge company directors, and many of the directors were insulated from local public opinion by living in Philadelphia or elsewhere at a distance from the bridge. The company kept the bridge closed to walking, seeming to flout the plain intent of the law. According to the *Kingston Leader*, "People are wondering why a railroad corporation should be allowed to set the laws of the state at defiance."[5]

In the next few years, even though the bridge was not open to public walking, some people walked on it anyway and found it safe. One summer evening in 1897, an eastbound local train failed to arrive as scheduled at Pratt's Mills station, just west of Highland. The train failed to arrive because, as the waiting passengers learned only later, its engine had been derailed. Eventually the passengers waiting for this train, having become, according to the *Poughkeepsie Eagle*, "weary," and some even "a little mad," gave up waiting for it, and considered getting to Poughkeepsie by walking across the bridge. They seemed to feel they had a right to walk over it since it was the railroad which owned the bridge which had let them down. The more timid passengers, "afraid to venture on the big structure at night," walked through the village of Highland and down the long slope to Highland Landing, a walk of three miles, to take the ferry across the Hudson to Poughkeepsie. However, most of the passengers decided to walk along the tracks to the bridge and then walk across it, a walk altogether of about the same distance as the walk to the ferry, but a walk which would take them directly to Poughkeepsie. When these adventurers reached the entry to the bridge, a watchman there refused to let them enter. He "drew his club, threatening dire vengeance on all who tried it." But he was "overcome" (how the *Eagle* left unsaid), and they walked "safely" across the bridge in the dark to Poughkeepsie.[6]

For decades thereafter, according to the state's Public Service Commission, controversy over the right to walk on the bridge "raged." Meanwhile, the bridge managers continued to refuse public walking. They did not build the public entrances to the bridge which had been recommended to make the bridge available for public walking. They did not either strengthen the flimsy outside railings that would protect walkers from falling off the bridge or erect interior railings that would protect walkers from being struck by passing trains.

By 1920, the Rev. J. F. Sheahan of St. Peter's Catholic Church, which was close to the bridge in Poughkeepsie, made a legal complaint on behalf of people who continued to believe they had a right to walk across the bridge. With the assistance of such Poughkeepsie lawyers as John J. and Philip A. Mylod, Sheahan complained to the state Public Service Commission which regulated railroads, including railroad bridges.

The commission held hearings on the complaint. Commission members personally inspected the bridge. The Central New England Railway, the owner of the bridge at the time, arranged to run a train onto the bridge to take them, along with lawyers for both sides, to see for themselves how safe the walkways were. On August 24, 1920, the commission issued a ruling on the issue.

The commission acknowledged that the laws of both 1891 and 1893 were clearly intended to require the bridge owners to permit public walking on the bridge. The commission thanked the Rev. Sheahan and his associates for their "public spirit" in asking why then walking was not permitted. However, the commission determined that if the bridge were to be opened to public walking, alterations in the bridge were necessary to make it safe for that purpose, at significant cost—the CNE claimed it would cost $250,000. The commission concluded that because the construction of the bridge was completed at a time when the bridge company charter did not require it to permit public walking, the state could not constitutionally later add the requirement that it do so, unless the state gave the company "due compensation," as it had not done.[7]

For many years thereafter the commission's decision kept the issue quashed. It was not until the bridge was over a hundred years old that the question of allowing pubic walking on it re-emerged as a pressing public issue.

23. A BRIDGE FOR AUTOMOBILES? (1919-1930)[1]

"For modern engineering the impossible does not exist."

AFTER WORLD WAR I, when automobiles and trucks were becoming common, motorists were impatient with Hudson River ferries—they ran infrequently and were unreliable. Except for the Poughkeepsie Railroad Bridge, there was still no bridge of any kind across the Hudson all the way from the ocean up to Albany. By 1919, motorists were campaigning to add a highway onto the Poughkeepsie Railroad Bridge, so they could drive across it.

When the New Haven Railroad seemed reluctant to respond to the campaign, motorists appealed to the Pubic Service Commission to compel the New Haven, through its subsidiary, the Central New England Railway, to add a highway to the Poughkeepsie Bridge. In response, New Haven officials argued that, while originally the bridge charter had allowed the bridge to be built for highway as well as railroad use, changes in the charter while the bridge was being built relieved them of any responsibility to open it to highway traffic, the same as it had also relieved them of responsibility for opening it to foot traffic. They also argued that from the engineering point of view, adding a highway to the bridge would add too much weight for its safety, no matter where on the bridge such a highway might be added. In August 1920, the Public Service Commission, agreeing with the railroad, decided it was not obliged to rebuild the bridge for highway use.

Meantime, a rumor arose that the New Haven Railroad would be amenable to having its Central

CLIPPINGS ON PROPOSALS TO CONVERT THE BRIDGE TO HIGHWAY USE (*Poughkeepsie Eagle News*, Dec. 7, 1920, Jan. 27, 1921)

New England subsidiary sell the bridge to the state, giving the state the opportunity to turn it into a highway-only bridge, and freeing the New Haven to build a new railroad bridge. CNE Superintendent G. W. Clark dismissed the rumor of such a sale as "tiresome." However, both the New Haven management and its Consulting Engineer Ralph Modjeski, while believing it would not be wise for the state to buy the bridge, came to believe that if the state did so, it would have advantages for the railroad. At the time, Engineer Modjeski was pushing the New Haven Railroad's president, E. J. Pearson, to regard the life of the bridge for railroad purposes to be short. But Pearson came to doubt that even if the New Haven sold the bridge to the state, the New Haven could afford to build a new one to take its place.

Late in 1920, the Poughkeepsie Automobile Club invited Engineer John F. O'Rourke to come to Poughkeepsie to speak on the feasibility of remodeling the Poughkeepsie Bridge for highway use. By this time, O'Rourke was well known for having figured in building not only the Poughkeepsie Bridge but also the Pennsylvania Railroad's passenger tunnels under the Hudson River at New York. On his visit to Poughkeepsie, O'Rourke

HUDSON RIVER BRIDGES (SELECTED)				
Opened	Location	Name	Traffic	Design
1889	Poughkeepsie	Poughkeepsie Railroad Bridge	Rail	Cantilever
1924	Upriver from Peekskill	Bear Mountain Bridge	Auto	Suspension
1924	Selkirk-Castleton	Alfred H. Smith Bridge (sometimes called Selkirk-Castleton Bridge)	Rail	Truss
1930	Poughkeepsie	Mid-Hudson Bridge (in 1995 renamed the F. D. Roosevelt Mid-Hudson Bridge	Auto	Suspension
1931	NYC	George Washington Bridge	Auto	Suspension
1935	Catskill	Rip Van Winkle Bridge	Auto	Cantilever
1955	Tarrytown (Thruway)	Tappan Zee Bridge	Auto	Cantilever
1957	Kingston	Kingston-Rhinecliff Bridge	Auto	Truss
1959	Castleton (Thruway Extension)	Castleton Bridge	Auto	Cantilever
1963 (second span, 1980)	Newburgh (I-84)	Newburgh-Beacon Bridge (in 1994, renamed Hamilton Fish Bridge)	Auto	Cantilever

inspected the bridge and expressed his pleasure in renewing old acquaintances. In his talk, O'Rourke argued that a two-lane highway could be added to the bridge underneath its railway tracks. But it would be expensive, he maintained, particularly because of the cost to construct the highway approaches.[2]

As agitation for some form of highway bridge continued, in 1921 the Poughkeepsie Chamber of Commerce invited another eminent engineer to speak about the possibility of adding a highway to the Poughkeepsie Bridge. He was Gustav Lindenthal, the well-known Austrian-born designer of the Hell Gate railway bridge over the East River—the bridge whose opening had deprived the Poughkeepsie Bridge of its most famous passenger train, the Federal Express. With a "halting" German accent, Lindenthal reviewed for a large audience how the Poughkeepsie Bridge had been built cheaply, as "a desperate effort to get the bridge done on any terms." Although the public did not know it at the time, he said, "it was well known to engineers . . . that in building the bridge you did not get the structure that would take care of the future." Like other bridges, it was "built too light." Of course when it was finished, "everywhere it was known as one of the great bridges of the world," Lindenthal said. But since then, the weight of trains has kept increasing. The CNE has already had to strengthen the bridge more than once, and, he predicted, will have to strengthen it again.

Nevertheless, Lindenthal agreed with O'Rourke that the railroad bridge could be remodeled for highway use. "For modern engineering the impossible does not exist," he said. Bridge building "has progressed farther probably than any other science of building in the history of mankind." But some kinds of bridge building, he explained, are more suitable than others.

Adding a highway to the bridge on the same level as the railroad tracks, next to the tracks, would be expensive, Lindenthal argued, and would make the highway lanes narrow, so they would be easily choked if an auto broke down. Adding a highway under the tracks would require strengthening the towers, and it would be a question if their foundations were strong enough. Besides, it is not wise to build a highway under a railway track, he said, because "everything trickles down" through the "open space between the ties"—doubtless meaning "everything" like coal, cinders, brine, and oil.

To prevent the trickling down, rebuilding the track level with a solid, fireproof flooring would be both "a considerable expense and a considerable weight." If the highway level were built with a cheap wood floor to carry the traffic, such as was originally built on the Brooklyn Bridge, it would add little weight, but it would quickly wear out from the traffic. Moreover, wood can catch fire, and a fire on a bridge "cannot be put out quickly," and could mean the bridge's "destruction" from heat causing the "twisting up of the steel"—a prophetic comment to make about the Poughkeepsie Bridge.

With both the New Haven Railroad and respected independent engineers pointing out difficulties in remodeling the railroad bridge for highway use, if in varying degrees, local boosters turned instead to calling on the state to build a new highway bridge at Poughkeepsie. Franklin D. Roosevelt, of nearby Hyde Park, the recent Democratic candidate for the U.S. vice presidency, became "intensely" interested in the proposal.[3]

In 1922, an association of advocates of a new highway bridge at Poughkeepsie asked the engineering firm of George W. Goethals, the engineer of Panama Canal fame, to propose the design of such a bridge. The Goethals firm proposed to place the bridge about half a mile south of the railroad bridge, and to design it, as federal navigation authorities had suggested, like the railroad bridge, with four piers in the water and the same spacing between them, so that water traffic could pass smoothly in a continuous course under both bridges. By 1923, the New York state legislature voted to build a new highway bridge at the site Goethals had proposed. But by 1924, the state asked not Goethals but Modjeski and his associate Daniel E. Moran to design the new bridge, at the same time that Modjeski continued to be a consulting engineer for the railroad bridge.

For the new highway bridge, Modjeski and

ELEANOR ROOSEVELT CUTTING THE RIBBON TO OPEN THE NEW MID-HUDSON BRIDGE, AT ITS HIGHLAND END, AUG. 25, 1930 (FDRL). With left, STATE SUPT. OF PUBLIC WORKS FREDERICK S. GREEN, and right, THE DEISIGNERS OF THE BRIDGE, RALPH MODJESKI (a long time consultant on the Poughkeepsie Railroad Bridge) and, farthest right, DANIEL E. MORAN.

Moran abandoned the Goethals design, and developed a more pleasing design of their own, a suspension bridge, with only two piers in the water. As we have already had occasion to observe, in 1930, Franklin D. Roosevelt, by this time New York State's governor, along with Mrs. Roosevelt, opened the graceful new Mid-Hudson Bridge, providing the Poughkeepsie Railroad Bridge a worthy companion.[4]

24. Hoboes (1890s-1940s)

"Where they boiled in oil the inventor of toil."

QUESTIONS about hoboes abounded: Why did anyone become a hobo? Why did hoboes reject home and family? Why were they willing to put up with discomfort and danger? When railroaders and others came across hoboes, why did they sometimes want to hinder hoboes, sometimes want to help them, and sometimes want to become hoboes themselves?

About when the Poughkeepsie Bridge opened, a veteran railroader claimed that so many "tramps"—as hoboes were usually called at the time—were coming by train to Poughkeepsie, that if he stood at Poughkeepsie's Hudson Line station in the early morning hours when the trains came in, and swung a baseball bat, he would hit one every time. They ride, he said, between cars and under cars, and "notwithstanding they get killed, or lose an arm or a leg, their number never seems to decrease."[1]

Not long afterward, on a dark winter evening with the temperature almost down to zero, a New York & New England passenger train was stopped at the Willimantic, CT, depot, when a tramp, dressed in a shabby, thin coat, approached its engine. Timidly, he asked the engineer if he might ride with him in the cab to Boston. Engineer Potter replied, no, the rules of the road forbid it. The tramp shivered and moved on.

Engineer Potter had brought his train through Hartford, where it connected with the Bridge Route, and had stopped it in Willimantic to let passengers out for a hurried supper at the depot restaurant. When their time was up, the locomotive bell clanged, the passengers rushed to climb aboard, and the train chugged off.

After the train had resumed its normal speed, according to the *New York Sun*'s account, Engineer Potter and his fireman, leaning out of a cab window, "saw the wretched tramp," the same one who had begged for a ride to Boston. Trying to stand on the cowcatcher, he was shaking with the shaking of the engine. Then, as the fireman and engineer both watched, the tramp "made a desperate effort" to climb higher on the engine nose. "In doing so he grasped a live steam pipe. The pipe was blistering hot, but the tramp dared not let go, for he would have fallen to the roadbed."

"The train was still thundering on at fifty miles an hour, and the engineer dared not apply the brakes, for he knew the sudden jar. . .would fling the wretch under the wheels." But Engineer Potter saw a way he might save him. He crawled out along the side of the locomotive, with the fireman following him, "and yelled to the man to hold up his hand. The engineer then fairly pulled the tramp, with the aid of the fireman, into the cab." The tramp fell to the floor. Though burned and exhausted, he had found a ride to Boston.[2]

Like Engineer Potter, the Bridge Route's Conductor Nate Blodgett tried to keep tramps off his trains. One night in the 1910s, Blodgett was climbing into his CNE caboose as it pulled out of Simsbury, CT, on its way toward the Poughkeepsie Bridge, when he found two tramps—or "hoboes" as they were beginning to be called—on the rear platform. Blodgett promptly "pulled the air [brake]," he remembered, "stopped the train and put them off." The next evening, Blodgett was returning east on another train when, at Highland, just before crossing the Poughkeepsie Bridge, he "saw the same two guys in an empty gon [gondola car] on a west bound extra."

In dealing with hoboes, Blodgett recalled, he often had the help of Brakeman Johnny Kepler. He "was about 6 feet 3, weighed about 220, liked to fight, and could." With Kepler's help, Blodgett had "very little trouble" with hoboes.[3]

During the Great Depression of the 1930s, when probably more than a million hoboes were riding America's railroads, Americans often sang about hoboes. When they sang "Big Rock Candy Mountain," they seemed to identify with hoboes who refused to work:

> There's a land that's fair and bright,
>   Where the hand-outs grow on bushes
> And you sleep out every night,
>   Where the box cars are all empty
> And the sun shines every day. . .
>   Where they boiled in oil
> The inventor of toil
> In the big Rock Candy Mountains.

When Americans sang "Hallelujah, I'm a Bum," they seemed to relish playing at being hoboes themselves:

> When springtime does come,
>   oh won't we have fun!
> We'll throw up our jobs
>   and we'll go on the bum.

In Highland, at the western end of the Poughkeepsie Bridge, hoboes arrived on trains of both the Bridge Route and West Shore lines. But according to the recollection of Arthur Lyons who grew up there in the 1920s and 1930s, more hoboes arrived on the Bridge Route. One reason was that the Bridge Route had a larger proportion of freight rather than passenger trains, and hoboes preferred to ride on freights. Another reason was that unlike the West Shore freights, Bridge Route freights had to slow in Highland for crossing the Poughkeepsie Bridge, and when they slowed hoboes had a better chance for jumping on and off.

Whether the hoboes came and went by the West Shore or the Bridge Route, they often camped in a glen in Highland almost underneath the bridge's approach viaduct. Lyons does not recall going into the camp when hoboes were there. He didn't go there, he explains, because his father wouldn't allow it. But Lyons remembers seeing the camp fires glow at night. He also remembers going to the camp site when the hoboes were gone and finding cans of Sterno—cooking fuel—which the hoboes had left behind. Hoboes drank the Sterno, he explains, for its alcohol.

Dorothy Phillips grew up near the same hobo camp. She called it "Hobo Heaven." Her house was high on Mile Hill Road, and she remembers the camp as downhill, along a clear stream, in a low place near the bridge viaduct.

Phillips, Lyons, and many other Americans found themselves fascinated by hoboes. How did men become hoboes? Was it because, as songs about hoboes suggested, they couldn't find jobs, or were disappointed in love, or were weary of repetitive trivia like washing dishes? Or more fundamentally, was it because they resisted growing up? Or were in pursuit of a private dream which was nobody's business but theirs?

Hoboes often drank, both Phillips and Lyons knew, but they were seldom aware of hoboes becoming obnoxious. A few of the hoboes settled in the Highland area, even if their drinking persisted. One, Phillips recalls, became a house painter, another an electrician.

Railroad men were not usually worried about hoboes, it seemed to Phillips, though occasionally she heard that the "bulls"—the railroad police—would be rough on them, to keep them under control. Her father, Tom Phillips, as a track foreman for the West Shore Railroad, did not seem afraid of them; to her, he did not seem afraid of anything. However, he did not allow her to visit the hobo camp. When hoboes came to her house for food, her mother, with her father's consent, would feed them, as Arthur's mother did too. But at both houses, the mothers fed them not inside the house, only on the porch.[4]

Antonio Marano, whose father was a Maybrook yard worker, recalls that when hoboes came to his family's door in Maybrook, they never asked for money, only food, and they seemed grateful when they received it. His mother would feed them the same spaghetti and woodchuck that the family ate.

Marano remembers that in the 1930s, there was a hobo camp along the track which led from Maybrook toward Campbell Hall. It was in a ravine, in a woods. The hoboes made fires there, and cooked over those fires.

Marano recalls that a black hobo often came to Maybrook, especially by the Ontario & Western line. One of his eyes seemed blue, perhaps because it was an artificial eye. This hobo sometimes stayed in the Maybrook region for weeks at a time, whether in a hobo camp or elsewhere, but by winter he was always gone. He regularly carried two packs with him, one for his clothes, the other for his simonize polish equipment. The boys called him "Simonize Sam." If railroad police found occasion to bother him, he would try to appease them by giving their automobiles a free simonize job.

When Marano was in his teens, if he was hanging around with other boys on the railroad tracks and they saw hoboes, they would talk with them and give them cigarettes. They never had trouble with hoboes, and did not know anyone who did. Later, after Marano had become a brakeman on the Bridge Route, if hoboes asked him what freight to take to ride to New England, he would tell them. Sometimes he would even given them a ride in his caboose.

By the 1940s, the number of hoboes on the Bridge Route was declining, as they were on other rail routes. More opportunity for jobs was a factor in the decline. So also were better government programs to help the poor and elderly. So also were faster trains, with fewer stops, making it less feasible for hoboes to ride them.

Nevertheless, in the 1940s, while Poughkeepsie Bridge maintenance man Ferris Davis recalls that hoboes were no longer camping under the western end of the bridge, he still saw hoboes riding the trains that crossed the bridge. Moreover, Brakeman Alexander recalls often still seeing hoboes in the 1940s, everywhere along the line from Maybrook over the bridge to New Haven. When his train stopped and he walked the tracks beside it, as he regularly did, checking for hotboxes, Alexander often found hoboes. Sometimes he found them "sitting up" in empty box cars.

We brakemen were not afraid of hoboes, Alexander explains, even though when we were at work on a train we usually carried a gold watch in the bib of our overalls, and fifty to one hundred dollars in our pockets. We knew that hoboes "could easily attack" us and take our watches or money and run. "But they wouldn't do that. They figured that they were privileged to be on that train," and "they didn't want to spoil the privilege." Hoboes would attack other hoboes to prevent them from interfering with the hoboes' privilege of riding a train. But they would not attack trainmen.

As a track foreman, THOMAS J. PHILLIPS (shown on a Hudson River dock) was not afraid of hoboes. But as a father, he would not allow his daughter Dorothy to visit the hobo camp near his house (Poks. New Yorker, June 6, 1948).

Although both Engineer Potter and Conductor Blodgett were capable of being severe with hoboes, both Alexander and Marano, as they looked back on their years on the Bridge Route as brakemen or conductors, found it natural to express their pleasure in their railroad work by comparing themselves to hoboes. Their work, they said, made them feel like hoboes, like "paid" or "glorified" hoboes.[5] Evidently on some level, both of them, like many other Americans, yearned to be hoboes, if only in imagination.

25. POUGHKEEPSIE REGATTA (1895-1949)[1]

"Casting anxious eyes at the railroad bridge, waiting for the smoke bombs to tell us who won."

FOR FIFTY YEARS, college students gathered at Poughkeepsie for the annual rowing races known as the Intercollegiate Regatta. These races came to be called, next to the Henley Royal Regatta in England, the most famous rowing races in the world.

Before coming to Poughkeepsie in 1895, the intercollegiate races had been held at a variety of sites, as on Cayuga Lake near Ithaca, NY, and on the Delaware River near Philadelphia. Poughkeepsie's Hudson River site offered advantages. It was easily accessible by train as well as by Hudson River boat. Its course was wide, nearly straight, and somewhat protected by highlands from wind. According to *Harper's Weekly*, one of the site's "best points" was that from the start of the race, the coxswains could "get a line" on a "particular pier in the huge railway bridge. . .and steer direct for it, without fear of losing sight of it," freeing them to concentrate on the rowing. Moreover, the bridge provided a convenient location from which to signal to spectators who had won the race. The famous bridge engineer, Gustav Lindenthal, once quipped that the Poughkeepsie Bridge was "more beneficial for the regattas running under it than for the trains running over it."[2]

Crews participating in the regatta were all men, and they participated by invitation only. They came from such prestigious colleges as Cornell, Columbia, Pennsylvania, the Naval Academy, Wisconsin, and Stanford. In the early years, the crews arrived by steamer or train. They brought along their own shells—they were usually 8-man shells, each 60 feet long, only 2 feet wide, with pointed bows.

Rowing might demand grueling practice even in winter, as with weights or hydraulic rowing machines. However, as the artist Thomas Eakins, a rower himself, suggested in his paintings, rowing was an elegant sport, emphasizing skill and grace, and the rowers were likely to be of only moderate weight.

The rowers often arrived in Poughkeepsie early, sometimes more than a week early, to familiarize themselves with the site. To honor the visiting rowers, friends of the regatta held receptions, concerts, and dances, occasionally providing an opportunity for them to meet Vassar girls. Among the sponsors of such events at various times were such local bridge promoters as the Platt, Innis, and Roosevelt families, and among those attending were Bridge Engineer John F. O'Rourke and his wife.[3]

Before the races were held, usually in late June, thousands of spectators converged on the Poughkeepsie region, many of them by special trains on the three rail lines which served the region, the Hudson Line, West Shore Line, and Bridge Route. In 1897, the Bridge Route ran four special trains to the regatta, two originating in Hartford, one in Bridgeport, and one on the Ontario & Western line. In other years the Bridge Route offered special excursion rates on regular trains, or altered the schedule of regular trains for the convenience of the spectators, as by making special stops at the western end of the bridge for spectators to detrain near prime viewing sites.

Spectators crowded onto both sides of the river, but especially onto the Highland side, one reason being that the race course was laid out closer to that side. On both sides of the river, some spectators chose sites near the water, others chose to climb up rocky bluffs overlooking the water.

While the spectators watched, rowers practiced their strokes on the river, and trains kept crossing the great bridge above them. In the 1930s, some of the rowers, as a Syracuse University rower recalls it, placed bets on the number of cars in the trains crossing the bridge. They tossed coins into a hat to create a stake. Whoever guessed the number of cars in a particular train most accurately won the stake.[4]

On racing day, boats maneuvered in the river for favorable viewing positions. Among such boats on occasion were the yachts of John Jacob Astor, Frederick W. Vanderbilt, and J. P. Morgan, as well as ferry boats sent upriver from Newburgh, and Day Line steamers making special excursions. In 1938, six destroyers came up the Hudson at once, bringing the Naval Academy's entire second class from Annapolis to view the races. When Franklin D. Roosevelt could, he tried to watch the racing from his presidential yacht, the *Potomac*.[5] Roosevelt rooted for the Academy's crew, the "Navy."

In the earliest years, some spectators rode a special train that "steamed slowly out on the bridge," and parked there so the spectators could view the races. Many more spectators, however, including journalists and regatta officials, rode an "observation train" on the West Shore Railroad along the Hudson River shore, a train which became a regular feature of the racing. This special train, having two locomotives, one at each end, facing in opposite directions, could move easily up or down the river shore, but was especially intended, during each race, to move downriver in pace with the straining oarsmen and their rhythmically dipping oars. The train included 25 to 50 cars. They were often ordinary flat or gondola cars, but were fitted with tiers of wooden bleacher seats, open to the air. To reduce the smoke which might annoy the passengers in their open-air cars, this train's locomotives burned hard rather than the usual soft coal. On the train, the passengers were sometimes packed so tight that it was said that there was hardly room for the ladies

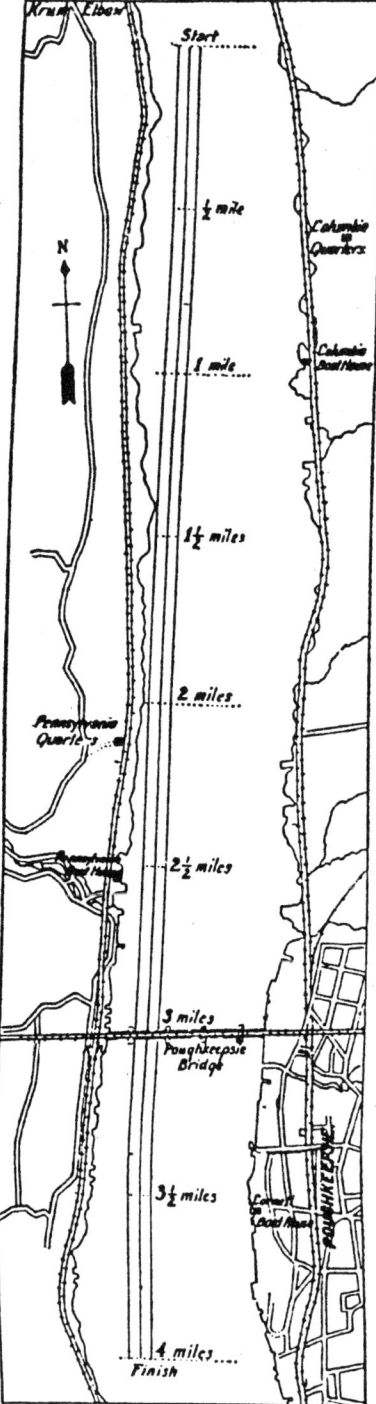

THE RACE COURSE (adapted from *Harper's Weekly*, June 22, 1895). While the location of the crew quarters and boat houses changed from time to time, according to this map those of Pennsylvania were on the west shore in Highland, those of Columbia on the east shore in Hyde Park (on the estate of John A. Roosevelt), and those of Cornell on the east shore in Poughkeepsie city.

to open a fan, but they managed to wave flags and burst out now and then with such yells as "Cornell, Cornell, I yell, yell, yell."[6]

In the regatta's early years, while pedestrians were not usually permitted on the railroad bridge, newsmen were stationed there to report the racing, which they did over special telegraph and telephone wires strung on the bridge. By the 1930s, when radio broadcasting had become well developed, radio commentators were stationed there to broadcast the regatta spectacle live. Among the commentators was Graham McNamee, who, in his enthusiasm, would talk breathlessly and tear off his necktie.

The shells raced in lanes, sometimes as many as ten or eleven of them depending on how many colleges the Regatta Association had invited to race that year. The lanes were marked both by buoys and by conspicuous signs on the railroad bridge, signs that numbered the lanes.

The start lines were always placed north of the railroad bridge and measured from it, but they were placed differently for different races. For the longest race, the four-mile Varsity Race, the start line was usually placed on the river opposite the southern end of the F. D. Roosevelt family estate in Hyde Park. The other two races being shorter—three miles for the Junior Varsity and only two miles for the Freshmen—their start lines were placed closer to the railroad bridge. The races always passed under the bridge. The finish line was usually about a mile south of it, in nearly the same location year after year.

To mark the locations for the races, from about 1923 on, the West Shore Railroad's track-section foreman, Thomas J. Phillips—the father who allowed his family to feed hoboes on their porch—would marshal the help of West Shore Railroad surveyors signaling from shore. They would locate the start and finish lines. They would also locate where the stake boats were to be anchored to mark the head of each lane. When a race was about to begin, each crew rowed its shells to the appropriate stake boat, at the head of the appropriate lane.

SHELLS PASSING UNDER THE BRIDGE IN POUGHKEEPSIE'S FIRST INTERCOLLEGIATE ROWING REGATTA, 1895, with spectator boats crowding the Poughkeepsie shore. (Drawing, *Poughkeepsie Daily Eagle,* June 25, 1895)

As starting time approached, spectators tended to wax noisy, but regatta officials called out with megaphones asking them to be silent, to help them hear the opening ritual. If the wind was right, some spectators might hear the referee starter calling out three times, "Are you all ready?" Then they might see his pistol puff out smoke, hear its crack, and see the shells leap forward.

In the shells, each rower's feet were tied into shoes which remained in a fixed position, attached to the shell. However, each rower sat on a seat which moved on rollers as he rowed. He rowed by pulling with both hands on one oar.

The shells being delicately balanced, if the wind or currents were rough, they sometimes tipped over. To avoid rough winds, the races were sometimes postponed even at the last minute, to the frustration of the crowds. To avoid turbulent currents, the races always went downstream, and were timed so that the tide was flowing out, with the current, toward the ocean.

As the shells approached the finish line, spectators strained to see which shells were leading. Each shell flew a flag showing its college colors, Cornell's red and white, Columbia's light blue and white, Pennsylvania's dark blue and red. But especially if many crews were racing, the spectators sometimes could not see well enough to be sure which colleges were ahead. For authentic information, they often looked for signals from the great railroad bridge which towered over the race course.

Beginning in the regatta's early years, by arrangement with whatever railroad owned the bridge at the time, regatta officials placed signalmen on the bridge to indicate which shell passed first under the bridge, and then again which shell first reached the finish line. The signalmen did this by setting off fireworks — that is, smoke and noise bombs. As a sports writer recalled in the 1940s, "We can remember casting anxious eyes at the railroad bridge, waiting for the smoke bombs to tell us who won the big event. A puff of smoke, then a muffled boom that echoed off the distant hills. Slowly —one—two —-three!"[7] If the signalmen set off three bombs, that would mean that the crew racing in lane three had won.

Another way the signalmen did their signaling was to let down flags from the bridge—at first only from the railroad bridge, later, after the companion Mid-Hudson highway bridge had opened nearby, from it too. Once in the 1930s, when Phillips was in charge of arranging the signaling, he hired a nineteen year old, Arthur Lyons, to help let down those flags from the railroad bridge. Since it was depression time, Lyons had seldom held any job at all. He was delighted to have this job, even if it was only for one day.

While Lyons had grown up in Highland, he had never been out on the bridge before. He felt the walkways along the edge of the bridge were narrow and the railings alongside them were flimsy. He found it scary to look down to the water, whether he was looking down through the railroad ties or over the edge of the bridge.

As the shells slid down river, passing under the bridge and continuing on toward the finish line, some of the shells were close together, but Lyons and the experienced signalman he worked

THE STARTER HAS FIRED HIS PISTOL TO START THE RACE—A PUFF OF SMOKE IS VISIBLE (Intercollegiate Rowing Asso., *Official Program*, 1910). The stewards' boat shows at the left. The stake boats which mark the starting line show left center.

SYRACUSE UNIVERSITY VARSITY CREW. Their oars are so tall they reach above the photo. The coxswain is kneeling. (Intercollegiate Rowing Asso., *Annual Regatta*, 1941)

with could see from their vantage point on the bridge which shells were ahead. They could identify which college each shell represented by both the lane it was racing in and the colors it displayed.

The two men worked as fast as they could. They hooked college flags to a rope. First they hooked the flag of the college that was ahead in the race at the top of the rope, then the flag of the college that was second, and so on. Then they payed out the rope over the edge of the bridge, displaying the flags. As each bomb or flag made its message known, spectators screamed and boat whistles screeched, together lifting a great roar over the river, from shore to shore.

The two men were to hang out the flags for each of the three races in succession, Freshmen, Junior Varsity, and Varsity. In the intervals between races, Lyons found himself hanging onto the railing, insubstantial though it seemed. In one interval, the unexpected happened. A train huffed onto the bridge. As it moved farther out over the river, it seemed to Lyons that the movement of the train caused the bridge not to sway from side to side as he expected it might, but to bounce up and down. He grew queasy. As the engine came closer, roaring and hissing, Lyons felt it was growing into a monster who might demand all the room on the bridge. When the engine came up onto Lyons, he cringed against the railing. He felt that the engine crewman, watching him from high up in his cab, was laughing at him. Lyons vowed he would never go onto the bridge again. He never did.[8]

Because of the open setting of the regatta, it was not practical to charge spectators for watching. This helped to make it difficult to pay the regatta expenses, even when the sponsoring committee included Vincent Astor, President Henry N. MacCracken of Vassar College, and President Franklin D. Roosevelt. Besides the difficulty in raising funds, feuding among the regatta promoters and concerns about the vagaries of the river currents were among the reasons for regatta officials preferring to leave Poughkeepsie. A proposal to send an observation train onto the Poughkeepsie Bridge in addition to the traditional one on the West Shore rail line was not enough to stop the regatta from moving away.[9] In 1950 it moved to the Ohio River at Marietta, later to Lake Onondaga near Syracuse.

After the regatta left Poughkeepsie, other boat races were held at about the same location on the Hudson. But none of these other races had the panache of an observation train running along side the racers, passing with them, race after race, under the great railroad bridge. Also none of them had the excitement of having the winners announced ritually, by the firing of bombs and the lowering of flags, from the great bridge.

Railroaders at Work

26. CONDUCTOR BLODGETT, STORYTELLER

(1906-1910s)

"Cat, if you could cross that bridge in the shape it's in, you have a home for life."

As a young railroader handling freight, Nathan ("Nate") Blodgett could lift a 350-pound barrel of sugar. He could lift it onto his back, and then step backwards to load it into a freight car door. Once he took turns loading barrels this way with another crew member, Matt Kane, who was taller and stronger than he was. Kane loaded the first, third, and fifth barrels, while Blodgett loaded the second and fourth. Afterwards, Blodgett was pleased when Kane told him, "You're a good little man for your inches."

Blodgett began working for Bridge Route-related railroads in 1906 when he was twenty years old. He worked for several different railroads, but especially the Central New England. He worked in a variety of roles, including that of brakeman, conductor, and yard crane operator. At various times he was based in different rail yards, including Boston Corners and Maybrook, NY, and Canaan, Hartford, and New Haven, CT.

Blodgett grew up near Waterbury, CT. His father was an engineer on the Hartford, Providence, & Fishkill Railroad (later part of the New York & New England). His father, even after he quit railroading, stayed close to railroad workers. Blodgett's mother, observing this, liked to say, "Railroad men stick together."[1]

Like his father, Nate Blodgett enjoyed the forthright style of railroad men. He respected their ability to handle heavy and dangerous work. He

CONDUCTOR BLODGETT AND CREW WITH THEIR CABOOSE, AT THE CNE'S CANAAN, CT, YARD, 1910s (FULLR). Blodgett is third from the left, with arm outstretched.

felt positive about many of the particular men with whom he worked. However, Blodgett was not always politic in what he said to his fellow railroaders or even to his bosses, as the following story, which he told about himself, illustrates.

THE RIGHT SOLUTION

While Blodgett was the conductor for the CNE on a Hartford to Winsted local freight, some CNE officials suspected that he and his crew were deliberately working slowly. Blodgett and his crew did local pickups and deliveries along their route. At Winsted they swapped cars with other trains, and did the switching necessary to head their own train back to Hartford. The whole round trip, Hartford to Winsted and back, was only seventy miles, but because it usually took about fifteen hours, the CNE had to pay the crew overtime.

One day, one of Blodgett's superiors, J. W. Cuineen, "rode the job" to see how Blodgett and his crew did their work. While the train was stopped at Simsbury to take on water, Blodgett recalled, he and Cuineen were together on the freight house platform when Cuineen said, "Blodgett, I consider this job can get done in twelve hours."

Blodgett replied, "Not and do the work, Mr. Cuineen."

Cuineen insisted, "Yes and do the work."

At the time, as Blodgett recalled later, Blodgett had in his pocket the waybills for cars to be dropped and freight to be unloaded. He took out the bills and offered them to Cuineen, saying, "You run the train and show me how." Cuineen seemed affronted. He "turned and walked away."

After that, Blodgett suspected that Cuineen "was getting even" with him. But meantime, G. W. Clark, the CNE's general superintendent, instead of telling Blodgett what to do, asked him why his local freight job took so long. Blodgett answered, "There is just too much work for one crew."

Clark asked, "What would you suggest?" and Blodgett replied, "If another crew was put on, to handle the drop cars and do the switching at Winsted, both crews would make it without overtime and the work would be done." Clark "put on another crew," and according to Blodgett, "it proved to be the right solution."

Blodgett, as he looked back at it later, explained the difference between Cuineen and Clark. Cuineen "started at station work." He moved up to dispatcher, chief dispatcher, trainmaster, eventually to a division superintendent, all without direct experience in train service. That made Blodgett distrustful of him. Clark, however, "started on the Erie as a brakeman, then a conductor. He came to the CNE as general yardmaster at Maybrook." Clark "knew the train service because he had worked at it."[2]

As suggested by the above example, Blodgett liked to recall railroad life in story form. Three more of his stories, taken from his letters of recollection, follow, ranging from the grotesque to the humorous.

THE HEADLESS CORPSE

One night, Engineer Fred Woodin had stopped his freight train at Salisbury, CT, near the New York State border, waiting for a westbound train to pass. While waiting, he was fussing over his locomotive, oiling it, when he noticed there was blood on its driver wheels.

Woodin notified the dispatcher in Hartford, explaining that he could not recall seeing or hitting any object on the tracks that would put blood on his drivers. The year being 1910, Woodin's headlights were likely to be lighted with oil, allowing him only an indistinct view from his cab into the night.

The dispatcher issued an order to the next train which was to take the same route that Woodin's train had just taken, which was from Poughkeepsie eastward into Connecticut. The next train heading east was a work train, and its conductor was Nate Blodgett. He and his crew were to run their train slowly and watch for whatever could have put blood on Woodin's driving wheels.

When Blodgett's train was approaching Salt Point, NY, Blodgett saw "an object looking like a bundle of clothes." Blodgett and his crew stepped off their train. They examined the bundle. They decided it contained a human body, "a headless and footless" one.

The crew searched further along the tracks. They found a head "propped up on its neck against a telegraph pole."

One of the crew, as Blodgett remembered it, bravely "picked up the head by the hair," and peered at the face. "Nobody I know," the crewman said, and put the head back down on the ground.

CNE rules required the crew to report finding a human body, which meant they had to take their train to the Salt Point station, to telegraph their report. The rules also required, however, that someone stay with the body until a coroner arrived.

Conductor Blodgett designated the crewman who had "so bravely picked up the head" to stay with the body. However, when the crewman faced staying with the body alone, he "put up such a squawk," that Blodgett designated a second crewman to "keep him company."

Blodgett's train went on to Salt Point, and as it was too early for anyone to be on duty at the station, Blodgett went to the station agent's house. Blodgett persuaded him to go to the station to telegraph to the dispatcher in Hartford explaining what Blodgett's crew had found.

Eventually, investigators discovered the deceased was John Mackessey, a laborer from Poughkeepsie. The CNE had hired him to help build a water tower at Stanfordville. When he first came to work he was intoxicated, and as the day went on, he became more so. Before the day was over, he had been fired.

To get home, Mackessey had taken a CNE passenger train from the Stanfordville station. However, because he had no money to pay his fare, the conductor had put him off the train at the next stop, which was Clinton Corners. Then, as Blodgett figured it out later, Mackessey must have "set out to walk to Poughkeepsie." He had walked about three miles, to near Salt Point, when, "getting tired, he had laid down on the tracks." Whether intentionally or in an alcoholic stupor, he had put "his head on one rail and his feet on the other." Woodin's freight train had "put an end to his career."[3]

COUNT YOUR MEN

According to another Blodgett story, one day a CNE work train was made up in West Winsted, CT. The train included a flatcar carrying a steam shovel ditcher. It also included other flatcars intended to haul away the loads of dirt the ditcher would dump onto them, and an old coach for the section men who were to assist in the ditching.

When the train was nearly ready, its conductor, Nate Blodgett, went into the coach to ask the section foreman Mike McMahon if his section men were ready. McMahon counted the men with him in the coach, and said they were ready.

En route westward through the Litchfield Hills, the train picked up other section gangs, as conductor Blodgett recalled afterward, one at Norfolk, and still others at West Norfolk, East Canaan, and Canaan. One more gang joined them at Salisbury, and then the several gangs began working together near State Line, on the New York State border, clearing out ditches.

At the end of the day, the work train headed back to West Winsted, dropping off the various gangs at their respective home stations. After arriving at West Winsted, with only McMahon's gang being left in the coach, the train crew "were about to cut off the engine," Blodgett recalled, "run [it] around the train, [and] couple on behind" to "push the train in on a spur track for overnight storage," when there was a crash." An engine had "hit our rear end." It was a pusher engine from Canaan, on its way to Simsbury to help a train climb back west over the mountains.

"We were in yard limits," Blodgett remembered. So the pusher's engineer, Frank Marcy, should have been going slowly enough to "have stopped within half his range of vision." Unable to stop quickly enough, the pusher engine slammed into the ditcher train, forcing its "rear caboose platform to raise up." It "slid over the front coach platform," and as it did so, one of the men who had been riding with Mike McMahon in the coach "was caught by the leg." His leg was mangled. He could not free himself. The train crew could not free him either.

Nearby there happened to be men repairing a bridge who were equipped with saws. Conductor

CNE ENGINE #200 (FORMERLY #4) ON TURNTABLE AT THE ROUNDHOUSE IN THE CNE'S HARTFORD, CT, YARD, ABOUT 1905 (FULLR). Blodgett as a conductor was sometimes based in this yard.

Blodgett asked one of them to saw a piece out of the caboose platform, to free the injured man. Meanwhile, Blodgett called an ambulance. Once the injured man was freed, the ambulance took him to Winsted Hospital where his leg was amputated above the knee.

The CNE's assistant superintendent, G. W. Clark, investigated the wreck and held Engineer Marcy responsible. The CNE gave Marcy thirty demerits.

Meanwhile, questions arose about the injured man. Blodgett discovered that his name was Joe Trowbridge. On the morning of the accident, when the work train had set out from West Winsted, and Mike McMahon had counted his section gang, he had included Trowbridge in his count.

Later, Conductor Blodgett learned that Trowbridge, on the day of the accident, after riding with the section men in the morning to their work site near State Line, had not worked with them there, but rather had walked from there to Millerton, NY, about a mile, to look for work. Unable to find it, he had walked back to where the section men were busy at their ditching work, and had "mingled" with them again, as if he were employed with them, seeming to be accepted by them. He may have worked with the section men at this time, though Blodgett afterwards doubted it. At any rate he stayed with McMahon's section men until the end of the day, when he rode back with them to West Winsted.

During the day, Conductor Blodgett seemed not to have noticed Trowbridge at all. Later Blodgett reflected that he knew the foremen of the various gangs which his work train picked up, but "didn't know all of their men." Perhaps Foreman McMahon did not know all of his men either. Perhaps he did not know Trowbridge. Or perhaps McMahon, along with the men in his section gang, empathizing with Trowbridge's attempt to find work, chose to tolerate his presence on the work train that day.

Blodgett did not comment on what the story meant to him, and it is a question what that tells us about Blodgett. He did not comment on how ironic it was that Trowbridge should be the only one injured, the unemployed Trowbridge, the adept Trowbridge who seemed to insinuate himself among the section men so smoothly that they scarcely seemed to know that he was not one of

THE "DINKEY" ENGINE IN WHICH CONDUCTOR LAKE CARRIED HIS CAT OVER THE POUGHKEEPSIE BRIDGE IN ORDER TO DISPOSE OF IT, with engineer Hicks (right) who stymied Lake's plan. Photographed in Poughkeepsie (BEAUJ). The American Bridge Co. used this engine to rehabilitate the bridge, 1906-1907. It was built by Baldwin 1881, originally used by the Manhattan Rwy. as #284, and sold to the American Bridge Co., March 1906.

them, the only one accepted into the section gang without having been properly employed. Blodgett also did not even hint whether McMahon was disciplined for treating Trowbridge as part of his gang, or indeed, how Trowbridge coped with his legless life.[4]

THE CAT ON THE BRIDGE

According to another Blodgett story, when the Poughkeepsie Bridge was being strengthened in 1907, Theodore ("Dory") Lake was the conductor for a little work train. Pulled by a "dinkey" steam engine, the train moved materials about on the bridge, as needed.

One day Lake brought a cat to work with him. Although it was a cat which he had kept at his house in Poughkeepsie, he now wished to dispose of it. Lake brought it with him on his little work train, in its caboose. Then once when the train was at the Highland end of the bridge, the end away from the cat's home in Poughkeepsie, Lake "turned the cat loose."

Lake's engineer, Norman Hicks, saw him do it, and considered that it provided him and his fireman an opportunity. When they had a chance, they caught the cat, Blodgett recalled, "and put it in the seat box" inside the dinkey engine, without Lake seeing them. Then once when the train was back at the Poughkeepsie end of the bridge, they "let the cat go," also without Lake seeing them.

When Lake arrived home that evening, he found his cat sitting on his back steps. Lake was impressed. The Poughkeepsie Bridge was long. It had many openings in it through which a cat could fall. As a result of the reconstruction of the bridge, it was cluttered, making it especially difficult to walk across.

Lake addressed the cat: "Cat, if you could cross that bridge in the shape it's in, you have a home for life."[5]

27. STUDYING FOR EXAMS (1909-1946)[1]

"Yellow lights to front & sides, red lights to rear."

WHEN NATHAN BLODGETT was twenty-three years old, on July 1, 1909, he took a Central New England Railway exam. He took it to qualify to be promoted from brakeman to conductor. Part of the exam was physical, including a check on his ability to distinguish colors. The main part of it, however, was oral, including nearly 500 questions, and it took over four hours. The oral part was based on the CNE's book of rules. Although the New Haven Railroad controlled the CNE at the time, the CNE had its own book of rules, which was somewhat different from the New Haven's.

After Blodgett passed his exam and was promoted to conductor, he assisted others in preparing for similar exams. In 1913, when Blodgett was a conductor on a Canaan-Poughkeepsie local, Jack Beebe, who, like Blodgett, lived in Canaan, was fireman on that local and was studying for the exam to qualify to become an engineer. When their train "tied up" in Poughkeepsie over night, Blodgett recalled, they both slept in their caboose, and Blodgett tutored Beebe: "I put him through the book of rules." Soon afterwards, Blodgett also tutored Frank Doolittle, another fireman who worked with him on the same local, also to qualify to become an engineer. By this time, the CNE and the New Haven had issued a new rule book together, so that both railroads used the same rules.[2]

What was it that the various railroads which operated the Bridge Route from time to time wanted their men to know? Two practice exam booklets are available to help answer this question. The earlier booklet is not dated, but since it was published jointly by both the New Haven and Central New England railroads, it was likely to have been prepared between 1914 when they began publishing joint rule books, and 1927 when the CNE ceased to exist.

In this booklet, the questions were printed. The answers, however, were written out in ink. They were written in a meticulous, clear hand, with good spelling, by Winne Veeder Stover. He filled out his name on the booklet's cover, and explained that he was already a telegraph operator, and was retaking the exam, as was required periodically for him to keep his position. In accordance with the requirements, Stover used the rule book to help him fill out this exam booklet. He filled it out essentially for practice, to prepare for taking the oral exam.

The later practice exam booklet was filled out in 1946 by Albert Alexander, a brakeman in the Maybrook yard. At that time, Alexander was looking to be promoted to conductor, and like any Bridge Route railman looking for promotion, he was required to fill out an exam booklet in preparation for taking an oral exam on the rules. By that time the CNE having long since been swallowed by the New Haven Railroad, the rule book was exclusively the New Haven's book.

The format of Alexander's exam booklet was essentially the same as Stover's, with printed questions followed by spaces to write out the answers. Requiring railmen to write out the answers, Alexander explained, enabled the railroad, in the event of an accident, to say to any employee involved, you wrote out in your own handwriting what you were supposed to do.

During World War II, it seemed to Alexander

that the New Haven Railroad made promotions easily, without serious testing on the rules, because it had a pressing need for men. However, it was after the war, in March 1946, soon after Alexander had been discharged from the army, when he filled out the blanks in his exam booklet and then went to New Haven to be examined individually, in an oral interview. At that time, it seemed to Alexander that because the railroad no longer felt a pressing need for men, the examiner tested him severely. In fact, the examiner reported that he failed the exam. It was long before Alexander felt willing to try the exam again.[3]

A few sample questions and answers are given here from the two exam booklets, the first ones from Stover's booklet, the later ones, as is indicated, from Alexander's. The first samples are on the means of communication which railroaders used every day on the job:

HAND, FLAG, AND LAMP SIGNALS:

Q. What is the hand, flag, or lamp signal for stop? A. Swung across the track.

Q. For proceed? A. Raised and lowered vertically.

Q. For back? A. Swung vertically in a circle at half arms length across the track.

VISIBLE TRAIN SIGNALS:

Q. What signals must be displayed on the rear of every train by day? A. Marker lamps without lights.

Q. What by night? A. Yellow lights to front & sides, red lights to rear.

Q. What signals must be displayed on the front of an engine to indicate an extra by day and by night? A. White flags by day and in addition white lights by night.

DEMONSTRATING LAMP SIGNALS
(*The American Railway*, 1889)

AUDIBLE TRAIN SIGNALS:

Q. What is the whistle signal, when a train has parted? A. Three long, repeated.

Q. To back when train is standing? A. Three short.

Q. Approaching stations, junctions, and railroad crossings at grade? A. One long.

Q. For fire alarm? A. One long, three short repeated.

Underlying several of the samples taken from Alexander's 1946 exam booklet, was the assumption that to prevent accidents, extreme care was required not only in sending train orders but also in receiving, understanding, distributing, and following them.

TRAIN ORDERS:

Q. What is required for [train] movements not provided for by time-table? A. Train orders will be issued.

Q. By whose authority? A. The superintendent.

Q. How must those for a train be addressed? A. To the conductor and engineman, also to anyone who acts as its pilot.

Q. To whom must a copy be supplied by the operator? A. To each employee addressed.

Q. What is required of the conductor or engineman and others to whom a train order is addressed? A. They must read it to the operator and then sign it.

MOVEMENT OF TRAINS:

Q. Who are responsible for the safety of the train? A. Both the conductor and the engineman.

Q. Must conductors and enginemen compare watches with each other before commencing each day's work? A. Yes.

Q. How is a train superior to another train? A. By right or class.

Q. How is right conferred? A. By train orders.

Q. How is class conferred? A. By time table.

Q. Is right superior to class? A. Yes.

Q, Are extra trains inferior to regular trains? A. Yes.

Q. How will extra trains be governed with respect to opposing extra trains? A. By train orders.

In the late 1950s and early 1960s, on the line from Maybrook across the bridge to New Haven, the New Haven Railroad was introducing the Centralized Traffic Control (CTC) system. By the new system, switches and signals were electrically controlled directly from one control panel, located in New Haven. Train orders, whether sent by telegraph or telephone, were declining in significance, for exams or otherwise.

TRAIN ORDER, 1891, giving instructions on which train has the right to use the Bridge Route track from Silvernails, Columbia County, NY, to Hibernia, Dutchess County, NY (STICK). Telegraphed by the dispatcher in Hartford, CT, to the station at Silvernails. Delivered by the station's operator Wheeler to Train No. 5's conductor Potter and engineman Riley, and signed by them as having been read and understood.

28. "THE CNE BOOMER": A FOLKSY TALE

(Ca. 1920)[1]

"Boomers have a way that's mighty taking."

THE FOLLOWING humorous, poetic tale was written by an unknown author about 1920. Saturated with railroad slang, it hints of the low life on the margins of the Bridge Route.

On the trip described, the engineer was the legendary Casey Jones. The conductor was Sap Connors, probably a fictitious name. The head brakeman was a "boomer," that is, a railroad worker who was too restless to work steadily at any job.

In the story as printed below at the left, certain expressions which readers may find puzzling have been italicized. Then at the right, on the same line, possible interpretations of those expressions have been listed.

Tho I am not much of a spieler, the story I tell to thee	
Is the story of a boomer, a *shack* on the CNE:	Brakeman
He hired out at Maybrook, a God-forsaken spot,	
But he didn't intend to linger, so to him it mattered not.	
He got a place at *Stormses'* (a good place by the way),	Rooming house
And told *Frank* how he'd come across as soon as he got his pay.	Roomg house operator
The *caller* knocked upon his door, 'twas a little after eight,	Crew caller
And he signed up on the call-book, for a Waterbury freight.	
He dropped in on Mrs. Johnson to sting her for a feed,	
And he wasn't the first boomer to do this little deed.	
He filled his *shirt*, then crossed the road,	Stomach
To bum Frank for the *making*,	Makings of a cigarette
It's funny how these boomers have a way that's mighty taking.	
He got a *sack*,	Sack of tobacco
And rolled a *pill*, for the roundhouse made with groans.	Cigarette
They'd *boarded* 3213 with an engineer named Jones.	Listed on a board
He *bent the iron* for the main, and Casey done the honors.	Threw the switch
They stopped at *"XC" office*,	East end, Maybr'k Yd.
For an old *rail* called Sap Connors.	Railroad man
Sap handed up the clearance, and said, "Come in on *eight*.	Track 8
We're going to get 50 cars of Waterbury freight."	

"The CNE Boomer": A Folksy Tale

The boomer coupled on ahead, and then *tied up the air*,   Connected brake hose
Let off the brakes and murmured that everything was fair.
Sap checked them and swung his lamp.
The wheels began to turn;
The boomer was making another start, an honest living to earn.

They pulled up to the *plug* at Highland,   Water plug
The fireman swung the spout,
The boomer snored up on the seat—he'd been two hours out.
Casey drifted out on the bridge, by that *shanty in the sky*,   Phone shanty on
And called up the dispatcher to let 3213 by.   Poks. Bridge

"You got the *board*," the fireman cried,   Clear signal
   and Casey opened her up.
They rattled by the depot,
   like a can on a poisoned pup;
They cleared the crossing at Hopewell,
   and roused the boomer out.
   "*Pin ahead*," said Casey, "we'll pull down to the water spout."   Uncouple

They unloaded and went into Jimmie's.
The boomer began to *chin*,   Ask for credit
'Twas quite a job to stand him off,
But his *chin music* won again.   Clever talk
The next stop made was Brewster, the order board was red,
"*Set your train off* at Fairgrounds   Drop the freight cars
An' *turn*," was what the order read.   Turn the engine around
They woke the boomer at the switch;
He got down and *lined the rail*,   Set the switch
And pulled down through the middle,
While Sap *cut off her tail*.   Uncoupled the rear cars

They coupled onto the *crummy*,   Caboose
And started for Danbury and the *loop*   Danbury's circular track
With thoughts of George in the restaurant,
Irish stew and beans and soup.
The boomer climbed down at the station;
A sign 'cross the street he spied.
"The Edelweiss Cafe" it said,
And that's where the boomer hied.

That night Sap lost his *head man*, the CNE lost a shack,   Head brakeman
Mrs. Johnson lost a feed, and Storms his room rent and sack,
Jimmie at Hopewell is hopeful,
But his face has a worried look,
As he thinks of the CNE boomers
Who have said, "Put that on the book."

TONY DI ROSA AS HE APPEARED WHEN HE ARRIVED IN THE U. S.
by himself, aged 15 (DiROS)

29. WATER BOY DI ROSA (1927-1939)[1]

"I looked at the river below and realized that I had just better stay still."

IN POUGHKEEPSIE one morning, a sixteen-year-old boy got up early. From where he was staying, in the shadow of the railroad bridge, he walked a few blocks to the New Haven's Bridge Route station on Parker Avenue. There he saw a crowd of workmen, and figured they were the men he was to join.

The boy, Tony Di Rosa, could scarcely understand English, and so at first he just listened to what was going on. Then when Di Rosa heard one of the crowd speaking Italian, he approached him, and explained he was looking for Mr. Giambu who was to be his boss. The man pointed out Giambu, an Italian-speaking foreman who weighed over 200 pounds. Giambu asked Di Rosa to join the other workmen in stepping into a waiting work train. Just after 7:00 AM, the train took off, heading over the Poughkeepsie Bridge.

Riding over the bridge for the first time, Di Rosa found it exhilarating. He felt, as he later recalled, "that the engineer who had the vision to build such a long and beautiful bridge, that would hold a heavy load such as a railroad train, surely deserved to be congratulated."

Anthony Di Rosa had grown up in a family of Italian immigrants in Buenos Aires—before World War I, Buenos Aires and New York City had the largest concentrations of immigrants from Italy in the world. Tony had become appalled that his family lived in one wretched room, with a roof so leaky that night after night they had to shift their beds. He had dreamed of earning enough money to buy his parents a decent house of their own. He had heard enough about New York City to know that he wanted to go there so that he could help his parents. Struggling to convince his parents that he was determined to go, he had sometimes refused to eat. Eventually he had won their permission to take a ship heading for New York, by himself, at the age of fifteen.

In New York, not impressive in appearance, not tall or strong, he had looked for work. Though he could hardly speak English, neither could many of the other job-seekers who had crowded with him into a labor office in Grand Central Station. Eventually an Italian-speaking recruiter, even though he suspected that Tony was not as old as he claimed, had offered him a job as a water boy for a work gang on the New York Central Railroad. He had arranged for Tony to take a train out of Grand Central, and Tony had gradually discovered while riding on that train that he was to live with his work crew in a train car in a Hudson Line rail yard at New Hamburg, and would work on a section of track along the Hudson River between New Hamburg and Poughkeepsie.

His parents had told him that he had relatives in Poughkeepsie who came from his parents' hometown in Italy. On his days off, he gradually became acquainted with them. They were the Accelaro and Giummaro families, several of whom were railroad workers and proud of it. They were among the several thousand Italians already living in Poughkeepsie, railroad jobs being one of the major attractions which brought them there. His relatives had hugged Tony. In accordance with the Italian extended family custom, they had invited him to their various homes, all in the neighborhood around Dutchess Avenue and Delafield Street, close to the Poughkeepsie

Railroad Bridge. He had discovered that they and others he met in their neighborhood cooked their food much the same as his mother did, and spoke Italian with much the same accent as his parents, and he was delighted. They had soon invited him to live with them in Poughkeepsie, and with their help he had secured the promise of another job as a water boy, this time on the New Haven Railroad.

After the New Haven work train carried Di Rosa and his work crew over the Poughkeepsie Bridge, it continued on, often passing through farm country, until it finally dropped them off as they neared Maybrook. There the crew gathered picks, hammers, and barrels of spikes, and began to lay tracks. Foreman Giambu gave Di Rosa two pails and explained that he was to carry them a mile to a well, to fill them with water. It was August 1927, and it was hot. There was scarcely any shade for the workmen and they were sweating. They kept calling, "Water boy, water boy." Di Rosa, who had turned sixteen only that month, kept running to the well and back as fast as he could.

The next day some of his crew called him "Grease Ball." They taunted him in English that they knew he could not understand. The foreman told the men to lay off.

Preparing for work every morning, Di Rosa would walk to the Barone grocery near his house, almost under the railroad bridge, to buy his lunch to carry with him, a sandwich and perhaps a fruit or onion to nibble on.[2] Then he would run to catch his 7:15 AM "work train." Riding it across the bridge day after day, Di Rosa dwelt on the Hudson River as "majestic," on the bridge as an engineering "wonder," and then farther on, on the beauty of the orchard fruit he saw ripening in the sun. Concentrating on such positives helped him forget the cruelty of some of the workmen toward him. On the job his feet grew sore from so much walking. At home at night he often soaked his feet. Still every day, as he remembered it, at some time or other he would stop what he was doing to thank God for his work, his family, and the friends and relatives he had found in America. In gratitude, he planned to apply to become a citizen as soon as he could. Every Saturday when he was paid, he would write his mother, telling her he was doing well, and would include with his letter a money order, to help her buy a house.

On weekends—he always had Sundays off and sometimes at least part of Saturdays too— he occasionally worked at odd jobs, like shoveling in a gravel pit. He also might walk up Main Street, looking in the store windows. Or he might walk north along the river with some of his cousins, boys about his age, a mile and a half north of the bridge to Woodcliff Pleasure Park, where they would watch the girls.[3] On Sundays if he attended Our Lady of Mount Carmel Church—the church just south of the railroad bridge which Italians had built for themselves—various Italian families might invite him to join them for dinner.

As winter approached and it became too cold for his crew to continue their track laying, he was laid off for the time being. Then to pay for his room and board, he looked for part-time work. He was growing taller and stronger, and able to do heavier work. He helped the Barones at their store by unloading supplies that came to them by truck. He shoveled snow on the city's major shopping streets. He pulled ice cakes from the Hudson River with tongs, and loaded them into horse drawn wagons. In between jobs, he sometimes walked to the New York Central station only five blocks from his house, to take a train to New York, to visit friends. When the engines whistled and let off steam, it seemed to him they were "grand."

Meantime in Poughkeepsie, boys sometimes yelled at him in the street, "Here comes the wop," and "Hey, wop, why don't you go home?" Such taunts goaded him to keep learning English, if not passionately. He tried attending an English class for foreigners at a high school. But since he was already able to read and write in both Italian and Spanish, he was impatient when he found that many of the students in the class, a mix of nationalities, did not even know how to read or write in their native languages. He decided that the class was too slow for him. He picked up newspapers from trash cans—he was too chary of his money to buy newspapers. He discovered that by relating newspaper pictures to their captions he could pick up new words.

In early April, he was called to rejoin his railroad crew. He found himself pleased to ride over the bridge again, and to notice along the tracks that icicles sometimes sparkled in the trees. On the first days of the new season, when his crew was repairing switches along the line, it was still so cold that the men needed little drinking water, so he was anxious to run errands for the foreman, to prove his usefulness. By May, however, the crew had resumed their normal track laying work, driving spikes with long-handled hammers, and the men needed more water. He kept hearing their cry, "Water boy, where are you?"

Always conscious of his goal of helping his mother buy a house, he came to feel that his railroad job was not helping him to reach his goal fast enough, and he sought additional ways of earning money. He came up with a scheme to pick the berries and dandelion greens which he had noticed along the railroad tracks across the bridge in Highland. When he asked the Barones at their store if they would sell what he picked, Bartolo Barone promised they would if Di Rosa did his picking early in the morning.

One summer Sunday morning Di Rosa got up early, gathered some empty bags, and by 4:30 AM walked to the railroad bridge. The bridge guards allowed Di Rosa to walk on the bridge because he was a New Haven Railroad employee.

When he reached Highland, he picked along the tracks until he had one small bag of blackberries and four bulging bags of dandelion greens. Then, with berry juice smearing his face, he was walking back toward Poughkeepsie over the bridge, on the boardwalk along one edge of it, when he saw a train coming toward him. He was startled. All his reasons for uneasiness bubbled up in him—his youth, his distance from his parents, his foreignness, his crazy dandelion picking scheme. Of course he had walked on the bridge before, and knew, at least in one part of his mind, that there was room enough for the train to pass him if he huddled close to the railing. But this time he was carrying fat bags, and felt there was hardly room. "I looked at the river below," he recalled afterward, "and realized that I had just better stay still and let the train go by." As the engine grew closer, he felt that its engineer was disturbed by his being on the bridge. "The engineer on the train saw me and started to blow his whistle a couple of times, and I think he also slowed down. When he came near, he yelled some words that I couldn't hear or understand and slowly the train passed me."

By the time Tony was twenty-one years old, he had married Marie, a Poughkeepsie girl. She being pregnant, he had married her secretly, in a hurry. It had not seemed to matter to him whether the ceremony was religious or merely civil. Like other Italian males, Di Rosa seemed to combine occasional church attendance and an easy use of popular religious language with aloofness from the church.[4]

After their baby arrived, Tony and Marie Di Rosa lived in a rented apartment on the corner of Clinton and Thompson Streets, not far from the New Haven's Bridge Route rail line. It was early 1933, and depression gripped the United States. By this time, Tony Di Rosa had helped to make it possible for his parents in Argentina to buy a house of their own, but he himself had lost his railroad job and found nothing to replace it. When it turned extremely cold, he lacked the money to buy the wood or coal necessary to keep his apartment stove going. Tony scoured the countryside for wood that he could cut up and carry home on his shoulders, but that was not enough. To keep warm, he and his family slept in bed fully dressed. He came to feel desperate.

One gloomy night, as he recounted it afterwards, he picked up an empty potato bag and walked to a nearby railroad yard. Di Rosa did not say whether the yard was the New York Central's or the New Haven's. In either case, it belonged to a railroad for which he had worked, but that did not stop him. He found several locomotives whose tenders were loaded with coal. He climbed onto one tender and filled his bag with coal. He threw the bag to the ground, picked it up, and was about to put it on his shoulders, when suddenly he heard someone call out, "STOP." He froze on the spot.

A watchman emerged out of the darkness and asked him if he had put coal in his bag. Di Rosa admitted he had, and, as he recalled it afterward, became so nervous he was shaking. "The watchman said he should call a cop and have me arrested for stealing," Di Rosa recalled, "but I begged him to consider that I had stolen it in order to keep my family alive. I told him everything about my family and about how I couldn't find a job, even though I was willing to work at any job at all. He listened, and looking at me very seriously, he told me to load the bag of coal on my shoulder and go home to heat the house. I thanked him with all my heart."

At home, Di Rosa got the coal burning in his stove. He and his family were soaking up the heat when he noticed that the stove was getting too hot and was smoking. He opened the windows. Soon he felt he had to ask Marie to take the baby out of the house. Then he heard a fire engine coming. People on the street had seen smoke coming out of his windows and reported it.

Di Rosa met the firemen on the second floor, and led them into his apartment. They looked at his supply of coal and told him that it was soft coal which was suitable for generating the "tremendous heat" necessary to produce steam, as for locomotives, but that it was not suitable for his stove. "I told them I knew nothing about that. I had to use it to keep from freezing." The firemen advised him not to use any more soft coal and left.

A couple of days later, a man came to the door to ask where to put 500 pounds of hard coal. "I couldn't believe it was for me, but the man insisted it was." Di Rosa was amazed to learn that the city of Poughkeepsie had made this gift to him even though he was not yet an American citizen. "A thing like that could only happen in America,"

BARTOLO BARONE BY HIS GROCERY, IN POUGHKEEPSIE NEAR THE BRIDGE (Rinaldi and Perretta, *History of the Italian Center*, 1981). When Di Rosa picked dandelion greens along the railroad tracks, this grocery sold them for him.

he recalled. "I hoped that someday I could help someone in need, the same way the humanitarian officials in Poughkeepsie had helped me."

By 1939, Tony Di Rosa had learned enough about coal to become a Poughkeepsie coal dealer.[5] Among his customers was President F. D. Roosevelt in nearby Hyde Park, whom he admired.

In the past, Poughkeepsie coal dealers had often imported their coal from Pennsylvania by the Bridge Route for which Di Rosa had worked. Now, however, Di Rosa, like many other coal dealers, was not bringing in coal by railroad. His gratitude to railroads did not dominate his business decision. Di Rosa was driving a truck to Pennsylvania to pick up his coal.

30. BILL FELL, PAINTER ON THE BRIDGE (1940s)[1]

"We looked after each other."

HE was used to dangling from the bridge, high in the air over the Hudson River. As he dangled, he sat in what he called a swing seat. It hung from the top of the bridge by rope. He was not belted into his seat, but he was fitted into it so that he was not likely to fall out.

Originally, Bill Fell did not know much about the bridge. He grew up in the Bronx. Before he began to work as a painter on the bridge, he had never walked on it or ridden across it on a train.

Fell first knew the mid-Hudson River region when his parents brought him along for summer visits to Red Hook. He first lived in the region during the depression of the 1930s, when he was in the Civilian Conservation Corps (CCC) camps which President F. D. Roosevelt had established to train unemployed youth. In 1937, he married Anita who had grown up in Poughkeepsie and was more aware of the bridge than he was. A favorite Sunday pastime of her family had been riding a streetcar down the slope of Main Street to the river, and then settling somewhere on the shore to watch the river scene, including the boats passing under the bridge and the trains passing over it.

Perhaps two years after he married, Fell began work on the bridge, but at first he told Anita only that he was working for the New Haven Railroad. He was afraid it would scare her if he told her he was painting on the bridge. A few months later, however, when she happened to see him climbing on the bridge, she began to understand, and became uneasy.

Fell knew that few people were willing to work hanging from such a dangerously high bridge. However, he was thankful for his job. The New Haven Railroad paid the bridge maintenance men, including its painters, better than it paid its track section men, and he considered himself well paid.

Initially, Fell concentrated on learning to do his job and do it safely. But later if he was feeling youthfully exuberant, he might allow himself to take unnecessary risks. By about 1940, when Fell and his wife were living in Poughkeepsie on Vernon Terrace, close to the bridge, if he was late in joining his work crew where he was supposed to join them at the entrance to the bridge, he would try to catch up with them later. He would do what his wife did not want him to do, climb a fence near his house and then climb up one of the bridge towers.

On the bridge, the more senior painters often worked from scaffolding where possible. The younger painters like Fell worked especially from swing seats, or in locations where seats were not suitable, they worked by crawling among the girders.

Preparing to paint, Fell and his fellow painters first scraped off the rust and old paint from the steel.[2] When they were ready to brush on paint, as Fell remembered it, they first sought out the badly rusted spots and gave them an undercoat which they called "spotting," and the paint they used for it was red. Then when they were ready for the finish coat, the paint they used was black.[3]

When Fell was painting from a chair, if he had finished painting one girder and needed to move his chair to paint another one, he might swing his chair until he was able to grab that girder. If he ran out of paint, he recalled, his buddies up above on the top of the bridge would lower a bucket of paint to him by rope. Sometimes, depending on where he happened to be painting at the time, to get the paint to him, they might have to swing the paint bucket. When it swung near enough, he would reach out, grab the bucket, and hook it to his chair.

If the painters felt a need to introduce a little hilarity into what they were doing, Fell recalled, they might swing their chairs out toward each other until they could grab onto each other's chairs. Then, though the foreman would scold them for it, they would tease each other, "horse around," and sometimes spatter paint on each other.

Once as a barge came by, the painters saw below them a man sprawled out comfortably on the barge deck reading a newspaper. To stir up fun on a dull day, the painters shook their brushes enough to flip a bit of paint down toward him, and watched. First he felt a few drops hit him, and he stirred, perhaps thinking they were rain drops. But then he touched his fingers to his head, looked at his fingers, looked up at the painters, and he knew. He stood up. He yelled up at the painters. He jumped up and down. He was still jumping up and down as his barge carried him off into the distance.

One of Fell's buddies, Ferris ("Davie") Davis, considered Fell to be agreeably feisty, the kind of person you liked to tease and in turn liked to tease you. Once, Davis recalled, one of the painters working in the steelwork above Fell, flipped paint down onto him to tease him. Fell thought he knew who was doing it. When Fell had a chance, he dumped a whole can of paint over this painter's head, only to discover later that he had dumped it on the wrong person.

Fell's crew understood they were expected to repaint the whole bridge every four to five years. There were periods, however, when to save money, the New Haven Railroad or its subsidiary CNE did not keep to such a schedule. In 1917, portions of the bridge had not been painted for ten years; in 1933, for eight years; in 1964, for thirteen.[4]

In Fell's time, the whole bridge maintenance crew sometimes consisted of as many as twenty men, only some of whom were painters. There were no blacks among the regular maintenance men, although later there were to be blacks both among the bridge watchmen and among the track section men who occasionally repaired tracks on the bridge. There were also no American Indians among the regular maintenance men, although once a group of Mohawks were brought in by contract to speed up the painting, and in accordance with their reputation for being adept in working in high steel, they did speed it up.

Maintenance men who had more seniority were kept working all year. But because painting was not done in winter, those who had less seniority, like Fell, were likely to work only seven months a year. In winter, Fell looked for temporary employment elsewhere.

When he was working on the bridge at lunch time, Fell recalled, he might climb to the bridge's top. But that would be only if he were working where he could easily reach there. If he were working low down on the bridge, near the base of a tower, it would take so much energy to climb up to the top that he would not want to eat anything after he got up there. In case he could not easily reach the top, his buddies on the top would let down his lunch to him by rope.

When Fell was chosen to take a turn at fetching drinking water for his crew, he was pleased. It could mean a relaxing walk on the bridge, perhaps a long walk, depending on where his crew happened to be working. Once, however, when he had carried two empty buckets to near the west end of the bridge, and had climbed off the bridge, and was walking into a wooded area towards a spring, suddenly out of the bushes came guards with guns pointed at him. They wanted to know what he was doing. "I had some explaining to do," he recalled. Eventually he learned that the guards were there to protect President F. D. Roosevelt who was expected at the West Shore line's Highland station, on the Hudson River shore below the bridge.

As he worked, Fell was sometimes aware that the bridge was moving under him. He knew that

BRIDGE MAINTENANCE CREW ON THE BRIDGE, ABOUT 1945 (DAVIS). The one of the crew who made this photo available, Ferris Davis, identified most of the crew as follows. In the back row, from the left, first an unknown; then Herbert "Hoot" Tyson; Bill Burger; Davis himself (in his late twenties); then an unknown; and finally Tom Gorden. Front row from the left, Pete J. Heady; then a man whose name is unavailable but who, instead of being a New Haven employee, was a Central Hudson electric utility employee concerned especially with the electric cables that crossed the bridge; then Frank Covert, the New Haven's Bridge Inspector who was in charge of the bridge's maintenance work until about 1945; Clair Charder (without a hat) who found working at heights more difficult than most of the crew; and finally Jack McMann (just home from war service). Fell was apparently not in the picture; he might not yet have returned from war service.

any such huge steel structure is likely to move somewhat, from the weight of the traffic, or the wind, or changes in the temperature. He knew that to allow the bridge to move without buckling or snapping, certain bridge joints were built to be flexible, as by the use of pins or sliding rollers.

When a train was crossing the bridge, if he heard it apply its brakes even though it was supposed to avoid doing so, he would feel the bridge shudder. If he was painting in a precarious situation at the time, he would stop what he was doing, hold tight onto some solid part of the bridge, and wait.

At some hazardous painting sites, Fell understood, painters had nets underneath them to protect them. He wondered why they never had them at the Poughkeepsie Bridge. Nets or no nets, some of the maintenance crew stayed on their bridge jobs a long time. Fell himself stayed for five years. Four of the group he worked with each continued for at least eighteen years, and Davis for thirty. Although earlier at least one maintenance worker had met his death at work on the bridge, during the time either Fell or Davis worked on the bridge from about 1939 to 1973, according to their recollection, no maintenance worker met his death while working on the bridge by falling into the water or otherwise.

Was Fell proud of having done such daring work? Looking back at it years later, he would not say so directly. What he preferred to emphasize was that he and his fellow painters, when they hung high in their chairs over the water, were dependent for their lives on each other and on those above who were handling their ropes. He insisted, "We looked after each other."

WHEN CONDUCTOR CANSAS WAS RUNNING A TRAIN ACROSS THE POUGHKEEPSIE BRIDGE AND HE LOOKED DOWN TO THE HIGHLAND SHORE, THIS IS APPROXIMATELY WHAT HE SAW: THE WEST SHORE STATION TO THE LEFT OF THE TRACKS, THE POUGHKEEPSIE FERRY SLIP TO THEIR RIGHT. Photo perhaps early 1900s (LLOYD). Behind the station what appears like a dwelling house with a white roof is Dean's Hotel.

31. LANTERN SIGNALING: THE CANSAS FAMILY (1930s-1940s)[1]

From the river shore, they would watch for his train to slide out on the bridge.

ROCKY CANSAS was a stocky man. He had a big smile. He smoked big cigars. For much of his life, he worked on trains that ran from Maybrook across the Poughkeesie Bridge and on into Connecticut.

Cansas was born into an Italian immigrant family in New York City. About 1925 he began working for the Bridge Route, that is, for the CNE. When he married, he was living in Orange County, near Maybrook. By the time he had become a conductor and had children, he and his family lived in Highland, near the western end of the bridge. At that time, his wife Lil, assisted by her children, ran a luncheonette at Highland Landing, on the Hudson River shore, just north of the bridge.

Cansas' family was well aware that he crossed the bridge by train almost every day, and they developed the habit of watching for him. Though his train schedule varied, they usually knew what it was. Sometimes they might miss his train. Often they saw it slide out on the bridge.

If they could, they would watch outside their luncheonette—sometimes all three of them would, the mother Lil and the children, Lola and Bob. They would wave up to Cansas' train, and were delighted if they could spot him way up there on his caboose, waving back. In the daylight, they would wave handkerchiefs to him, and he would

PEOPLE SEEM UNCONCERNED TO BE WALKING ON THE WEST SHORE RAILROAD TRACKS at Highland Landing, on the Hudson River shore. The West Shore depot is on the left of the tracks, the slip for the ferry to Poughkeepsie is on their right. Postcard, postmarked 1909 (SHUKR).

wave a handkerchief back. In the dark, they would wave railroad lanterns to him, and in response he would swing his railroad lantern to them. They were battery-powered lanterns, one of which the family still has.

Their luncheonette was located on the river front, where the steep, twisting Mile Hill Road came down to it, next to both the West Shore Railroad station and the slip for the ferry to Poughkeepsie. Occasionally people waiting for a train or ferry noticed that someone near the luncheonette was signaling up to the bridge. If they had time, they might walk to the luncheonette, perhaps to buy ice cream, or perhaps to ask the family about the signaling. If it was dark, they might join in the signaling, blinking their car lights at the same time that the family waved their lanterns. They might watch with the family for Conductor Cansas' answering lantern light, swinging back and forth, from the end of his train, as it rumbled high up across the hovering, black bridge.

After the nearby Mid-Hudson highway bridge opened in 1930, it drew automobile traffic away from the ferry. By 1941, the new bridge had reduced its tolls so that the ferry could no longer compete with it, and the ferry closed. In turn, within a year the luncheonette had lost so much patronage that the family felt obliged to close it. But they still lived nearby in Highland, and sometimes, when they knew that Rocky Cansas was coming over the railroad bridge, they would walk down to the river. They would again wave their handkerchiefs or lanterns up to him, and watch to see if he would respond.

A YEAR AFTER THIS FERRY SCHEDULE WAS ISSUED, THE FERRY STOPPED RUNNING (Moffett, *To Poughkeepsie and Back*, 1994). Soon after it stopped, the Cansas family closed their luncheonette.

Time Table
Poughkeepsie and Highland Ferries
Effective October 1st, 1940

Leave Poughkeepsie			Leave Highland		
a.m.	p.m.	p.m.	a.m.	p.m	p.m.
6:00@	12:20	5:20	6:15@	12:30	5:30
6:30@	12:40	5:40	6:45@	12:50	5:50
7:00	1:00	6:00	7:15	1:10	6:15
7:30	1:20	6:30	7:45	1:30	6:45
8:00	1:40	7:00	8:15	1:50	7:15
8:30	2:00	7:30	8:45	2:10	7:45
9:00	2:20	8:00	9:15	2:30	8:15
9:30	2:40	8:30	9:45	2:50	8:45
10:00	3:00	9:00	10:10	3:10	9:15
10:20	3:20	9:30	10:30	3:30	9:45
10:40	3:40	10:00	10:50	3:50	10:10#
11:00	4:00	10:30*	11:10	4:10	10:15*
11:20	4:20	11:00*	11:30	4:30	10:45*
11:40	4:40	11:30*	11:50	4:50	11:15*
12:00	5:00		12:10	5:10	11:40*

Notes:

@ Omitted Sundays morning.
\# Trips omitted Saturday and Sunday.
* Saturday, Sundays and Holidays only.
Departure of boats at times not guaranteed.

32. TOWERMAN BEAUJON (1930s-1950s)[1]

Pulling levers of the "armstrong variety."

GROWING UP IN CANAAN, CT, Leroy Beaujon often hung around its old CNE yard there, that is, the New Haven's Poughkeepsie Bridge Route yard at the west end of town. "A switcher tied up there every night," he recalls. In the early evening, when the switcher was through running for the day, he "would often walk over to pester the crew." They would be "servicing their locomotive," which was "invariably a 2-6-0 Mogul," getting it ready "for the next day's run."

Leroy Beaujon also remembers the freight train that arrived in Canaan early every morning on the Berkshire line, on its way north. Canaan was the junction of two New Haven lines, the Bridge Route running from Maybrook east over the Poughkeepsie Bridge to Hartford, and the Berkshire line running from South Norwalk north

CANAAN, CT, STATION (SHUKR), built in 1872 and still standing today. In Beaujon's childhood, it was a New Haven station where two active New Haven lines, both shown, crossed: the Bridge Route which ran east (middle right) toward Hartford, CT, and the Berkshire Line which ran north (middle left) toward Pittsfield, MA.

through Danbury into the Berkshire Mountains. On hearing the Berkshire line train approach sometime between 5:00 and 6:00 AM—he remembers the sound reverberating for miles—"I would quickly dress and hop on my bicycle, often dragging along a neighbor about my age, and race up to the towered Canaan station to see the 'big show.' " The train was always double-headed, being "pulled by two huge steam locomotives." The first, he reports, was a 4-8-2, Mountain type, and the second a 2-10-2, Santa Fe.

Leroy Y. Beaujon was already focusing on locomotives, as he has continued to do ever since. He would watch as the two locomotives maneuvered into position to take on water. Then as they began to move on north, he recalls, "the ground virtually shook and the sounds of their exhaust became almost deafening."

About 1936 and 1937, though the New Haven had reduced service in his region, Leroy remembers that it ran special Sunday ski excursion trains through Canaan, three of them. They were among the New Haven's seventy-five weekend ski trains which originated at the time in Grand Central Station in New York. The three which ran through Canaan reached there through Norwalk and Danbury on the Berkshire line. While two of them continued on the Berkshire line north into the Berkshire Mountains, the third one switched at Canaan eastward onto what had been the Bridge Route's main line, which Leroy remembers as the "old light CNE rails" running from Canaan upgrade to Norfolk. This train usually consisted of eight to ten coaches, and it took a fairly heavy locomotive, an I-4, class 4-6-2, as Beaujon recalls, to haul it up the grade.

Leroy's father was curious about this ski train. Several times he followed it as well as he could by automobile, "of course taking Leroy along." They followed the train by driving on Route 44 "which more or less paralleled the railroad." They followed it to "the ski area located near what today is the Blackberry River Farm," two or three miles west of the village of Norfolk. They watched as the skiers stepped out from the train, and "spread out over the snowy landscape like ants escaping from an anthill."

In this mountain location, there were no facilities to permit the train to turn around, to prepare for its return trip to New York. So once the train had disgorged its skiers, Leroy and his father would watch it back away. They would watch it back all the way down hill to the CNE yard in Canaan where the locomotive turned on the wye track. In the late afternoon, it would back its train up to Norfolk to pick up the now weary skiers. Then the train would be ready for its return trip to New York, locomotive first.

Much earlier, the New Haven had already begun to downplay the Bridge Route's original main line from the bridge into New England, that is, its northern claw from the bridge through Canaan and Norfolk to Hartford. Under the impact of the great depression, the New Haven had gone so far as to abandon parts of this line altogether. In late 1938, it abandoned the section from East Canaan to Norfolk Summit, making the New York to Norfolk ski excursion no longer possible.

Near the end of his last year in high school, on Sunday, May 4, 1947, Leroy made his first trip over the Poughkeepsie Bridge. Since regular passenger trains were no longer running across the bridge, the train he boarded was a special excursion train, for rail fans. While it originated in New Haven, Leroy boarded it in Danbury. It was to cross the bridge to Maybrook and return, and consisted of twelve "streamlined" coaches and a "baggage-observation" car whose doors were left open to allow fans to take photos. A Canaanite, Dennis Foley, who was a road foreman of engines based in Danbury, was assigned to help operate its locomotive which, as Leroy took note, was New Haven engine #1407, an I-5, class 4-6-4 Hudson — it was an unusual locomotive to run on the New Haven to Maybrook line. Leroy rode with Foley and others in the cab of the locomotive, but only as far as Poughkeepsie, where he climbed out to make room for other fans to take their turn in the cab. So Leroy was only riding an ignominious coach when he had his first ride across what the New Haven called at the time "our impressive 212-foot-high Poughkeepsie Bridge."[2]

When Leroy Beaujon graduated from high

BEAUJON WAS ON THIS EXCURSION TRAIN IN 1947 WHEN HE FIRST RODE OVER THE POUGHKEEPSIE BRIDGE (NYNH&H, *Along the Line*, July, 1947)

school in June 1947, his fascination with railroads had already developed so far that "almost immediately" he looked for work with the New Haven Railroad. By September, he recalls, he had completed "a couple of weeks of training in my hometown," and had passed "the rulebook examination given in New Haven." Then he was "put to the test" by being given an assignment. He was assigned to a station on the Bridge Route, seven miles west of the Poughkeepsie Bridge, at Clintondale, NY. He was assigned there "to relieve the agent, who was looking forward to going on vacation for a couple of weeks."

Beaujon did not have a car. Because there was no longer any passenger service anywhere on the Bridge Route, he took a bus from Canaan to Poughkeepsie, and then changed to another bus which took him across the Mid-Hudson highway bridge, within sight of the Poughkeepsie Railroad Bridge, and on to Clintondale. He planned to board somewhere in the village.

When he got off the bus in tiny "downtown" Clintondale, he went into a grocery store to ask where the station was. It was only then that he learned that the station was located about two miles out of the village.

He walked past woods. He walked past rows of apple trees and grape vines on rolling hillsides. When he finally walked down Station Road and arrived at the station, he felt it was a lonely site. There was a feed store nearby, but what had once been the busy fruit-growers' cooperative storage plant had just closed.

He met the station agent, William J. Kelly, an experienced agent, and told him he did not have a car. Kelly was astonished that the New Haven would assign anyone to this station who did not have a car. "He as much as told me that there were no places in the immediate area to stay, let alone eat. He said he wouldn't blame me for not working there under those circumstances. I agreed wholeheartedly. I decided then and there to head

LEROY BEAUJON (BEAUJ), in his early years a railroad worker, in his later years a railroad historian.

back home to Canaan."

Agent Kelly knew that an eastbound locomotive, a Santa Fe, would soon pass by the station. At the time, the New Haven still kept a few Santa Fes in Maybrook, in hump yard service. This one happened to be on its way from Maybrook to New Haven for maintenance, and it was by itself, not pulling a train. The agent decided this locomotive would provide an easy way for Beaujon to get as far as Poughkeepsie. They flagged it. When it stopped, Beaujon climbed up, and this time he had the exquisite pleasure of riding over the Poughkeepsie Bridge in a locomotive cab.

The New Haven Railroad did not seem offended that Beaujon declined to work at Clintondale. It employed him for about ten years thereafter, in a variety of locations, most of which were only tangentially related to the Bridge Route. He worked in a message center at Grand Central Station. After brief training as a tower operator, he worked at Tower #3 in the Bronx near the Harlem River yards, and at Tower #44 at South Norwalk. He also worked at stations on the Berkshire line as a relief station agent, performing all the miscellaneous functions station agents were likely to perform, including selling tickets, handling U.S. mail, handling Railway Express (he liked handling express baggage for the commissions it earned him), ordering freight cars for shippers, and controlling the passage of trains by operating manual signals as authorized by the dispatcher in New Haven.

Beaujon also worked as a relief tower operator three or four miles east of Danbury at the tower at the junction of the Berkshire line and the Bridge Route, or as this part of the Bridge Route was usually called by then, the Maybrook line. Here, as at other New Haven Railroad towers where Beaujon worked, the telegraph was on its way out. He himself was not trained as a telegrapher. Although in the late 1940s, certain towers and stations, including Canaan, still had working telegraph sets available "for the old-timers to use" if necessary, Beaujon's communication with the dispatcher at New Haven was always by telephone. Under the telephoned directions of the New Haven dispatcher, Beaujon, as the Berkshire Junction tower operator, was expected to "force switches open and shut manually." He did so by pulling levers which he called of the "armstrong variety." That is, he explained, they were heavy levers which required a strong arm. "They did not have any power assists attached to them."

Beaujon recalls that after crossing the bridge on return from his frustrating attempt to start work at Clintondale, he crossed the bridge again several times, all on rail-fan trips, all of them running from New Haven to Maybrook and back. On each trip he was aware that he was passing the strangely located Clintondale station, the lonely station, out among the woods and vineyards, way out of town.

THE LONELY STATION AT CLINTONDALE (postcard, LILL), where Beaujon declined to work.

33. MARY CARMODY, CREW CALLER (1939-1969)[1]

She was thankful the Lord had made her Irish so that she had guts.

NEAR THE BEGINNING of World War II, the increasing number of trains passing through the Maybrook yard pressured yard officials to hire more workers. Men being hard to find, the yard was hiring more women. It was then that Mary Carmody found a job at the yard.

Her job was to carry messages from place to place within the vast Maybrook yard, walking from one end of it to another, in all kinds of weather. It was not an easy job. Men usually did it, not women. Some men had tried it and given up.

At first the walking hurt her feet. When she came home she could do little but sleep. She was separated from her husband, and had only one hand—she was born that way. But she was determined to care for her son. She kept saying to herself that she was thankful the Lord had made her Irish so that she had guts. She kept going, regardless of wind, rain, or snow. Her feet grew less sore. Her legs grew firmer. She liked being outdoors. She gradually became so healthy that she seemed to become impervious to catching cold.

At the beginning her pay was only $19 a week, but it was equal to the pay of the men who did the same work. She learned to stretch her dollars. She did her own housekeeping, her own canning, and raised her boy largely by herself. As the war stretched on, her pay went up. After the war, unions helped keep the pay going up. Eventually she felt she was well paid.

Like almost everyone else at the yard, Carmody wore overalls at work. She also wore high top shoes. In rough weather she protected the waybills she was carrying by hiding them under her coat.

Wartime was a great time to work on the railroad, it seemed to her. Most railroaders, including herself, worked hard, did what they were supposed to do, and had a good time at it. She liked seeing an engineer fuss over his steam engine the way women fuss over babies—going over every inch of his engine, wiping the dirt off, putting a bit of oil here and there. She liked how the train crewmen, especially the older ones, wore white railroad caps, red checkered scarfs at their necks, and pinstripe overalls. Younger, newer employees dressed more carelessly, it seemed to her, as in T shirts, and were less enthusiastic about their work. She didn't like that.

While war pressures brought a few more women to work at the yard, still there were not many of them. A few women washed engines, more worked as clerks. Often it was women who typed the waybills that Carmody picked up at the main office—that is, records about railroad cars, what was in them, and where they were to go.

Over many years, railmen were usually respectful of her, she recalled. But some of them, partly because they were unaccustomed to having women moving around the yard, teased her. Once she was sitting on a platform in the sun, waiting for waybills to be made up, when someone turned a hose on her, soaking her. On other occasions, when locomotives were passing near where she was walking, she recalled, their crews might send a shower of soot out of their smoke stacks, making her look as if she worked in a coal mine. Some women, she believed, would wither under such treatment. But she found ways to tease back. Once she picked up some bubble-girl decals, sexy pictures meant to be pasted on walls. She pasted them on a locomotive. When an inspector discovered them, he made trouble for the locomotive's crew.

Ordinarily she did not ride on the trains which moved about the yard. But once when she saw a string of freight cars moving slowly in the direction she wished to go, she climbed between the cars to ride. A conductor passing by on the roof of a car moving in the opposite direction happened to see her. He scrambled down from his car, grabbed a ladder on her train, hooked her on his arm, and scared her by swinging her off that train. Then he explained to her that the cars she was riding were not just being shunted about the yard but were part of a train that would pick up speed and take her out of the yard on its way to the Poughkeepsie Bridge. Yardmen were always looking out for her, she recalled.

Carmody looked out for the men too. If men had sickness in their families, she would bake something for them. Everybody else would too, she recalled. When she took her lunch at the yard's YMCA, she did not especially huddle with the women there. She felt comfortable sitting at tables with either men or women.

Eventually the yard acquired a new communication system that reduced its need for office-to-office messengers. It installed underground pneumatic chutes, operated by air pressure, through which messages were whisked from office to office, a system such as many department stores used at the time to move cash and receipts. Meantime, Carmody was assigned a new job in the yard as another kind of messenger, a "crew caller," the only woman to be assigned to such a job.

For this job, she was expected to notify the "extra" men—the many train crew men who did not have enough seniority for regular jobs—when they were needed for work. She would notify the ones whose names came up on the top of the "extra board." At first, if an extra man did not show up for work at once when she notified him that he was needed, he lost out on that job. Later, the yard gave the men two hours notice. The men did not like to miss a call. They would often call her office to say where they could be reached, at home or elsewhere. She might reach them by telephone. Or if not, she might have to go out to find them, as at a barber shop, or she might slip a note under a door for them. She found that she could usually locate the men in time.

After Penn Central absorbed the New Haven Railroad, it cut back on its use of the Maybrook yard. In 1969, Penn Central asked Mary Carmody to work in New Haven. But she was well along in years now, and felt she had friends around Maybrook. Also, she was fond of her house nearby, outside of Walden, with its view across open fields toward the Shawangunk Mountains. She was fond too of the chance it gave her at night, when she was lying in bed, to listen for the familiar low hum of the Bridge Route's diesels going by. She declined to move to New Haven, and that meant that she lost her Maybrook job.

In her retirement, even in her seventies and eighties, Mary Carmody went regularly to the dinners which the old-time Maybrook railroaders held every year. She joined them as they greeted their friends, grieved that the yard was closed, remembered the good old days working there, admitted that they might be growing hard of hearing, and remembered those who had passed away. In January 1998, Mary Carmody herself died, joining those who will be missed at future reunions.

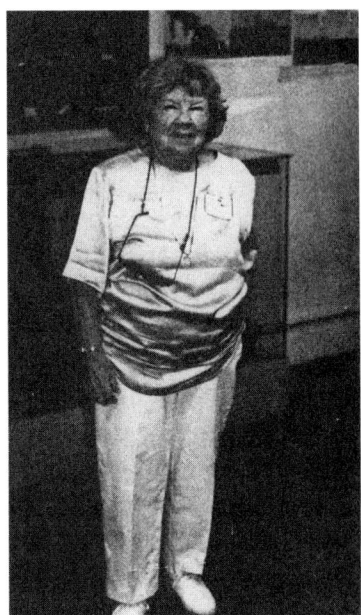

MARY CARMODY, RETIRED, VISITING THE MAYBROOK RAILROAD MUSEUM (MRM)

34. DISDAIN FOR A PUSHER (1960)[1]

Confessing your mistake, learning your lesson.

WHEN STEAM LOCOMOTIVES still ran on the Bridge Route, it was customary for special pusher locomotives to assist them to move heavy freights up steep grades. After diesel locomotives replaced the steam locomotives in the 1940s, because diesels were more powerful, at first pushers were seldom used. A decade later, however, the New Haven Railroad, under pressure to cut costs, sometimes reduced the number of diesel units it assigned to freight trains. Even on heavy freights going from Maybrook across the Poughkeepsie Bridge toward New Haven, it sometimes cut the number of diesel units, perhaps from four to three. Then at Hopewell Junction, much as in the steam engine days, the New Haven might find it necessary to add a pusher to get the freights up the mountains of the Stormville-West Pawling region, in southeast Dutchess County, on their way to Danbury.

One day in 1960, Pete McLachlan was a fire-

FOUR DIESEL UNITS, ALCO Fs, AT MAYBROOK, PREPARING FOR A FREIGHT RUN OVER THE POUGHKEEPSIE BRIDGE AND ON INTO THE MOUNTAINS EASTWARD, April 1957 (photo by Jim Shaughnessy, SHAU). If the New Haven kept four diesel units on a heavy freight as it moved into the mountains of southern Dutchess County, it probably would not need a pusher.

THREE DIESEL ENGINES PULLING A FREIGHT EASTBOUND PAST THE STATION AT HOPEWELL JCT., NY, 1946: ALCO diesels, two FAs and one FB (McEN). If trains were long and heavy, would three diesels be enough?

man on a diesel pusher waiting in the Hopewell Junction yard. The pusher's crew, under orders from the dispatcher, was waiting for an eastbound freight to arrive from over the Poughkeepsie Bridge, and expecting its crew to signal for the pusher to assist it to climb the mountains. However, when the expected freight came—it was a long, Boston-bound, diesel-powered Advance OB-2—the engineer in its head unit gave the waiting pusher a sign that he did not need its services, and kept his train moving disdainfully on.

As Fireman McLachlan recalled, the pusher's crew telephoned the dispatcher to report what had happened. The dispatcher, doubting that the OB-2 would be able to make it all the way up the grade without a pusher, told the crew to hold their pusher where it was and call him again in half an hour—the crew, having no two-way radio communication facilities available in their locomotive, were dependent on stepping out of their engine to telephone. Half an hour later, when the crew phoned the dispatcher again, the dispatcher told them that the Advance OB-2 had stalled trying to get up the mountain, and that as a result other trains were being delayed; he asked them to take their pusher to the rescue. The pusher climbed up the mountain and eventually got the stalled train moving again.

The New Haven's road foreman ordered an investigation of the incident. He asked the crews of both the pusher and the OB-2 freight to come to a hearing in New Haven. In preparation, the pusher crew, to protect the miscreant freight engineer whom they regarded as their friend, concocted a story. They agreed to say that his train would have made it over the mountain without help if one of his diesel units had not become overheated and thus lost power.

At the hearing, as Fireman McLachlan recalled, before the pusher crew had a chance to tell their concocted story, the road foreman surprised them. He winked at them and said he knew that to protect the engineer, they were ready to say that the train had stalled because one of its diesel units had lost power. He had often heard that story, he declared, and in fact, he explained, that's the story he himself had already told the superintendent. Now, he said, let's go for coffee.

As the pusher crew learned only later, the miscreant engineer had already admitted to the foreman that he had made a mistake in trying to go over the mountain without using the pusher. On the New Haven, it seemed to McLachlan, if you made a mistake and confessed it, it was usually understood that you had learned your lesson.

35. TOWERWOMAN COOPER (1956)[1]

"I can see on my model board when an east-bound train hits my signal near Clintondale."

FOR MANY YEARS, two signal towers controlled the passage of trains over the Poughkeepsie Bridge. One tower at one end of the bridge controlled trains moving in one direction, the other tower at the other end controlled trains moving in the other direction. From 1925, however, the signal system became sophisticated enough to enable one signal tower to do all the controlling.

During the mid-1950s, it was a signal tower at the east end of the bridge which did all the controlling of trains on the bridge and a considerable distance beyond it as well. In this tower, sometimes the operator was a woman.

Railroaders related to the Poughkeepsie Bridge had normally been men. Railroad company directors had been men. Maintenance gangs, track layers, and train crews had been men.

As early as 1873, however, on what was to be the bridge-related Wallkill Valley line, a woman was the station agent at Forest Glen, in Ulster County, NY. In 1892, the ND&C, which was to become part of the Bridge Route, chose Laura McCurdy as telegrapher at its station in Millbrook, Dutchess County, NY. Since telegraphy did not require heavy muscular work, it was considered more suitable for women than most railroad work. In McCurdy's case, the ND&C explained that the particular reason for appointing her was to relieve the station agent just before and after train time, so that he could concentrate on selling tickets, billing express baggage, and answering questions. However, in 1893, when a woman applied to the ND&C to be a telegrapher, the ND&C superintendent explained that on his line in small stations, station agents and telegraphers often interchanged duties, implying that telegraphers might occasionally be expected to do heavy lifting; therefore he had "but one or two places" which would be suitable for a "lady" telegrapher. Nevertheless, in that same year, of 206 station agents serving the New York Central, 26 were women.[2]

On the CNE itself, in the early 1900s, there were at least two women telegraph operators in Columbia County, NY, one at Ancram, and one at Ancram Lead Mines; both were daughters of the station agents there. From at least 1926 one of them, Mary Hoyt, had become the Ancram station agent. In 1941, the agent at the Matteawan station, on the New Haven's line (formerly ND&C) between Hopewell Junction and Beacon, was Florence M. Leion. Meantime in the early 1900s, it is doubtful if any of the Maybrook yard's office workers were women—a photo of the office workers outside their office building taken before 1907, does not clearly show any women. By World War II, however, when the demand for labor increased the opportunities for women, as we have already seen, not only did Mary Carmody work as a Maybrook yard caller, but several other women also worked there, some in the yard's office. Also especially relevant here, on bridge-related rail lines a few women had long operated signal towers, as in the notable example of a Pennsylvania Railroad tower operator who played a significant rescue role in the 1889 Johnstown flood.[3]

By the 1950s, the Poughkeepsie Bridge tower operators, whether men or women, were no longer significantly using telegraphy to communicate with the dispatcher. As the *Poughkeepsie New Yorker*, August 19, 1956, explained:

205

Did you ever notice how slowly trains cross the Poughkeepsie railroad bridge? That's because the New York, New Haven, and Hartford Railroad insists that east-bound trains must take [at least] ten minutes for the crossing and west-bound trains must take seven minutes.

Those figures come from Mrs. Elizabeth Cooper of Poughquag, the signal tower operator at the east end of the bridge from 8 a.m. until 4 p.m. Her husband, Edward Cooper, replaces her at 4 p.m. and carries on until midnight.

Mrs. Cooper sets the signals which bring the trains across the bridge. She sets them with both tall hand controls and electric switches. The first thing to do is close the "derail," she says. That puts the rails into position so that a train can go on the bridge.

There's a derail at each end of the bridge because it is now single-tracked—it was double-tracked originally—and the railroad isn't taking any chances of getting an east-bound and a west-bound train on the bridge at once.

Mrs. Cooper's circuit runs from Manchester Bridge on this [Poughkeepsie] side of the river to a point a short distance east of Clintondale on the west. That means that she sets the signals for the entire [twelve-mile] distance, in accordance with orders which she receives on a direct wire from the dispatcher in the main railroad office in New Haven.

"I can see on my model board when an east-bound train hits my signal near Clintondale," Mrs. Cooper said. "If everything is all right, I'll set my levers and clear him, give him a clear signal, a green light.

"When he comes down, I can follow him because lights go up on my board as he progresses. Once you give him a clear signal at Clintondale, you never take it away. There's quite a [down] grade on the west side of the river [coming toward the bridge]. It would be very difficult for an engineer to stop on that. Things might pile up.

"East-bound trains have preference over west-bound trains. Always. East-bound trains are loaded. Most of them are going to Boston. Most of the west-bound trains are empties going home.

"I wouldn't interrupt a west-bound train unless I really had to, but it can be stopped. After it hits the signal at Manchester Bridge it comes in slowly. From Manchester Bridge to here it's in what we call 'yard limits.' There's a speed restriction on all trains in yard limits, just as there is going over the bridge."

If a train does exceed the speed limit on the bridge, a special bell rings in the signal tower to call Mrs. Cooper's attention to the fact. She must report that to the New Haven [dispatcher's] office. In fact, she must time every train as it comes over the bridge and send a record of that to New Haven.

"The bell hasn't rung more than three times since I've been here," she said, "and I've been here since May 27, 1954."

The signal tower is actually a small one-story frame building, painted barn red, at the edge of the railroad tracks above the northeast corner of Washington Street and Parker Avenue [in Poughkeepsie]. The little building is long and narrow, with windows on four sides. A coal stove provides heat in winter. The model board Mrs. Cooper spoke of is above a north window. It is approximately eight feet long and two feet high. A schematic drawing of her circuit is on the board, with lights at intervals.

Mrs. Cooper uses several telephone lines to do her work. She has one switchboard with a city wire and two direct wires to the railroad terminal in Maybrook. She uses that setup when railroad workers call in and ask to speak to Maybrook.

Another smaller switchboard has one direct wire to the dispatcher's office in New Haven. There's also a message wire to New Haven on that board. She uses that when the agent's office in Cottage Street or the track supervisor in Parker Avenue ask her to send messages to New Haven, perhaps to have a car traced.

A third wire on this switchboard is connected with the telephones which are placed at intervals on the bridge. That's where a call would come from if a train should be in trouble out there, she said. She also has a regular outside phone.

She has never walked all the way across the bridge, Mrs. Cooper said, but she did walk out to about the middle early this summer. That was to ask the engineer of a train which was stalled there to phone to the dispatcher.

"I wasn't too scared," she said, "just a little maybe. I was working on the night shift at the time."

The dispatcher gives her the train orders over the dispatcher wire on the small switchboard, she said. In most cases that means telling her when a train will come in.

"If I had a train working in Hopewell Junction taking or leaving cars, and an east-bound out of Maybrook, I would get a 31 order," she said. "A 31 order is the most important. That means I'd give the east-bound train a red signal so that it would stop right here.

"There aren't many interruptions of service, but there are bound to be some. If the lights on my board don't show that a train is progressing, I can't send another one in."

While Mrs. Cooper was speaking, a [railroad] motor car operator came into the tower and asked for clearance to use the track to Hopewell Junction. She obtained the clearance by telephoning to the dispatcher's office in New Haven.

A motor car operator looks for breaks in the tracks, broken signal wires and similar things which must be repaired, she said. He also tests grade-crossing lights.

Only freight trains move on Mrs. Cooper's circuit. She has [on her shift] about five [regular] trains a day, with possibly two or three extras and two switchers, she said. The switchers take cars around the yards [in Poughkeepsie], to Western Printing and Lithographing Co. and to Hudson River State Hospital.

Mrs. Cooper was a secretary at a nursery in Manchester, Conn., before her marriage in 1932. However, she always has been fascinated by railroading. Her father was a conductor on the New York, New Haven, and Hartford for 49 years, and

ELIZABETH COOPER PULLS A LEVER IN HER SIGNAL TOWER TO ALLOW A TRAIN TO CROSS THE POUGHKEEPSIE BRIDGE (*Poughkeepsie New Yorker*, Aug. 19, 1956, by permission of its successor, the *Poughkeepsie Journal*)

her husband has worked for the railroad for 15 years, in signal towers, and as agent in Danbury, Berkshire Junction and Poughkeepsie.

"I like my job, like it very much, " she said, "but I won't be able to keep it. For the first year and one-half it was a steady job. Now I am on the spare board. That means I'm filling in, but doing it full time.

"I was put on the spare board because a lot of jobs were done away with. Men who have more rights—seniority—can take it. I haven't enough rights to keep it."

DIESEL ENGINES APPROACHING THE POUGHKEEPSIE BRIDGE, CROSSING OVER VINEYARD AVE., IN HIGHLAND. They are ALCO locomotives hauling a New Haven excursion train for the Connecticut Valley Chapter, National Railway Historical Society, from Hartford to Maybrook and return, May 5, 1956 (Photo by Jim Shaughnessy, SHAU)

36. DRIVING DIESELS OVER THE BRIDGE (1960s)
by Engineer Peter G. McLachlan

"I took another bite on the train air."

PETER MCLACHLAN drove locomotives across the Poughkeepsie Bridge, first as a fireman from 1959 to 1965, and then as an engineer into the early 1970s. In the following account, which McLachlan made available for this book, he explains what he was concerned about when he drove across the bridge:

On the New Haven Railroad, as fast as we sometimes ran our trains, we all respected the speed limit for the Poughkeepsie Bridge. None of us wanted to dive from the bridge 212 feet into the water.

Keeping to the bridge speed limit was complicated by the New Haven's forbidding us to use air brakes on the bridge. This rule sprang from the fear that air braking would set up dynamic motions which could shatter the bridge's steel structure.

Coming on a westbound freight from Hopewell Junction toward Poughkeepsie, I could do 50 miles per hour until within approximately two miles of the bridge. Then, preparing for the bridge, I would slowly shut off the throttle to idle. After drifting along for a minute or so, I would apply 10 lbs. of engine brake to bunch up the slack slowly, so as not to kill the crew in the hack [caboose].

Westbound freights were usually pulled by three to four diesel units. The freights hauled from 4500 to 6000 total tonnage, in from 90 cars to the full limit of 150 cars. Altogether, all the way back to the hack, we had 45 to 75 feet of slack to worry about.

After judging enough time had passed so that I had the slack in—in the 1960s we did not yet have any radio communication with the crew in the hack to help us decide — I would apply approximately 8 lbs. of train air, leaving the engine brake at 10-15 lbs., till I felt us slowing down. We were still doing approximately 40-45 miles per hour and were about one and a half miles from the bridge.

The approach on each side of the bridge was down grade, so the weight of our train was now pushing us along. I was now watching my speedometer and wanting to get down to approximately 25 miles per hour by the time we reached within about one quarter mile of the bridge. Near the old Poughkeepsie station on Parker Avenue, I took another bite on the train air of some 10 lbs. to slow us to approximately 18 miles per hour. Once by the station, with the air brake grabbing hard, I would release the train air completely, while increasing engine brake to approximately 25 lbs. By the time the train air began to release, my speed would be down to 10-12 miles per hour, ready to start across the bridge. Nevertheless, as the train's first cars moved out on the bridge, the rear cars would still be moving downgrade, so our tonnage would slowly push our train faster out over the bridge. We could not let our speed push up much. The New Haven's speed limit on the bridge was 12 miles per hour, and the time minimum to cross the bridge was 7 minutes. There were three machines on the bridge to record what we did. Their records were picked up once a week for reading.

When we were about two-thirds of the way across the bridge, our speed might be up to 15 miles per hour, but then we were starting to go up a hump. The Poughkeepsie Bridge is not quite level; toward its western end it has a slight rise.

The bridge was originally built that way. With the rear of our train still going downgrade east of the bridge, and our not being allowed to use the train brakes, going up the hump helped slow us and keep control of the slack in our train. We also held the train back a little with the engine brakes. Yes, we were allowed to use the engine brakes on the bridge.

As our locomotives pull off the bridge, we slowly, slowly, release the engine brakes, and notch out our throttle to no more than notch one or two, stretching out the train. The hump is still keeping our speed down safely.

Soon our engine is starting to climb cliffs. Presently we are stretching out with the throttle, pulling hard up into Highland.

By the time the caboose is coming off the bridge, my throttle has to be in the eighth notch, the engines pulling all they are worth. Many an engineer would stall a train because they were too timid to open up. They were too timid to pull the hack off the bridge faster than 12 miles per hour.

Between Gaffney's Curve, near Clintondale, and the Maybrook yard, we would often get up to 60 to 62 miles per hour. Approaching Maybrook, we would let the upgrade near the Berea "XC" tower slow us down naturally.

Coming back east over the bridge was basically the same but more difficult. The eastward grade down to the bridge was steeper. Each train carried greater total tonnage, up to 9500 tons, but fewer cars — eastward trains as a rule were not much longer than 95 cars. This meant we had fewer cars to provide the necessary air braking power. So we had to start the approach harder, with more air braking, and earlier, and hold air on to the last minute when we reached the bridge.

Oh yes, I misjudged sometimes, and would hit the bridge a little too fast. Sometimes a little more than a little! Whew! But only about three times. You learn.

Left, CONRAIL ENGINEER PETER McLACHLAN OF NEWTOWN, CT, the author of this chapter, with TWO RAILROAD HISTORIANS, center, LOUIS V. GROGAN OF PAWLING, NY, and, right, AUSTIN McENTEE OF POUGHKEEPSIE, in the Danbury yard, 1992 (Photo by Vic Westman, from McEN)

37. CONDUCTOR ALEXANDER, PROTECTOR

(1939-1970s)[1]

Being a team player, with difficulty.

WHEN WORLD WAR II began in Europe, Americans of Italian descent were sometimes under suspicion. After all, Italy's government was fascist and allied with Germany. Nevertheless, on October 1, 1939, the Maybrook yard asked Albert Alexander, then twenty-one years old, to begin work as a brakeman. Alexander was proud to consider himself one of the earliest children of Italian immigrants to have been hired at Maybrook to serve in such a responsible train service position.

Albert J. Alexander was born in Maybrook into a family who had anglicized their name. Albert's father, born in Italy, came to Maybrook as a laborer to help build the rail yard, and then stayed on, continuing to do railroad work, especially maintenance. When Albert himself became old enough to work, railroading was almost the only kind of work known in the town. But this was during the depression of the 1930s, and there were few openings in railroading. So, like many other young men, Alexander worked for a time for New Deal agencies created to employ the otherwise unemployed, including the Civilian Conservation Corps. Eventually, however, on his father's advice, he applied to Sam Scott, the Maybrook yardmaster, for work. At first, since Alexander had training in shorthand and typing, Scott offered him clerical work. But when Alexander insisted he preferred outdoor work, Scott put him on the waiting list to become a brakeman.

When Alexander began his brakeman's work, it seemed to him that he needed guidance. But the experienced railmen around him, perhaps because they were jealous of his securing such a good job, gave him little help. So he fell into awkward situations. Once when he jumped off a moving train, instead of jumping face forward, in the direction the train was moving, he jumped face backwards. He was jarred. On a more scary occasion, after Alexander had made a run from Maybrook to Danbury, he found he was not needed to work on any train that would take him home to Maybrook. His superiors arranged for him to "deadhead" home, that is, to hitch a ride home on a westbound train on which he was not to work. They told him to walk to where the rail line crossed a certain Danbury street, and then, when the train's locomotive came in sight, to signal to it. They assured him that by the time the caboose came by, it would be going slowly enough so that he could hop on.

When the train's caboose came by, it seemed to Alexander, inexperienced as he was, that the train was still moving rapidly. But he slipped his lantern handle over his arm, and reached out both of his hands for the "grab iron" just in front of the steps at the end of the caboose. He missed it and panicked. He felt that he must climb onto this caboose or die. In the next second, when, right after the steps, the only remaining grab iron came by, he lunged for it, and this time succeeded in seizing it. But he found that the force of the train's movement kept his feet flopping in the air at the end of the train until, with a supreme exercise of will, he pulled himself up onto the platform.

In time, Alexander felt he fitted well into railroading. Working with a group of men seemed to him "like being on a team," and he liked that. When problems occurred, even such a serious problem as a derailment, the bosses did not usually come around to tell the trainmen what to do; they usually let them handle such problems by

CONDUCTOR ALEXANDER (middle) ON A DIESEL SWITCHER IN POUGHKEEPSIE, with, left, fireman Art Malley, and right, engineer Bill Trepton (MRM)

themselves, and he liked that. Also he relished the view from his train, including the view as it crossed the Poughkeepsie Bridge—he liked to imagine what that view was like when Henry Hudson first came up the river. Of course being outdoors much of the time, he had to endure rough weather. He also had to accept the dangers inherent in railroading. But once railroading gets "in your blood," he recalls, you do not want to do anything else.

BRAKEMAN

As a brakeman, Alexander was expected to check his whole train for trouble. When his train was stopped, he would check it by walking along beside it. If it was dark, he would carry an oil lantern which might barely give him enough light. He would check the hand brakes on each car to make sure they were off. He would check for loose ladders. He would look for anything dragging underneath the cars. He would sniff for any journal boxes which were smoking from lack of oil.[2]

Once Alexander was a brakeman on a train making a run he often made, from Maybrook to New Haven. It was a long train, which limited his opportunity to discover any hotboxes by seeing or smelling their smoke. Moreover, it happened to be a dark, foggy night, Alexander recalls, which made it harder. After crossing the Poughkeepsie Bridge, his train was moving through Arlington when it passed a Maybrook-bound train, and the crew in that train saw that Alexander's train had a hotbox. However, by the time the crew tried to hand-signal the news to his train—the usual signal a trainman used to report a hotbox was to hold his nose with one hand and point to the ground with his other hand—the trains had almost passed, so Alexander and his crew missed seeing the signal. Two-way radio-phones not being available for Alexander and his fellow Bridge Route trainmen until the early 1970s, there was no other way for the passing train to warn Alexander's train about the hotbox. So Alexander's train continued on, with its crew still not knowing. Finally thirty miles farther on, in Brewster, the hotbox burned off its bearing, and its car derailed. Afterwards, railroad officials tried to hold Alexander and his crew responsible for the derailment, but Alexander argued that on a long train on such a dark night, they could hardly be expected either to see the signal from the passing train or to detect the hotbox on their own. In the end Alexander and his crew escaped punishment, even any loss of work time.

Another function of brakemen was to pass along signals, such as signals from a conductor to the engineer at the head of a train. Once Alexander's train, as it picked up and set off cars near Danbury, was backing around a curve while he sat on top of one of the train's boxcars passing signals along to guide the train. Because it was night, he was signaling with an oil lantern. But he happened to make "a quick move" which "jarred the light." It went out. For safety, he explained, these lanterns were made so that if they were disturbed by even "the least little jar," the light would "snap" out.

Alexander knew that with the lantern out, the engineer, looking back over his train around the curve, could no longer see any signal to guide him. In such a situation, it was against the rules for him to continue backing the train, but Alexander knew that the engineer would not want to stop. Dangerous though it might be, he would want to keep going until he could see a signal somewhere to guide him. Alexander still remembers feeling the pressure to get that lantern lit quickly. He felt in the dark for a match. He felt his way to "open up the lid of the lantern, turn the wick up real high, lift up the globe, strike the match and throw the flame in against that big wick."

Eventually during the war, unmarried as Alexander was, he was drafted. After the end of the war in 1945, he returned to Maybrook, and exercising his seniority rights, became a brakeman again.

Soon after his return, Alexander became eligible to take an examination for promotion to conductor, and, as we have already seen, decided to take it. To prepare for it, Alexander studied the New Haven Railroad's rule book by himself, and he considered that he studied hard. When he was examined, however, in an oral examination in New Haven, the examiner decided Alexander did not understand the rules well enough, and told him he had nerve trying to become a conductor.

Alexander felt that it was unfair that during the war, when the railroad needed to hire more men, men who were not in military service were passed through the test easily, but after the war, someone like himself who had done years of military service was tested severely. Alexander felt insulted. He blew up. He told the examiner he didn't need the promotion anyway.

Afterwards for several years, Alexander continued to work as a brakeman and refused to retake the exam. By refusing to take it, he allowed two other men of less seniority to become conductor ahead of him. Eventually, however, he decided he had allowed these two men "to go ahead of me on the roster just because I was stubborn and mad." He decided that he still wanted to become a conductor, and had made a mistake trying to study the rule book by himself. This time he studied it with the help of experienced railroaders who could give him realistic examples of what the rules meant. This time he passed the exam, and in 1954 was promoted to conductor.

Whether he worked as brakeman or conductor, Alexander often ran into danger. Once in the 1950s, he was on a train near Hopewell Junction which had set off some cars and had begun to back around a curve. While the crew at the head of the train was waiting for a signal from a flagman at the rear of the train to allow it to continue backing, Alexander, on top of a car in the middle of the train, was struggling to keep the flagman in view so he could pass his signals along. Trying to do so, Alexander ran "lickety split" over the top of several cars. This was easy to do on those cars that were regularly spaced. But suddenly he came to a space between cars which was much longer — as he learned later, it was longer because a new kind of hydraulic coupling had been installed there. If he had been running more slowly, he would not have been so foolish as to try to jump that space. But "I was running so fast I couldn't stop," he recalls. If he hadn't made the jump successfully, he could easily have fallen and been crushed. Some forty years later while lying in bed, floating between sleep and awake, he still recalls this incident and it still gives him chills.

LOCAL FREIGHT FROM BEACON CURVING INTO THE HOPEWELL JCT. YARD, MAY 14, 1956 (Photo by Jim Shaughnessy, SHAU). At the top, double tracks lead in from Poughkeepsie; to their left is the Hopewell Jct. station, with an automobile showing white beside it.

BALANCING

On another occasion, Alexander struggled to balance train safety with protecting his fellow railroaders' jobs. At the time, Alexander was crewing on a train in the Cedar Hill yard near New Haven, preparing for a return trip to Maybrook. A car inspector was completing his usual inspection of the train's air brake system when a switcher engine coupled twenty-five more cars onto the train. The car inspector asked a yard brakeman to couple the brake air hose from these additional cars to the rest of the train. Then the inspector, without having tested the air brakes again, walked to his office shanty and phoned in his report that the brake tested okay.

As the train pulled away from the yard toward Danbury, Alexander and his crew in the caboose could see that their air gauge was dropping slowly, not enough to turn on the brakes, but it continued to drop. Alexander suspected that the brakeman, after coupling on the air hose, had forgotten to open the valve which would allow the air to pass through the hose connecting the twenty-five extra cars, and that after the train climbed up the mountain to Botsford, there would be insufficient air left to brake the train properly as it went down the other side. Alexander wanted to assure the train's safety, but he also wanted to prevent both railroad officials and the engine crew from knowing that the inspector and brakeman had not done their jobs. Alexander knew that the train would be required to stop at Botsford to pick up orders. While the train was slowly struggling up the mountain to Botsford, Alexander walked from the caboose over the top of the cars up to the twenty-fifth car. When the train stopped at Botsford, he climbed down between the cars and turned the hose line's connecting angle cock gently, letting through the air slowly so the brakes would not grab, and so "the head end" would not know what he was doing. When the train started up again, "nobody knew the difference," and we "went on to Maybrook with no problem at all."

In another incident illustrating Alexander's concern to protect his fellow trainmen, he was riding in the engine of a long westbound train which had crossed the Poughkeepsie Bridge and was approaching the Maybrook yard, when the engineer had trouble handling his air brakes so as to stop the train in front of a certain switch. Although the train was already well slowed, it looked to Alexander as if the engine was going to slide through the switch before it was thrown as needed. Alexander jumped down from the still moving engine, and ran ahead of it to throw the switch before the engine got there. He threw his "whole body weight" onto the switch handle, and "the handle came right off the switch." Alexander fell to the ground, but fortunately he had already thrown the switch successfully.

When Alexander climbed back on the engine, the fireman said, "Fool, why did you get back up? Why didn't you lay right there? We would have called an ambulance and you could have had the whole winter off."

That, Alexander replied, would have gotten the yard's track walker in trouble. The track walker should not allow a switch handle to be ready to fall off. He "is supposed to check them switches."

However, there were limits beyond which Alexander would not protect other trainmen. Once when Alexander was working as a conductor, one of the crew who showed up to work with him appeared to be drunk. Alexander knew he had come to work drunk before, and was "always giving trouble to everybody that worked with him." Nevertheless, conductors were hesitant to take the initiative to refuse to work with him. They were hesitant for one reason because when there had been strikes, he would not join the strikers, so they "figured that he was one of the company's pets." But this time what tipped the balance for Alexander so that he refused to accept this man into his crew was that another conductor had recently told Alexander that when this man had been working with him and was drunk as usual, he had threatened him with a knife. Alexander asked the crew dispatcher to call another man instead, as a conductor had the right to do, and the dispatcher did so.

Risks, Accidents, Safety

38. TELEGRAPHY AWRY: A HEAD-ON COLLISION (1916)[1]

In an instant, three locomotives became twisted metal, lying beside the track.

WHEN the Poughkeepsie Bridge was first coming into use, the system of controlling the movement of trains by telegraph had already been long established. At that time, the bridge management and its railroad allies arranged for the Western Union Telegraph Company to install telegraph wires along their tracks from Campbell Hall over the bridge to the Connecticut border. By their contract, Western Union was to supply poles, wires, batteries, "Morse instruments," and a foreman to direct their installation and maintenance. The bridge-related railroad companies were to supply the necessary labor for the installation, keep the right of way clear of brush, and provide the telegraph operators. The telegraph lines were to be owned by Western Union and become part of its general system, but were to be used jointly by the railroads and Western Union, with the railroads having use of the line free, and having priority for their own urgent business, such as directing the movement of trains. The railroads would establish public telegraph offices in their stations and arrange that their personnel would transmit commercial telegraph messages for Western Union, using Western Union forms.[2]

The Central New England Railway, when it took control of the Bridge Route, continued the original arrangement with Western Union. However, at various times other railroads related to the Poughkeepsie Bridge Route made other telegraph arrangements. For example, on the west side of the Hudson River, the Lehigh & Hudson River Railroad owned and operated its own telegraph line, while on the east side of the Hudson, the Poughkeepsie & Eastern owned its own telegraph line but arranged for Western Union to operate it.

Meantime, while the bridge was being built, the telephone, which had evolved out of the telegraph, was already available to the public at each end of the bridge, limited and unreliable though its service was. Soon afterward, the local Highland telephone company, to make long distance connections, strung telephone wires across the bridge. By 1901, despite Western Union protest, two more telephone companies had strung their wires over the bridge. In the next few years, both telephone and telegraph wires were often strung along railroad rights of way, and bridge-related railroads increasingly used telephones for incidental purposes, such as notifying consignees of the arrival of freight. For three decades yet, however, it was more often the telegraph rather than the telephone which was used to dispatch trains.[3]

OPERATORS

In 1892, on the Bridge Route, the Central New England & Western employed 26 telegraph operators, by 1914, the Central New England (CNE) employed 60. Their average pay was more than that of trackmen, but less than that of engineers or conductors.[4] Regular telegraphers usually worked only in dispatchers' offices or in larger stations. If smaller stations had telegraph offices at all, to keep costs down, their station agents often doubled as telegraphers. A few years later, most CNE telegraph offices were single ones, one office at each station, but there were extra offices at major interchange points, including two offices at

Hartford, two at Danbury, three at Maybrook, and three at Poughkeepsie (one was in the signal tower at the bridge entrance).

If one of the public wished to send a telegram from a Bridge Route station to San Francisco, the station telegraph operator would customarily look up the appropriate fee in a Western Union reference book, charge the sender for it (fees were high enough to inhibit frequent telegraph use), and keep part of the fee as a commission. Such telegrams would be sent through a regional relay office, as at Albany.

Whether for public or railroad purposes, railroad telegraphers clicked out messages on senders. They clicked in Morse code (American Morse Code, not the International Morse Code used in shipping). They received incoming messages from receivers by listening to their clicking. As they listened, they wrote down the words they received, usually by hand. As they wrote one word down, they would be listening to the next word. If they were receiving orders directing the movement of trains and misunderstood them, or for any reason failed to deliver messages to the persons for whom they were intended, the result could be disastrous.

CONFUSING ORDERS

On the evening of Dec. 21, 1916, train number 195 out of Hartford, westbound, was delayed in leaving Hartford. By the time it departed at 5:10 AM the next morning, the delay had become so long that, as was customary in long delays, the CNE telegraph dispatcher in Poughkeepsie, who was in charge of train movements on the line, no longer called it by its normal numbering 195, but officially called it instead an unscheduled "extra."

The extra train's eastbound counterpart, train 194 out of Poughkeepsie, after it had reached eastward on the Bridge Route's single track line as far

THIS TOWER, CNE SIGNAL TOWER #194, WAS AT THE EAST END OF THE BRIDGE, AT WASHINGTON ST., POUGHKEEPSIE (STICK). Like all such CNE towers, it operated under the authority of the train dispatcher, who communicated with it by telegraph. In the 1910s and 1920s, this tower, along with the similar signal tower #193 at the west end of the bridge, controlled the movement of trains over the bridge. The train shown is coming east, off the bridge.

as the station at Winsted, CT, received from the dispatcher a telegraphed train order saying that it was to pass the extra at Canton, CT. With Engineer Everett Sisson in its cab, the train promptly took off eastbound for Canton.

Meantime, the westbound extra train from Hartford, when it reached Simsbury, CT, took on water and picked up more cars. One of the extra's engineers, Fred Woodin, went into the Simsbury station to talk to its telegraph operator, Downs. Although Downs was an experienced telegrapher, he had been on the CNE's Bridge Route only for about two weeks. While Downs had received a telegraphed order for the extra train, Downs later testified that he had not received a telegraphed order, such as he was supposed to have received, that train 195 had been annulled, and that an extra was taking its place. Downs recognized Woodin as from train 195, not from any extra train, and so did not deliver the telegraph order for the extra train to Woodin—the order that the extra train was to pass the eastbound 194 at Canton.

Downs watched the train leave, and then telegraphed the dispatcher to say mistakenly that train 195 had departed. He later testified that he did not receive any response from the dispatcher's office to indicate that his message had misidentified the train, as that office, if it were alert, should have understood. The result was that the westbound extra's Engineer Woodin believed the tracks ahead were clear for him, while the eastbound train 194's Engineer Sisson believed the same tracks were clear for him.

By about 8 AM, it was raining, and the extra,

TELEGRAPH CALL LETTERS ON THE CNE MAIN LINE
Arranged by Station, from East to West
From CNE, Time Table for employees, Oct. 5, 1913

DI	Hartford, CT	MO	Millerton, NY
BF	Bloomfield	BN	Boston Corners
CK	Clarkville	MN	Ancram Lead Mines
BD	Tariffville	B	Sissing Jct.
D	Simsbury	RD	Stanfordville
CD	Canton	RN	Clinton Corners
MI	Pine Meadow	DA	Salt Point
NH	New Hartford	VO	Pleasant Valley
WD	Winsted	CF	Poughkeepsie Jct.
WS	West Winsted	QA	Poughkeepsie
NK	Norfolk	HY	Highland
C	East Canaan	DO	Lloyd
A	Canaan	CD	Clintondale
WN	Twin Lakes	MD	Modena
CA	Chapinville	SM	St. Elmo
SA	Salisbury	WA	East Walden
AK	Lakeville	XC	Maybrook
		CH	Campbell Hall

with Engineer Woodin in the cab of the lead engine, hauling perhaps twenty-five cars, was puffing up "Horseshoe Bend," about a mile west of New Hartford, CT. Coasting down a long hill in the opposite direction, on the same single track, approaching the same curve, was the train from Poughkeepsie with Engineer Sisson in the cab. It was pulling only a caboose and one car, but the car was heavily loaded with cement.

Meanwhile, the agent at Canton came on duty and overheard the dispatcher telegraphing the agent at Simsbury to ask where the extra was. He cut in to tell the dispatcher that before he left home he had seen what looked like the extra passing through Canton. The dispatcher quickly understood. A crash was likely.

In a last minute effort to prevent a crash, the dispatcher telegraphed the operator at Pine Meadow, a station just east of New Hartford, asking if the extra had passed there. The operator clicked out his answer on the keys, "They are coming." The dispatcher clicked his reply, "Stop them." It took a moment for sending, receiving, and understanding. The operator clicked again, "I can't. They are gone."

A few minutes later, as Engineer Woodin in his engine was coming uphill around Horseshoe Bend, he suddenly saw Sisson's train coming downhill toward him. It was only about sixty feet away. Woodin believed it was too late for either train to stop quickly on the wet tracks. He jumped out. At about the same time, Engineer Sisson saw Woodin's train heading for him. Sisson also jumped out. The trains crashed head-on. In an instant, all three locomotives became twisted metal, lying beside the track. Telegraph poles along the tracks were shattered. Several cars buckled. One car split open, spilling its cargo of auto tires.

Engineers Sisson and Woodin had jumped free of the wreckage. Others of their crews were caught in it. Two fireman, Lewis Bennett of Hartford, CT, and William Dingee of Maybrook, NY, were burned and died. Other railmen were injured.

TELEGRAPH AND TELEPHONE LINES on CNE Tracks From CNE, "An'l Rep.," 1926	
Miles of Telegraph Lines: Owned by CNE Owned by others	112 888
Miles of Telephone Lines: Owned by CNE Owned by others	445 0
Miles of track over which CNE dispatches trains:	277
Percent over which CNE dispatches by: Telegraph Telephone	72.5% 27.5

The Interstate Commerce Commission investigated the crash. Its "direct cause," the Commission decided, was the failure of Operator Downs to identify the train at his station as an extra. Contributing factors were that the rules for handling telegraphed train orders were not followed. If the dispatcher had telegraphed the train orders in the same words to each station, as the rules required, and if the train orders had been repeated by the several stations which received them, to make sure they had been correctly received, as the rules also required, then Downs would have been required to repeat the message which he should have received that train 195 had been annulled. If he did not repeat it, then the dispatcher and other operators should have been alerted to the problem, and taken measures to correct it.

Directing the movement of trains on single-track lines by telegraphed train orders, while much superior to the previous system of following time tables and waiting for opposing trains to pass, had inherent weaknesses. It presented opportunity for human error and lacked a reliable system to detect the errors in time to correct them.

39. CARRIE'S COW (1920s?)

Fiction by John K. Jacobs

"The Lord hath taken away and now, the way I see it, He's fixin' to giveth back."

THE AUTHOR of this short story, John Jacobs, grew up about six miles west of the Poughkeepsie Bridge in Clintondale, NY, on a farm through which the Bridge Route passed. The rail line "was a dramatic presence in our quiet lives," Jacobs recalls. "We could set our clocks by the `Scoot'—the morning two-car train from Poughkeepsie that stopped along the way to pick up men going to work in the Maybrook rail yard." One winter his father himself rode the Scoot on his way to work in the Maybrook yard. He worked unloading coal from coal cars, Jacobs remembers, "knocking the frozen coal out of hopper cars and hoping that he himself would not slither along with the coal into the elevator below."

The basic tale in the story, Jacobs says, was one he heard as really having happened when he was a child. The "Million-Dollar Beef Train" out of Chicago, which plays a role in the story, really ran over the line near Jacob's house, heading across the Poughkeepsie Bridge. Margraf, the station agent in the story, really was a long-time station agent at Clintondale, though Jacobs changed his first name. But some details in the story, Jacobs admits, he "manufactured for fictional convenience." Carrie, he explains, "is based on a wonderful woman" who lived near Illinois Mountain beside the rail line. She was "tall, angular, flinty, totally independent, very poor, and not at all ashamed of it," a kind of person rare today. "Whether she was in actual fact the perpetrator of the strategy against the railroad which the story describes I cannot now say. But it would suit her character."

The story was first published by Jacobs when he was a student at Antioch College:[1]

Carrie kicked a shock of fodder in front of her cow. She took two burlap sacks and rubbed down the cow's flanks and udder, and then sat on the one-legged stool to begin her milking.

The children, Danny and Kate, visiting as usual, took off their jackets and squatted in the pungent animal heat that issued in wonderful abundance from the cow. "Aw, Carrie, why don't you give her a name?" asked Kate."

"Good Lord, you kids are ignorant; you don't name cows. S'pose I had sixty milking head like Pop did? Could I muss around with all them names? If it was bulls, it'd be different. Bulls got to have names. Pop had two bulls, Red and John L. Sullivan."

"But Carrie," objected Kate, "you haven't got sixty cows. You only have one."

Carrie milked on in tolerant silence. The children, understanding in a way that there was nothing more to be said on that subject, squatted, silent and solemn. Craning their necks, they watched the

JOHN JACOBS, the author of this story (JACBS)

long shadows cast by Carrie's squat, wire-faced railroad lantern, hanging on its hook. Carrie stripped out the remaining two teats and got up.

"Girls don't have much of a head for things like that anyway," she said. "Maybe Danny sees what I mean."

"Sure," said Danny confidently. "I know bulls is different."

The next day when the children walked over for milk, Carrie led the children through her pasture to the railroad fence, where they looked over at an immense dead creature lying by the tracks. Danny by himself walked a little farther along the fence to where there were rotten posts and rusty wires, and they could see that the cow had pushed through. Danny himself stepped through.

"Danny, don't!" shrieked Kate. But her fear instantly turned to envy for his bravado. Danny walked over to the cow and all around her. The frightfully broken beast, bloated, its great tongue lolling from its mouth, distended nostrils choked with blood—all of it fascinated him.

Then the two children walked back to Carrie and together they walked slowly back to the barn. "I told them railroad men that fence wanted mending," Carrie said. " I told 'em my deed says the railroad got to keep the fences in repair, and they said they'd tend to it. Never did, though."

"You could sue 'em," said Danny. "Maybe you could get a thousand dollars."

"I reckon I'll write 'em a letter. I don't only want what's right. Just enough to buy me a cow."

The children picked up their white enameled pail and trudged home, wondering. On the whole they accepted the cow's death with the indiscriminate skepticism of early life when many things are unfathomable. For five years there had been a cow, and now there was not, and that was that.

Behind the freight station, Clint Margraf, the station agent, searched among the sawed-up discarded railroad ties. Finding a chunk of reasonably straight grain, he took his hatchet and split it into kindling. Jimmy Gaffney, the father of Danny and Kate, had told him of the death of the cow, and he awaited the inevitable arrival of Carrie with some anxiety. He went inside the station to build a fire in the pot-bellied stove. It was soon smoking from the creosote in the old ties, his eyes were watering, and he failed to notice Carrie enter.

Carrie stood for a moment. "Put a capper on that chimney," Carrie said, "and she might draw better."

"Why hello, Miss Connor," stammered Margraf. "You give me quite a start comin' up so sudden like."

The two eyed each other for a moment, then Margraf picked up more kindling and started to put it into the stove. "Yes, sir, " he began in a loud voice. "I knowed a feller once, as stout an old feller as you'd want to meet. . ."

"Don't think I'd put any more wood on that fire," interrupted Carrie. "You'll smoke yourself out."

"Smoke myself out," laughed the agent uncomfortably.

"Like a woodchuck," said Carrie.

Margraf straightened, and took a deep breath into his great stomach.

"Now, Miss Connor, there ain't a thing I can do. The cow's dead, and you know the railroad ain't what it used to be. I just work here in my station. I don't have nothing to say up to Maybrook."

"I didn't hear nobody askin' nobody for anything," said Carrie. "But since you speak of it, I was looking for somebody to write me a letter, and I guess it might as well be you."

When the letter was written and sealed Carrie walked slowly home along the tracks.

Two days later, after she had washed her scant breakfast dishes and scoured the frying pan, she went again down to the tracks and trudged the two long miles to the station. She found the mail trains had brought no letter for her. On three succeeding days she made the same trip, noticing as she passed the carcass, that the stench of its decomposition had increased. Still no letter had come.

On the fourth day as she approached the station, she saw parked there a large, gleaming automobile. Inside the station, she found agent Margraf talking to a big, red-faced man.

"Morning. That letter come yet?" called Carrie.

Margraf stirred officiously and the two men

came towards her. "Miss Connor, this here is Mr. Sheldon from the main office down to Maybrook."

"Please to meet you, Mr. Sheldon," said Carrie.

The red-faced man stood before Carrie. He planted his thumbs in the pockets of his vest and glared. "Miss Connor," he said, "Mr. Margraf here tells me that a cow of yours broke through our fence and got hit by one of our trains." He paused to bite the butt from his cigar and spat on the floor.

Carrie waited.

"He tells me that this cow is still on our right of way. That's a dangerous thing, Miss Connor, and its illegal."

"What's left of it lays there, and I'll get it off—soon as I get sixty-five dollars to get me a new cow," said Carrie. "I know my rights, Mr. Sheldon, and I know that fence was just ready to fall down of itself."

Mr. Sheldon drew deeply at his cigar. "The law says we got to put up a fence and there was one there. The law says too you got to get that cow off from our property."

"You don't have to tell me the law, Mr. Sheldon! I know what my rights is!" flared Carrie.

"All right, all right," said Mr. Sheldon. "Go ahead and sue. But I tell you right now, it costs a heap of money to go to law. And we got the best lawyers in New York State. I'm telling you to get that cow off from our right of way. That's all I got to say. Good day, Miss Connor." He turned and walked out of the station.

Margraf ran his fingers through his thin grey hair. "I don't know, Miss Connor," he said. "I guess the law ain't writ for the likes of us."

"Oh, hush up!" sighed Carrie. Margraf's telegraph key was clicking, and he turned to it as Carrie walked out of the station and onto the tracks with long, angry strides.

Anger burned in her stomach. It was the futile anger that always came upon her when she was defeated because of her womanhood and her poverty—that was the real injustice. The very ties she walked on, spaced with no consideration for the human stride, irritated her. Eventually the fine drizzle that was falling began to soothe her. By the time she reached her own gap, she was calm once more. As she sat on the track to remove a cinder from her shoe, the cyclopean snout of a freight locomotive appeared far down the track, heading east toward the Poughkeepsie Bridge. Unhurriedly she replaced her shoe and walked towards her barn, occasionally watching the great locomotive labor up the grade. Black smoke fulminated from its squat stack and steam hissed from its under parts as it slowed and finally came to a stop.

"Never seen her work like that before," she thought. And then she saw. The tracks along the grade had been ground to a sheeny smoothness by innumerable, heavy freights such as this, and now, wetted by rain, they afforded scant traction for the great drive wheels of the locomotive.

The piston rods clanked once more. The train backed slowly down the grade, and then started up it again, this time with more success. Carrie stood in the rain and watched until the caboose disappeared over a slight rise in the direction of the Hudson River.[2] Still she stood, trying to grasp something that had flitted through her mind and had disappeared into the oblivion of a thousand other thoughts. She looked long at the cow, at the tracks, and then she found it.

* * * * *

All afternoon Carrie worked, cutting through the putrefied flesh with a large pig knife, hacking through the bones with an axe, and then carrying heavy bushel baskets full of the pieces up to the barn. On her last trip she took a spade with her and buried the head and hooves and entrails that were no part of her scheme.

After supper she took the railroad lantern and went out to the corn crib. She rolled out the iron pig cauldron from where it had been rusting ever since her father had died. She slung the cauldron to the charred beam that stuck out from the corner of the smoke house, and underneath it she built a fire.

All through the night Carrie sat on a stump before the fire, dozing fitfully and awakening to thrust logs into it. Periodically she would draw the steaming, yellow residue from the pot, and pour it into a tub. And when at long last, the morning crept murkily abroad, she was done. Her tub was brimming with thick, yellow tallow. Carrie got stiffly up from the stump and walked toward the house.

When she had finished breakfast and was wringing out her dishcloth, there came a knock at the kitchen door. Carrie opened it to find James Gaffney, carrying his dinner pail under his arm.

He smiled a bit uncertainly. "I was set on comin' to the wake, Miss Carrie, but you didn't send me word," he said.

"It was last night," said Carrie with a tired gesture towards the still smoldering fire, "and a fine time was had by all."

Jimmie shifted his feet on the door mat. "Miss Carrie," he said, "I been talking some with the men down to Churchill's store, and it ain't any use to go to law. It ain't only the money. It takes more than that. It's the same as an Act of God."

"Umm?" said Carrie.

"Well, we were talking like I say." Jimmie kicked a piece of mud off the porch. "And as we seen it, it was only fittin'. We took up a collection and we're gonna buy you a heifer. Now there ain't no charity in this, Miss Carrie, it's only fittin' and right. That's the way we seen it an———."

"Thank you, Jimmie, thank you," quietly interrupted Carrie, "and tell your friends I appreciate their kind sentiments, and you can tell 'em for me too that the Lord hath taken away and now, the way I see it, He's fixin' to giveth back."

"I'm sorry you see it that way, Miss Carrie. We only done this because ————."

"It's ten to eight," broke in Carrie. "Jimmie, you'll be late to work."

Jimmie walked back down the path and through the sagging gate. He was troubled about Carrie's refusal, but after he had walked a distance he knew how he would have felt had she accepted. He shook his head and grinned broadly to himself.

Early that night Carrie hooked two pails of the tallow to her old bucket yoke and made her way down into the pasture. She waited and listened at the foot of the embankment. Hearing nothing, she climbed up through the shifting cinders to the railroad bed. The two long steel ribbons caught the pale glimmer of the moon, making the tracks visible all the way down the grade to the first crossing; beyond that she could see the signal—was it lighted? Probably.

Moving almost stealthily, Carrie took the whitewash brush she carried, dipped it into one of the pails, and swabbed the grease onto one rail. Working slowly at first, then faster and more smoothly, she developed a quick rhythm of motion, dip, swab, step; dip, swab, step, first from one bucket then the other to keep the balance of the yoke. She kept wondering how much time she had before the refrigerator freight would come through from Chicago. Then she was under the signal without knowing it and she looked up in surprise. She could hear the metallic click of its mechanism. She sat down on the tool chest beneath it, slipped the yoke from her shoulders and rubbed the muscles of her neck. A few minutes later she replaced the yoke, got to her feet, and started to work back up the other rail. When she finally reached her own crossing, she placed her ear against the track and heard it humming with the approach of a distant train.

Carrie squatted beside the track and waited until finally a point of light showed far down the track. She strained her eyes toward the light. On and on it came, eating up the intervening darkness.

The train passed the signal and began to climb. At first it seemed unaffected, but Carrie knew her job was well done and she knew that jobs well done succeed. When she saw the train slow perceptibly, it seemed to her but the inevitable operation of a certainty. When finally the train hissed and ground to a stop and the ant-like creatures that operated it climbed down out of its innards, she turned and walked slowly home, swinging the empty buckets to the rhythm of her gait.

* * * * *

Early the next morning, the rumble of a heavy vehicle passing beneath Carrie's window awakened her. Throwing up the sash, she saw a truck, heavily loaded with cakes of ice, bump through her barnyard into her pasture.

When she had dressed and was going down the stairs, she heard another truck grinding up the lane to her house, and snatching her cap from its hook she went out into her barnyard. The truck, loaded with sand, came towards her, and stopped reluctantly a few feet from the spot where she stood in front of it, feet planted and arms akimbo. Jimmie Gaffney climbed slowly out of the cab and came towards her, with a grin that cleaved his face.

"James, " she said gravely, "I can't understand why a strong man like yourself would take advantage of a helpless woman like myself."

"How come?" asked Jimmie, still chuckling.

"Ain't you heard of the law of private property? This here is private property, James, and I can't understand why you'd come roaring through here getting me out of my bed and cuttin' my yard to ribbons. It ain't legal, James."

Jimmie's laughter rang out to the barn and back again. "Now look here, Miss Carrie," he said, "this here is the biggest order of ice and sand that our company ever had, and that's the Million-Dollar Beef Train from Chicago down there gettin' ready to thaw, and there's two more trains behind that."

"The law is the law, James. That's what they told me." A hint of a smile flickered across the corners of Carrie's mouth as Jimmie, staggering with mirth, climbed back into his cab and backed laboriously down the long lane. Far down across the pasture she could see puny men swarming about the locomotive, carrying cakes of ice from a truck, and lifting them into the ice compartments of the freight cars.

A week later after the check had come, and after the new cow, just freshened of a bull calf, had been installed in a newly bedded box stall, Danny and Kate came though the fields with their milk pail to the rambling farm house. Finding no one there, they ran out to the barn.

"No milk tonight, kids," said Carrie. "Look over there in the stall. He gets it all."

The children set down their pail and rushed to look over the stall rail at a wobbly new arrival. "Oh, Carrie, it's *beautiful*," Kate gasped. "And look," yelled Danny, "it's a boy! You *have* to give *this* one a name!"

40. SPECTACULAR SPILL (1943)[1]

Could it have been caused by sabotage?

A New Haven Railroad train was on its way from Maybrook toward the Poughkeepsie Bridge with seventy cars, when about two miles before the bridge, at a curve at Pratt's Mills on the outskirts of Highland, it hit a rock. The locomotive's forward wheels left the track and plowed into the track bed, bringing the locomotive to an abrupt stop. Its stopping buckled many of the cars it was hauling, including tank cars of gasoline and oil.

Some of the cars tumbled down an embankment and split open. They spilled out oil and gasoline which exploded into flames. Explosions followed explosions. They shot flames toward nearby houses. Some occupants rushed out so fast that they could scarcely take anything with them.

The accident occurred at 11:48 A.M., Saturday, Oct. 30, 1943. Since it was during World War II, naturally some observers suspected it was caused by sabotage. During the previous year, a German submarine had landed a group of men on a beach on Long Island, with instructions to destroy such key targets as the Hell Gate railroad bridge. The FBI had caught these men before they had done any damage.[2] But why not suppose that submarines might have landed more such saboteurs, and that putting the Poughkeepsie Bridge Route out of use might have been one of their objectives? Observers reported the fire to both firemen and the FBI.

Firemen could do little more than try to keep the fire from spreading. They often aimed their hoses at the nearby rail cars and houses that had not yet caught fire. Doing so, they lay on the ground, trying to avoid the heavy, oily smoke. When they crawled back out of the smoke, they were sometimes so darkened that they were hard to recognize.

Oil kept leaking out of crumpled tank cars onto the ground, spreading the flames. Oil ran nearby onto an old mill pond, setting the pond's surface ablaze, even as firemen were pumping water from the pond.

Smoke rose so high that it was seen not only across the Hudson River at Poughkeepsie but also as far north as Kingston and as far south as Newburgh. The Highland fire chief, William H. Maynard, called the fire "the most spectacular blaze in the history of the area."

Eventually more than 300 firemen, police, and civilian defense workers came to the scene. They came from as far away as Arlington on the east side of the Hudson, New Paltz and Modena to the west, and Lake Katrine to the north. The Red Cross trucked sandwiches to the fire site. Other groups mobilized clothing and ration cards for the families driven out of their homes.

As FBI agents arrived, they took control of investigating the accident, including looking for possible sabotage. They identified pieces of the rock that caused the accident, and sent them to a laboratory. They questioned witnesses, including the train crew.

The New Haven Railroad brought in wrecking cranes by rail from Maybrook, Danbury, and Hartford. That night, as the cranes worked to clear the wreckage from the tracks, pockets of oil continued to burst into flames here and there, but the cranes kept working. By eight o'clock the next morning traffic had been restored on one track, and by afternoon restored on the second track.

Three houses burned down: the homes of Ernest Robinson, Frank Brescia, and Joe Brescia. Although cars at the end of the train were saved, including two loaded with dynamite, more than thirty cars were lost, with about 261,000 gallons of precious gasoline and oil.

Several days later, the FBI announced that they had found the cause of the wreck. Two Highland boys, aged nine and fourteen, had confessed that they had put a rock on the track "just for fun."

41. FOLLY ON THE BRIDGE (Ca. 1960)[1]

Jumping from car to car.

ONE MORNING before dawn, a freight train of about one hundred cars was made up in Maybrook to head across the Poughkeepsie Bridge. Loaded on the train were several heavy steel girders which were so long that they sprawled over three flat cars, their weight resting on the first and third of the cars, with an idler car between.

At a Maybrook yard office, a crewman picked up two orders for the train, each one in duplicate. By mistake, he delivered to the train's engineer, not as he intended one copy of each order, but two identical copies of one order, both copies ordering the train to drop off one car in Poughkeepsie and several at Hopewell Junction. Correspondingly, he delivered to the train's conductor, not one copy of each order, but two identical copies of the other order, both of these copies ordering the train not to exceed twenty-five miles per hour on the whole trip because it was hauling hazardous steel girders.

It was only after the train had pulled out of the yard and was picking up speed that Conductor Albert Alexander, in his caboose, realized that the engineer up front had not received the speed limit order. But being in the caboose, Alexander had no way of communicating this information to the engineer far forward in the engine.

After the train had passed the Clintondale station and was nearing Gaffney's curve, the sharpest curve before the Poughkeepsie Bridge, Conductor Alexander grew upset that the train, considering its special load of girders, was going too fast for the curve.[2] Alexander felt torn over what to do. He imagined that the girders might shift on the curve, and fall off onto the westbound track, where he knew a train was soon due. On the other hand, he was anxious for this train to complete its trip to New Haven quickly, so that he could return to Maybrook within the time limits that federal regulations allowed for crews to work continuously. Alexander was passionately averse to being stuck overnight in New Haven.

Alexander decided to use an option that he was only supposed to use in an emergency. He decided to turn on the air brakes from his caboose at the same time that the engineer, as he brought his train around the curve, was also likely to be turning on the air brakes from his engine. This was dangerous to do because, if the two men were braking at significantly different rates, they could split the train in two. Nevertheless, in the caboose, Alexander opened the conductor's valve to release air from the air brake hose. He did so gingerly, only slightly turning on the brakes on each car of the train.

Alexander hoped his gentle braking might send the engineer, Chet Loop, a message to slow down, but the engineer, headstrong as he could be, seemed not to notice. After the train had passed the curve, Alexander tightened the air hose valve again, and the train seemed to resume its reckless speed.

ALBERT ALEXANDER, recently (ALEX)

227

When the train neared the Poughkeepsie Bridge, Engineer Loop slowed it to twelve miles per hour, the speed limit for all trains crossing the bridge, a limit that was regularly checked. Conductor Alexander knew that when the train stopped at the far end of the bridge near the signal tower, as it was required to do, he would have a chance to deliver the twenty-five miles per hour speed-limit order to Loop. He could be sure to be able to do this by letting enough air out of the brake system to force the train to stay there until he had time to walk forward from the caboose to the head end to deliver the order to the engineer. But walking that far on the bridge would be slow. He was too anxious to get to New Haven and head home again to use such a slow method.

He chose a faster method. Alexander climbed from his caboose onto the top of the next car. Day was breaking, helping him to see. He moved forward to the sound of the cars clicking over the bridge, balancing himself on the top of one car after another. He scrambled over the coal in open coal cars. He climbed over the sprawling steel girders. He jumped between cars, which was always risky to do, and normally not recommended whenever a train was moving, and less likely to be recommended when it was moving on a high bridge.

When the train stopped at the tower in Poughkeepsie at the eastern end of the bridge, Alexander had moved far enough forward toward the head of the train so that he was able to climb down from the train and walk the remaining distance to the locomotive quickly. He finally handed the engineer the speed-limit order.

Looking back, Alexander believes the Lord looked after him in his folly. He suspects that few other crewmen would be so foolish as to jump from the top of one car to another on a train crossing high over the Hudson.

AS TRAINS CROSSING THE BRIDGE GREW LONGER AND HEAVIER, QUESTIONS ABOUT THE SAFETY OF THE BRIDGE CONTINUED TO ARISE. Three New Haven diesels, U 25Bs, haul a long freight westward, July 13, 1967 (photo by J. W. Swanberg, SWANB).

42. BRIDGE SAFETY AND MAINTENANCE

(to 1974)[1]

Snapping rivets, dripping brine, rusting steel.

THE POUGHKEEPSIE BRIDGE managers knew that heavy trains could set the bridge shaking, putting pressure on its joints. They knew that the thrusting power of steam locomotives could contribute to such shaking. They knew that applying locomotive brakes could also contribute to such shaking.

Because of such dangers, the bridge managers warned engineers driving trains across the bridge to avoid braking "except when absolutely necessary and then only in the most careful manner." Moreover, in the bridge's early years, trains crossing the bridge were limited to about 7 miles per hour, with some variation for which direction a train was going. By at least 1913, however, the bridge speed limit was raised to 12 miles per hour, no matter which direction trains were going, and despite pressure to raise it further, for as long as trains continued to cross the bridge, it remained at 12.[2]

It was easy for trains to enter onto the bridge too fast because from both directions the tracks sloped down to the bridge. Moreover, it was tempting for engineers driving heavy, through eastbound freights to pick up too much speed as they left the bridge in Poughkeepsie because they were preparing for the upgrade ahead. This meant, as the inspectors who regularly checked the bridge explained, that while their locomotives might stay close to the speed limit, by the time their "tail ends" left the bridge, they were sometimes traveling up to 30 miles an hour. Such speeds caused a "great" vibration in the east approach viaduct over the Poughkeepsie streets, the inspectors said. The approach viaducts being of only "light construction," a New Haven engineer explained, it was especially because of them that the bridge speed limit had been set.[3]

During World War II, according to the recollections of one of the bridge painters, crews who brought the troop trains across the bridge seemed to consider themselves above the usual bridge speed limits. At that time, if the various railroad men who took turns timing the trains as they crossed the bridge could see that the trains were going too fast, the timers would wave and jump up and down to try to get the trains to slow down before they could set up vibrations which could weaken the bridge.

One winter, when an engineer was approaching the bridge from Highland too fast, he severely applied his locomotive brakes—brakes which were independent of the general train air brakes. He applied them severely enough to overheat the steel tires on his driver wheels. As the locomotive came onto the bridge, the tires expanded. They curled off the wheels. They caught in the locomotive's side rods. The engineer shifted to braking his train by its general air brake system, trying to do so slowly enough to avoid shaking the bridge or buckling his cars. When he brought his train to a stop, it was half way across the bridge. His cars had not buckled and his locomotive had not tipped over, either of which on the bridge could have been a disaster. But his locomotive had derailed.

To handle this wreck, it was impossible to move a big wrecking crane out onto the bridge to haul the locomotive away because such cranes were heavier than bridge weight restrictions allowed. Instead small equipment was brought in to take care of the locomotive on the spot, with an emergency crew. The crew worked at night in a cold, howling wind, trying to keep warm by drinking from pocket flasks. To put the steel tires back onto the engine wheels, they used a bottled gas-fed ring to heat the tires, expanding them. Then they sweated the tires back onto the driver wheels, hammering the tires as needed. As the tires cooled, they tightened in place.[4]

NAVIGATION AIDS

By terms of the bridge charter, the owners of the bridge were required to keep the bridge piers lighted during the navigation season. From 1877, when the first bridge pier rose out of the water, they employed Captain W. H. Sweetman, an Irish-born sea captain, to care for the lights on the piers. Going out to the piers by boat, he kept two oil lights shining on each pier. Sweetman, according to an engineer who knew him from his youth, "was a man of few words, obeyed orders to the hair," was "afraid of nothing," and was "just the man for the place."

By the 1950s, when the pier lights were lit by electric cable, many of the lights were placed about forty feet below the deck of the bridge. To tend these lights, maintenance men climbed down from the top of the bridge, and walked out to them on catwalks. There were eight wooden catwalks available for the purpose, each about 4 feet wide and 60 feet long, resting on steel supports, hanging high over the river.[5]

Though it was not required in the charter, bridge operators also assisted passing navigators by keeping a fog bell ringing on the bridge. In 1918, this bell was kept ringing by storage batteries which were charged by jars of acid. When the maintenance men replaced the acid in these jars, they poured the spent acid over the edge of the bridge. Inspectors found it difficult to persuade them to stop doing it, even though they explained to them that as they emptied the jars, the wind often blew the acid onto the bridge structure, causing the "rapid deterioration" of the steel. Later, in the memory of the bridge maintenance man Ferris Davis, when the bell was powered by electric cable, the bell was brass, 18 to 20 inches high, and it hung on the pier in the water which was closest to the Poughkeepsie shore. You could hear it, Davis recalled, all the way across the river.[6]

By terms of the bridge charter, during the navigation season the owners of the bridge were required to keep a tug ready to assist "tows"—that is strings of barges—to pass safely under the bridge. Tows often passed under the bridge, carrying such cargo as coal, sand, or bricks. Tows could blow a whistle to call the tug to come to

TOW WITH TUGS, as seen from the bridge through three cables attached to it (HOUST). Tows threatened the safety of the bridge, and in turn the bridge threatened the safety of tows.

their assistance which they were more likely to do if it was dark, stormy, or foggy.

In 1923, the Central New England and its parent New Haven Railroad, chafing against the expense of providing the tug, helped introduce a bill into the state legislature to eliminate the charter requirement that the bridge owner keep a tug ready. The bill passed the lower house, but failed in the upper house. In the early 1930s, when the cost of providing the tug had gone up, the tug was assisting more than 160 tows a month during the busiest months, and each tow might be pulling 12 to 14 heavily laden barges in a row, with three barges abreast. At that time, the New Haven's engineer of structures, P. B. Spencer, reviewed the question whether the New Haven should continue to provide the tug. It seemed to him that if a tow struck a bridge pier and there was no tug available to provide assistance, the tow's owner's might sue the New Haven, even if the bridge charter no longer required it to make a tug available. Also if a tow struck a bridge pier, it might damage a pier enough "to interrupt railroad traffic," which was certainly a circumstance to avoid. Spencer, like other New Haven engineers before him, recommended continuing to provide the tug. In the 1940s, the New Haven was providing a tug by renting one for $1,000 per month.[7]

On a hazy dark night in 1950, a tow of fourteen barges was passing north under the bridge, carrying sulphur to a chemical company in Buffalo. The bridge's navigation lights were on, its fog bell was ringing, and the New Haven's tug was at the rear of the tow, helping to guide it, but for unknown reasons the tow jackknifed. The tow's head struck one pier. Then its tail end struck another pier. Two of its barges sank just north of the bridge. Two other of its barges were damaged, began to leak, and, assisted by the New Haven's tug, were beached. New Haven engineers arranged for a diver to come from Boston to inspect the piers under water. With the help of his report, they decided the piers were still sound.[8]

STRENGTHENING

Over the years, the locomotives crossing the bridge kept getting heavier. Many of the early locomotives were Baldwin-built 2-8-0s weighing only 83 tons. After 1900, however, newer ones available to come on the Bridge Route were Rogers-built F-3s weighing 86 to 91 tons. To prepare for them, in 1906-1907, the New Haven Railroad, through its subsidiary Central New England Railway, strengthened the bridge. Contractors inserted along the whole length of the bridge a new center steel truss, in addition to the two outside trusses already in place. The new supports, which the new trusses required, changed the support pattern of the bridge framework as seen from shore from being a spidery double pattern into a more substantial triple. Also because certain piers had been found to have weak or empty spaces in them, contractors reinforced them, both above and below the water line, by drilling holes in them, inserting six-inch pipes into the holes, and then grouting them.[9]

In the next few years, locomotives continued to get bigger. Alco Schenectady was building huge L-1s, 2-10-2s, called Santa Fes, which weighed 280 to 295 tons, and the New Haven Railroad wanted to allow them to cross the bridge. Also the railroad was concerned that trains crossing the bridge eastward were regularly more heavily loaded than trains crossing westward. Since the bridge was double tracked, this kept putting extra pressure on the down-river side of the bridge, the side on which the eastward traffic moved. The uneven weight put a strain on the bridge.

To deal with these problems, in 1917-18 the New Haven Railroad again strengthened the bridge. To direct the work, the New Haven chose the Polish-born engineer Ralph Modjeski, the son of the great Polish actress Helena Modjeska known for her Shakespearean roles.[10] While trains continued to run over the bridge almost as usual, Modjeski arranged for workmen to jack up the tracks onto oak blocks, to allow them to install a new steel floor system. To distribute the weight more evenly on the bridge, he arranged for workmen to gauntlet its railroad tracks — that is, to overlap its double tracks so that in effect the bridge was made single tracked, thus concentrating the weight of the passing trains in the middle of the bridge. On the approaches, workmen bolted new steel columns into certain towers—the viaducts were gradually coming to consist less of iron and more of steel. By enabling the heavy new Santa Fe engines to run on the bridge, this strengthening paid for itself; it did so, New Haven officials decided, in less than two years.[11] The Santa Fes remained the mainstay locomotives on the Bridge Route until diesels began arriving during World War II.

INSPECTIONS

Once Modjeski was involved in the bridge, the New Haven Railroad kept him and his engineering firm involved, asking them to inspect the bridge annually. Over many years, Modjeski inspectors, along with the New Haven's own inspectors, not only viewed the bridge from the ground and from the water, but also scrambled over it assisted by ladders, by swing seats, by "wire hand rope," or even occasionally by scaffolding especially erected for the purpose. In their annual reports, the Modjeski inspectors made hundreds of recommendations, most of them for the routine care of the bridge, including painting, replacing rivets, tightening rods loosened by the bridge's swaying, sealing cracks in pier masonry, and shimming up girders which had settled.

In 1924, the Modjeski firm made a test of the pier foundations in the water, a test they did not usually perform. They did so by boring holes in

AS PART OF THE BRIDGE STRENGTHENING IN 1906-1907, DERRICKS LIFT A LONG TRUSS FROM THE WEST END OF THE BRIDGE, ALL IN ONE PIECE, in what was considered a "rather notable achievement" (photo and quote, *Poughkeepsie Eagle*, Dec. 14, 1907). The derricks, designed by Ray Murray, the American Bridge Co.'s resident engineer at the bridge site, were mounted on rail cars. The trusses, after being lifted, were sent to Athens, PA, to be reinforced, and then were reused. On the east end of the bridge, where the spans were longer, this method of removing trusses was not employed; rather they were removed more conventionally, piece by piece, with the aid of false-work.

LOOSE SPIKES ON A RAIL ON TOP OF THE BRIDGE (photo from Modjeski & Masters, Inspection Report, Oct. 15, 1942, PBEN) The author has added an arrow to point to one loose spike.

each river pier, to discover, as Modjeski explained, "the character of its interior construction." According to Modjeski, in a disturbing report, they found that at certain locations where they thought the piers should have been filled with concrete, they were "filled almost entirely with gravel, sand and sometimes even soft mud." However, Modjeski admitted that since the piers had shown no appreciable settling over the years, the piers at present must be carrying their loads "in a satisfactory manner." At the same time, Modjeski warned that the bridge, even if carefully maintained, might continue to function for only eight to ten more years.[12]

Responding to Modjeski's pessimistic predictions, the New Haven Railroad put maintenance men to work year round, as they had not earlier, and pushed for quality maintenance—in both 1928 and 1929 Modjeski inspectors called the maintenance work "first class." The New Haven also created a retirement reserve fund for the bridge, either for "renewing or replacing" it, whenever one or the other should become necessary. By 1935, the fund had accumulated over $2.5 million, but the New Haven president believed that the bridge would last ten years more, and "it is not at all certain...that retirement will be necessary at that date."

By the mid-1940s, when the bridge had continued to serve far longer than the Modjeski engineers had expected, they believed that several factors were reducing stress on the bridge, giving it extended life. One factor was that diesel rather than steam engines were beginning to be used to pull trains over the bridge; because diesels exerted their power more smoothly than steam engines, they caused less stress on the bridge. Another factor was that the "roller nest" bearings, which were supposed to allow the deck spans to expand and contract during changes in the weather but had rusted so they were "frozen" stiff, were being replaced by new bearings of the "permanently lubricated sliding type."[13] Still another factor was that less corrosion-causing brine was dripping from passing refrigerator cars, brine produced by the salt which was mixed with ice for cooling the cars. Less brine was dripping because refrigerator cars were changing from being built of wood, which could easily rot and leak, to being built of steel. In addition, whenever painting was behind schedule, as it often was, inspectors worried that this speeded the corrosion of the bridge's metal. But in 1946, after painting and other maintenance work had considerably recovered from wartime shortages, the Modjeski inspectors declared the bridge's condition was "better than it has been at any time since the year 1932."[14]

In consultation with Modjeski inspectors, New Haven engineers initiated orders for necessary supplies, such as rivets, pins, ties, planking, steel plates, steel eye bars, steel girders. When appropriate, they specified their size, and where holes should be drilled in them. They kept track of their arrival and storage at the bridge site. Under their guidance, maintenance men adjusted tie bracing. They greased pins. When necessary, they removed smaller worn pins, hammering them out if they could, and if not, employing an acetylene torch to cut them out. They replaced bent lattice bars and tightened the bolts on the sway rods, employing swing seats or scaffolding if necessary. In the masonry of both viaduct pedestals and river piers, they sealed cracks. On the bridge deck, they replaced railroad ties and walkway planks which were split or decayed. On the ground around the bridge approach viaducts, they cut back brush and trees, trying to keep the structure as dry as possible to reduce rusting.

REPLACING RIVETS AND PINS

When rivets were missing, worn, loose, or corroded, maintenance men replaced them. In the 1950s and early 1960s, Tom Houston Sr. would sometimes be on top of the bridge at a forge, presiding at a rivet ritual which his son, Tom Houston Jr., when he was a youth, found fascinating to watch. On the forge, Houston Sr. would heat rivets, some of them up to six inches long, until they were red hot. Houston would pick up red hot rivets, and throw them down, one by one, to the workers who were waiting scattered about the steel supports below, perhaps trying to stand or sit on a girder, or on a board stretched 10 or 12 feet between two girders which could bounce. Houston would throw the rivets through the openings between the ties or over the edge of the bridge, as needed. The men would catch them in funnel-shaped buckets, watching sharply to avoid being burned, insert them into the appropriate holes, and hammer them in.

In summer if you were riding a caboose across the bridge with the windows open, the train brakeman Christian recalled, you could hear the railing on the edge of the walkways rattle, and know the train was shaking the bridge, putting pressure on the steel pins that held much of it together. Similarly, on a cold morning, as crewman Houston Jr., recalled, if you were on the bridge as the sun began to shine, you could hear the bridge bang, and know that the sun's warmth was expanding the steel, also putting pressure on the pins.

Once a key steel pin, about ten inches in diameter, three feet long, and weighing a thousand pounds, needed to be replaced. This job was too big for the regular maintenance men to handle by themselves. This pin joined several of the vertical steel girders of the tower to a cantilever arm, holding it in place. As part of the job, it was necessary to hold the cantilever arm down, because without the pin being in place, that arm would naturally press up. In what maintenance man Ferris Davis considered a "neat" operation, a locomotive with two cars full of coal sat on the arm's deck to weight it down, while a heavy crane removed the old pin and inserted a new one.[15]

GIRDER DETERIORATION, east end of bridge, Poughkeepsie (Penn Central, Inspection Report, Oct., 1970, PBEN)

CRACK IN METAL PLATE (Penn Central, Inspection Report, Oct 4, 1973, PBEN)

UNDERWATER INSPECTIONS

Divers inspected the bridge river piers not only in 1924 and 1950, but also in 1969 and 1972. The bridge owners contracted for divers from specialized firms, and they came, in teams of two or three, from Massachusetts, West Virginia, or Maryland. The divers prepared themselves by studying the engineering records of the bridge. They arranged to dive from various boats, such as an outboard motor boat, an inflatable raft, or the

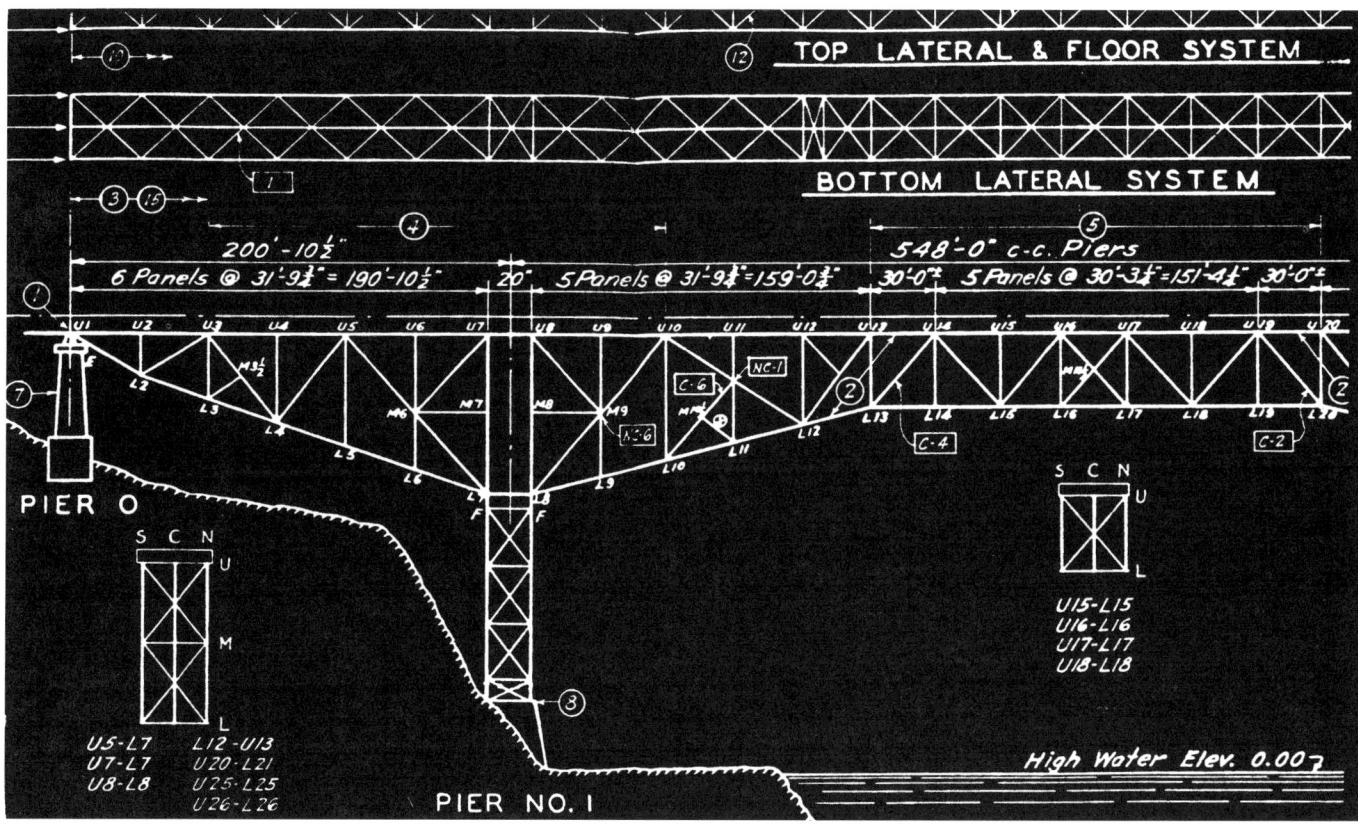

DRAWING INCLUDING THE WEST SHORE ANCHOR PIER (PIER NO. 1) (Modjeski, Masters, & Chase, Inspection Report, Oct., 1938, PBEN). Among the inspectors' recommendations, keyed to numbers marked on the drawing, were: (4) Clean, paint, and seal tops of corroded chord eye bars; (5) Clean and paint rusty lacing bars; (7) Clean out and seal the cracks in the west face of pier O; (8) Remove spalling concrete in base of pier No. 1.

tug which usually guided tows under the bridge. For the 1950 diver, the bridge maintenance man Ferris Davis recalls that his maintenance crew used a hand-pumped air compressor which the diver had brought with him, taking turns twenty minutes at a time, to pump air through a hose to the diver so he could breathe. The diver, an experienced man of about forty, wore a brass helmet which was fastened to his diving suit by bolts. The 1969 divers, however, were scuba divers, employing, as they explained, "lightweight diving equipment," including their own "high pressure" air supply.

Divers dropped down into the water only when tides were favorable. In the water, they searched along the piers' outer timbers for handholds by which to propel themselves down the sides of the piers, all the way to their bases. Finding the water was so turbid that they could scarcely see, they were forced to probe, as needed, with their fingers, knives, or sticks. They probed on several days—the 1950 inspection was by one diver diving on sixteen different days, the 1969 inspection was by three divers diving on three days—to discover whether any crib timbers were torn or missing. They probed for any timber joints which had split open. In the concrete, they checked for cracks and holes. At the base of the piers, they checked for "scour" by currents which might undermine the piers' stability.

The divers reported that the riprap placed around the pier bases was successful in preventing scour. They reported that while a few of the outside timbers of the pier cribs were missing, in general the timbers were in place and "hard." They found the concrete had developed some cracking and a few large holes. In a surprise, the 1969 team reported that in piers 2 and 3, the concrete had developed such large holes that the bridge should be "closed to traffic until emergency repairs are made." In response, bridge management, on the advice of engineers, considered that the holes in the cribs were only in the exterior weighting pockets and not in the central concrete portions of the piers which were designed to carry the loads, that they were not growing larger, and that only minor repairs were necessary. Taking a very positive view, the 1972 diver team reported that the condition of the piers was "a tribute to its builders."[16]

> (1) Place clamps at intersection of I-Bars to prevent clashing.
> (4) Replace badly corroded angle.
> (5) Replace cracked angle.
> (8) Insert shims or reshim floorbeam bearing.
> (10) Tighten sagging rod.
> (15) Tighten bolts in floorbeam bearing.
> (17) Repair spalled concrete areas.
> (22) Add collars on pin.
> (24) Tighten pin nuts.
> (25) Drive rivet which is missing, or rivets that are loose.
> (28) Reconstruct pin bearing of expansion chord.
> (30) Repair expansion detail.
> (33) Reconstruct electric wire supports on top chords.
> (32) Provide drain holes in transverse strut.

NOTES OF WORK TO BE DONE NEAR EAST ANCHOR PIER, from Modjeski (deceased) & Masters, Inspection Report, Sept., 1945 (PBEN)

PERSONAL SAFETY

Over many years, boys were fascinated by the bridge and persisted in climbing on it. One boy, Sal Guimaro, who climbed on the bridge where it crossed over the streets of Poughkeepsie, made a habit of doing stunts up there. He would attract a crowd, including his father who boasted of his acrobatic skill. Once when someone brought a camera to photograph him doing his stunts near Dutchess Avenue, the boy was outdoing himself for the camera high up in the bridge lattice work when he failed to complete a swing, fell onto rocks below, and died.

Once in 1919, a recently employed bridge maintenance man, Oliver Goring of Poughkeepsie, only twenty-four years old, was working with his crew out on the bridge, in the center of the river. After having had his lunch on the deck of the bridge with the other workmen, he was returning to his painting at a location slightly below the track level, when he lost his balance and touched an electric transmission line. It flashed. It knocked him into a net of electric lines which ran along the bridge—Central Hudson's long-distance transmission lines. The wires caught him, preventing him from falling to the water below, but burned him. He yelled for help. His fellow crewmen found a rope, tied a loop in it, and then gingerly, trying not to let themselves touch the wires, caught the loop around Goring's legs and pulled him off the wires. They lifted him, horribly burned and unconscious by this time, up onto the bridge. They set him onto the handcar they used to transport their equipment. They moved him to the Poughkeepsie end of the bridge, and then by ambulance to the Mansion Square Hospital. The next day Goring died. As the years passed, his wife, son, and grandson grew to feel that for them, the bridge had become his monument.

Train crew seldom walked on the bridge, but if they did, they might be wary of how they did it. Once around 1950 when a train was about to cross the bridge, its caboose derailed, and the conductor in the caboose applied the air brakes slowly, succeeding in stopping the train gradually without buckling it, but only after it had stretched out across the bridge. To explain to the engineer at the head of the train what had happened, the conductor asked Brakeman Frank Doolittle Jr., to walk from the caboose at the back of the train much of the way across the bridge. As he looked down to the water below, it "looked awful far" to him, and he was suspicious that some of the wood in the walkway was rotten. "Believe me," he recalled, "I examined every plank."[17]

Once an eighteen-year old began painting on the bridge, and he seemed, according to the recollection of maintenance man Ferris Davis, to be doing well, scrambling deftly from one steel beam to another, high over the water. But then while on a beam where he could hold on by his legs but could not reach out with his hands to grasp anything else, he looked way down to the water and suddenly froze. He could not get himself to move any farther. Davis and other co-workers talked to him, trying to persuade him he could not stay on that beam forever. Finally they let a rope down to him, which allowed him to steady himself enough to move off the beam. Once back on top of the bridge, he quit the job at once, without even asking to be paid.

The New York State's Department of Labor, after inspecting the bridge in 1951, recommended that bridge maintenance men working below deck, whether in swing chairs or not, wear safety belts with a safety line connecting to the bridge deck. It also recommended that a manned boat be kept on the water near where the men were working, so that if any of them fell to the water, the boat could assist. A New Haven division engineer endorsed these recommendations, and they were put into effect.

One summer soon afterwards, John M. Reed, a future lawyer studying at Fordham University, worked along with other students as a painter on the bridge. At first he found the work "hairy," but eventually it became "second nature." Whether he was painting in a "bosun's chair" or crawling on steel, he recalled, he was protected by the required "safety harness." He also took turns with other painters in manning the required boat, a rowboat, which was kept on the water underneath where the painters were working. If they were painting near a pier, he would tie up the boat to that pier and wait. If they were painting away from a pier, he would have to keep rowing to hold the boat in the appropriate position.

However, maintenance man Davis recalled, when workmen were in the swing chairs, the extra safety line became so tangled with other lines that it hampered the men doing their work. As for the manned boat, according to Davis, the maintenance men believed that if they ever fell into the water, the boat would be too slow to reach them.

As he worked on the bridge, Davis kept reminding himself, as one of the more experienced workers had told him when he was starting in on his bridge work, that familiarity with danger breeds contempt for it. But what Davis came to emphasize more than the danger was that whatever your work is, you should love it or not do it. Davis loved working on the bridge. Despite its dangers, he felt comfortable doing it. Long after he had retired from working on the bridge, at the age of eighty-one he joined a group to walk out on the bridge deck. When others around him grew queasy from the height, he seemed to chat easily, even as he stood at the edge of a hole through which with a slight slip he could fall down to the water.[18]

POLICE DOGS GUARDING THE BRIDGE DURING WORLD WAR II (NYNH&H, *Rider's Digest*, Jan. 1943)

WORLD WARS I AND II

During World War I, when the bridge was a major transportation link for fuel, munitions, and troops, soldiers were stationed near the bridge to guard it from sabotage. A boy recalled long afterwards that he was fascinated by these soldiers. They carried guns. They wore uniforms, hard hats, and tight pants. In bad weather they stayed in little huts built under the bridge approaches. The boy gladly ran errands for them.

During World War II, the Coast Guard stationed armed vessels by the bridge. Also along the right of way under both the east and west approaches of the bridge, the New Haven built a chain linked fence topped with barbed wire, and stationed forty policemen there. When a stray dog at the Highland end of the bridge by chance chased a trespasser up a tree, the head bridge policeman, Captain Fred Rozelle, decided to adopt the dog and train it to recognize trespassers. Gradually he adopted other dogs and trained them too. He posted the dogs at each end of the bridge, especially at night. They sniffed, cocked their ears, and patrolled wide areas.

At about this time, when troop trains passed over the Bridge Route—the trains that the poet Alvin Greenberg remembered as "the long, thrilling, khaki freight of the war"—the trains were guarded by armed soldiers. In one of these troop trains, among the soldiers chatting, sleeping, singing, agonizing, was one soldier, Bill Fell, who had been a painter on the bridge. He had trained at Fort Benning, Georgia, and was being sent from there on this train to an unannounced overseas embarkation port.

As Fell recalled this train, it traveled especially at night, and, as was customary, to hide trains as much as possible from any enemy airplanes, its shades were drawn. Armed guards, placed on each car of the train, were supposed to prevent the soldiers from looking out the windows, one reason being to prevent them from knowing where they were going. Fell wondered if the guards were also there to prevent those soldiers who were terrified of the unknown ahead of them from escaping the train.

When Fell knew from the persistent hollow rumble of his train that it was crossing a long bridge, he managed to peek out enough to recognize that his train was crossing his familiar Poughkeepsie Bridge. Soon afterward he also recognized, below the bridge in Poughkeepsie, his own neighborhood where his wife and child were still living. But his train clicked and clacked along, taking him inexorably away from his family, toward an unknown port where he would take ship for an unknown fate. Would he ever return?[19]

THE 1950s AND 1960s

From the 1940s into the early 1960s, there continued to be guards on duty on the bridge, working in eight-hour shifts, twenty-four hours a day—in 1954, two men on each shift. At certain times they walked from their stations at each end of the bridge to the middle, punched a clock, and walked back.

While the Modjeski inspectors often warned that there were not enough guards or maintenance men, those who were there did what they could. If the bridge's navigation lights went out, a guard might telephone maintenance man Tom Houston Sr., even waking him, his son recalls, in the middle of the night. Houston would go to the top of the bridge, locate the tower on which the navigation lights had gone out, and then in the dark climb down the tower's ladders, and sometimes where there were no ladders, climb down steel "lacing bars." He could not carry a flashlight because he needed both hands to hold on. He went down by feel. Some of the navigation lights were placed close to the bottom of the towers, so he might have to climb far down.[20]

While the main structure of the bridge was of steel, and would not burn, its railroad ties and walkways were of wood, and oil from passing trains tended to drip on them, so they could easily burn, and often did. Under the New Haven's orders, its employees, whether guards, maintenance men, or train crew, watched for fires and kept fire-fighting equipment ready. A telephone line connected six call boxes on the bridge directly to rail officials and to the Poughkeepsie Fire Department, and maintenance men kept the line functioning. Water barrels were located at intervals along the top of the bridge, more than thirty of them. If they leaked, the men replaced them; if they needed filling with water or adding calcium chloride so the water would not freeze, the men would do it. Water pipes, made of steel, also ran the length of the bridge deck; they brought in water from the Poughkeepsie city system. If they leaked, section or maintenance men welded the pipes, and before winter, they drained them to keep them from freezing. There were canvas fire hoses in boxes at intervals on the bridge deck, ready to attach to the water pipes when needed; occasionally the men checked the hoses under water pressure to see if they leaked and needed to be replaced.

On May 1, 1964, Albert Alexander was the conductor on an early morning train out of Maybrook heading across the Poughkeepsie Bridge. The train included about twenty boxcars containing new automobiles to be delivered in Hartford and Springfield.

BRIDGE MAINTENANCE CREW, on the western approach of the bridge, Nov., 1950 (DAVIS). Left to right, front row: Fred Healy (didn't stay long; he later became Poughkeepsie's Supt. of Highways), Joseph Dyseven (stayed long). Back row: Charles Chapman, Homer Odell, Roy Baker (foreman), Ferris Davis (stayed long). Not shown is Tom Houston Sr.

After his train had moved most of the way over the bridge, Alexander, in the caboose, noticed that the train had stopped. He got out of the caboose to see what was happening. It was daylight. He could see that ahead of the train, smoke was coming up out of the bridge. It was a fire, like many fires on the bridge, probably caused by sparks from the brakes or exhaust of a passing train, in this case, from the train ahead of his, a fast OB-2, a refrigerator train on its way to Boston. The fire was on the bridge's eastern approach viaduct, over Poughkeepsie's Delafield Street. Fortunately only a moderate wind was fanning the fire.

In the train's engine, the engine crew had taken out a fire extinguisher and played it on the fire. Alexander and his crew also walked forward to the fire. They gathered at nearby fire barrels where buckets were ready and organized a bucket brigade to pour water on the fire. Before any other help arrived, the trainmen had put out the fire.[21]

This minor fire was a foreshadowing, as it turned out, of what began ten years later as a similar minor fire, but which, without anyone being quickly available to extinguish it, grew into a disaster.

THE BRIDGE AS SEEN FROM THE CAB OF NEW HAVEN DIESEL LOCOMOTIVE U25B, AS IT PULLED A FREIGHT OUT OF POUGHKEEPSIE. Sept. 10, 1966 (SWANB). Although the New Haven Railroad was bankrupt, J. W. Swanberg, the locomotive's youthful fireman who took this picture, considered that New Haven's train crews were self-respecting.

Transition

43. DECLINE (1960-1974)[1]

Shrinking, bankrupt, negligent.

THROUGH much of the twentieth century, the Poughkeepsie Bridge and its connecting rail lines, like most American rail lines, were hurt by the shift from heavy to light industry. They also were hurt by the inflexibility of railroad management, railroad unions, and government regulators. They also suffered from governments on many levels subsidizing autos and trucks by providing free improved highways, but not comparably subsidizing trains by providing free improved tracks. Governments, with public support, chose to let railroads decline. Moreover, new bridges gradually opened over the Hudson which competed with the Poughkeepsie Bridge. While most of the new bridges were highway bridges, one was a railroad bridge, opened in 1924, the New York Central's Selkirk-Castleton Bridge, seven miles south of Albany near the Selkirk rail yard.

During the depression of the 1930s, the New Haven cut back many of its more unprofitable lines. With the permission of government regulators, it abandoned portions of what had once been the main Bridge Route line into New England, through Canaan to Hartford. It tore up tracks and hauled them away. Thereafter trains from Maybrook passing over the bridge heading toward New England went by the only other major route available, the more southerly route through Danbury.

After American rail freight traffic reached a temporary high during World War II, it resumed its long decline. By 1950, the New York Central was shrinking its Hudson Line tracks from four to two. Two railroads which funneled traffic through Maybrook onto the bridge stopped running altogether, the Ontario & Western in 1957, and the Lehigh & New England in 1962. About 1961, when the New Haven Railroad went bankrupt again, it was shrinking its Maybrook-over-the-bridge-to-Danbury tracks from double to single.

ON-THE-JOB MORALE

At various times in the 1960s, the youthful J. W. Swanberg, while he was perfecting his skill as a railroad photographer, served as a firemen on the New Haven Railroad. When he could, Swanberg avoided working on switchers and locals, preferring to work on the more dramatic through freights which went from Maybrook over the Poughkeepsie Bridge to New Haven.

Although the New Haven was running fewer trains over the bridge, for efficiency it was running what through freights it still ran as longer trains. Swanberg found them hauling as many as 150 cars, strung out well over a mile long. They were sometimes pulled by as many as five diesel-electric units, often Alco FA's lashed together, the middle ones being without cabs.

Swanberg came to admire the FA's as his all-time favorite locomotives. In much of his work on them, however, Swanberg felt he was compensating for the New Haven's neglect in caring for them. As he walked about inside them, unsteadily in their oil slick passageways, where the roar of the engines came to seem so painfully loud that he took to wearing earplugs, he manually adjusted the speed of the once-automatic engine-cooling fans. He banged on plugged sand-pipes to allow the sand to spray onto the tracks more freely. He bashed jumper cables into better connections.

Yes, there was some distrust between the railmen based in Maybrook and those based in New Haven. New Haveners considered the Maybrookites to be tobacco-chewing hicks while the Maybrookites carped at the New Haveners for drinking and playing cards on the job. Yes, there were occasional questions as to whether the New Haven was financially sound enough to back its paychecks. Still, Swanberg sensed train crews were holding up well. He found the trainmen self-respecting and their supervisors professional.[2]

PENN CENTRAL

In 1968, two of what had been America's greatest railroads, the Pennsylvania and the New York Central, combined to form the Penn Central Railroad. It was called the largest private railroad in the world. But since the two railroads forming it had themselves long been ailing, from the first it was a question if the new Penn Central could survive.

Wobbly as it was, Penn Central did not want to add to its problems by absorbing any other troubled railroads. But by December 1968, the Interstate Commerce Commission, casting about for some device to keep the bankrupt New Haven running, had forced Penn Central to absorb the New Haven, including its Poughkeepsie Bridge.

With these mergers, the traditional patterns for routing freight over the bridge were upset. As Penn Central saw it, for traffic from the West and South heading toward New England, it was better not to route it by Maybrook and the Poughkeepsie Bridge. It was better instead to route it through the Selkirk yard, the ex-New York Central yard near Albany, and then across the Hudson by the nearby Selkirk-Castleton Bridge. It was better because the yard at Selkirk was more efficient than the Maybrook yard. It classified cars electronically, braking the cars as they came down the yard's hump not by hand but by automatic retarding. It had up-to-date facilities for rail-truck container interchange. Routing through the Selkirk yard was also better because long distance traffic coming east through Selkirk would likely arrive on Penn Central tracks. Traffic coming east through Maybrook would reach there by other, independent railroads, such as the Erie Lackawanna or L&HR, and only from there move by Penn Central tracks, for a shorter haul.

In accord with this pro-Selkirk, anti-Poughkeepsie Bridge policy, Penn Central kept abandoning unprofitable Poughkeepsie Bridge-related lines. It also delayed Poughkeepsie Bridge traffic. Moreover, even though the Bridge Route sometimes offered cheaper and more customer-friendly service than other Penn Central routes offered, Penn Central discouraged shippers from specifying a choice of the Bridge Route on their waybills. The result was that Penn Central squeezed both the bridge and such independent bridge-related railroads as the L&HR and the Erie Lackawanna. The Erie Lackawanna complained to the ICC, and it eventually forced Penn Central to adjust its time tables to be more favorable to the

DOWNWARD STEPS BY BRIDGE-RELATED RAILROADS	
1935	New Haven Railroad bankrupt
1938	Erie Railroad Bankrupt
1957	Ontario & Western stops running
1960	Erie and Lackawanna Railroads, both ailing, merge into the new Erie Lackawanna
1961	New Haven bankrupt again
1962	Lehigh & New England stops running
1968	The Pennsylvania and NY Central Railroads, both ailing, merge into the new Penn Central. Soon afterward, the Penn Central takes over the New Haven
1970	Penn Central bankrupt; Lehigh Valley bankrupt
1971	Reading bankrupt
1972	Lehigh & Hudson River bankrupt; Erie Lackawanna bankrupt

Poughkeepsie Bridge, but not fundamentally to alter its anti-Poughkeepsie Bridge policy.

In 1970, Penn Central, unable to meld its sprawling lines into a coordinated, efficient system, and receiving insufficient government support, itself went bankrupt, the largest bankruptcy ever in U.S. history. In its fall, Penn Central not only hurt Bridge Route traffic directly, but also indirectly by dragging down other neighboring, still independent railroads which fed traffic to the Bridge Route by not paying them what it owed them. These railroads, if they were not already bankrupt, were heading there, including the L&HR, Erie Lackawanna, Reading, Lehigh Valley, and Central of New Jersey.

Soon after Penn Central took over the New Haven, it began tearing up portions of the Maybrook yard's track. It let go some of the yard's employees. By the early 1970s, Penn Central made clear it wanted to shut down the Maybrook yard.

By 1971, Penn Central had reorganized its personnel so that the trainmaster Willard B. Rogers, with his headquarters in Poughkeepsie, was responsible for both the ex-New York Central's upper Hudson Line (from Beacon to near Albany) and the ex-New Haven's "Maybrook Branch" (from Maybrook over the Poughkeepsie Bridge to Hopewell Junction). Maybrook, the region's only major yard, still had car repair shops and round-the-clock crews handling interchange of freight traffic. Personnel assigned daily to Maybrook was 118, to Poughkeepsie 23, to Hopewell Junction 4. The average number of carloads Hopewell Junction received per month for local delivery was 105 (industries served there included IBM and Bry-Dain Lumber), while Poughkeepsie received 240 (Western Publishing—paper, Smith Brothers Cough Drops—syrup, Effron Baking—flour, Dutton—lumber, Hudson River State Hospital—coal, Poughkeepsie city—salt). Smaller locations within twenty miles of the bridge also being actively served included Manchester Bridge (Miron—lumber), Stormville (Green Haven prison—coal, flour, feed), Modena (Croce—fruit boxes), and Clintondale (Central Hudson—lumber).[3]

ESTIMATED ANNUAL FREIGHT TRAFFIC AT SELECTED PENN CENTRAL SITES, IN MILLION GROSS TONS PER MILE Adapted from a Penn Central map for 1969		
Line	Site	Tons
Wallkill Valley Line (ex-NY Central)	Campbell Hall to Kingston	0.1
Poughkeepsie Bridge Route (ex-New Haven)	Maybrook, NY, to Danbury, CT ("Maybrook Line")	10.5
Hudson Line (ex-NY Central)	Approaching Poughkeepsie from south	19.8
West Shore Line (ex NY Central)	Passing under the Poughkeepsie Bridge	24.3
Ex-NY Central Main Line	Utica to Selkirk	64.0

In the early 1970s, Penn Central, pinched though it was, was introducing some improvements. On the Maybrook-over-the-bridge-to-New Haven line it was introducing what had long been in use on some other railroads, two-way radiophones for its train crews, at last providing crewmen with instant direct communication between locomotives and cabooses, and between trains and dispatchers.[4] At the same time, Penn Central was also rehabilitating one of the rail lines it used to help bypass the Bridge Route, that is, the Beacon-Hopewell Junction line, preparing it to better handle the considerable traffic which came to Beacon from the Selkirk-Castleton Bridge south on the Hudson Line, on its way to Danbury and New Haven.

In the early 1970s, while most of the trackage of what had once been the CNE had long since been abandoned, perhaps four to six freight trains a day were still going over the Poughkeepsie Bridge, and observers remembered them as being as long as the bridge itself. However, as Penn

Central continued what a member of the Connecticut Transportation Authority called its "deliberate policy of accelerated obsolescence" for the Poughkeesie Bridge route, traffic on the bridge continued to drop. By spring 1974, according to a railroad executive, only one train crossed the bridge daily, round trip, and it was "poorly patronized." It was run, he said, not as much because Penn Central wanted to run it, as to "satisfy an Interstate Commerce Commission order."5

POUGHKEEPSIE BRIDGE MAINTENANCE MEN From Penn Central's Bridge Roster, Sept. 19, 1969, PBEN		
	Entered Railroad Service	Entered Service on the Bridge
Dyseven, J. J. (Foreman)	5-11-28	5-11-28
Houston, T. F. Sr. (Foreman)	8-3-31	11-16-45
Davis, F. W.	6-12-39	12-1-47
Heady, P. J.	5-3-46	5-21-51
Dietter, E. S.	7-1-31	9-19-69
Hanford, W. L.	4-20-36	9-19-69
McMeel, F. J.	1-12-37	9-19-69
Grattan, J. K.	11-27-50	9-19-69
Heddin, E. D.	5-5-58	9-19-69
Munson, G. E.	6-22-64	9-16-69
Hughes, A. A.	1-30-64	9-19-69
Decker, F.	4-1-68	9-19-69

Those whose bridge service began 9-19-69, the date the list was created, had been Penn Central employees at either Maybrook or Great Barrington. When Penn Central eliminated their jobs at those sites, it offered them jobs on the bridge instead, plus lodging and meals nearby. However, in Nov. 1973, Penn Central let all its bridge men go.

THE REDUCED BRIDGE MAINTENANCE CREW, 1973, SHORTLY BEFORE THEY WERE DISMISSED, LEAVING THE BRIDGE WITH SCARCELY ANY PROTECTION (DAVIS). At the eastern end of the bridge, over Washington St., Poughkeepsie, are, left to right, Pete J. Heady, Ferris W. Davis, J. Kenny Grattan. Not shown was Tom Houston Sr. The shed in the background housed telephone equipment.

NEGLECTING BRIDGE MAINTENANCE

The Modjeski engineering firm, which had been inspecting the bridge for about fifty years, had been complaining for several years that the bridge management was no longer doing the repairs which the firm's inspectors considered essential to keep the bridge in service. Early in 1969, just after Penn Central took over control of the Bridge Route, the firm decided that under these circumstances it was "pointless" to continue its inspections, and stopped them. From then on, Penn Central relied on its own inspections of the bridge, and it did less maintenance than ever. Later in 1969, when Penn Central was cutting back on personnel everywhere, it kept only twelve men on the bridge for both maintenance and watch.

Fires continued to occur on the bridge as they long had, especially occasioned by sparks from passing trains. Penn Central employees, like the New Haven employees before them, were usually able themselves to spot fires and contain them. However, in 1971, Penn Central cut the hours of the watchmen who had walked across the bridge around the clock, and then in 1972 eliminated the watchmen entirely.

By summer 1973, Penn Central, seeing no way out of its financial morass, asked a federal judge

ONE OF THE LAST TRAINS TO CROSS THE BRIDGE: WESTBOUND PENN CENTRAL FREIGHT NE-97, MAY 1, 1974 (photo by Heyward Cohen, COHEN). At the time, regular service over the bridge had been reduced to this train, which made one trip daily each way, called CB-2 eastbound, and NE-97 westbound. The train was a joint Erie Lackawanna and Penn Central service. The diesel locomotives shown are three striped Erie Lackawanna ones at the head, followed by two black Penn Central ones, the total of five units suggesting that at least when it was eastbound, it was a heavy train. The train had originated at Port Jervis, NY, on the Erie Lackawanna line. At Maybrook it had picked up a crew from Penn Central and cars from the Lehigh and Hudson River. It had crossed the bridge and reached the Cedar Hill yard in New Haven. As shown, it was returning westward across the bridge, with some cars which were to be switched at Maybrook onto the Lehigh & Hudson River, and other cars to be taken to Port Jervis, where the train was to terminate.

for permission to stop all its rail service and sell off all its assets. By that time, the bridge maintenance crew had been reduced to five which was the minimum number Penn Central's insurance company demanded that it keep. At the same time, the Coast Guard was complaining that Penn Central was neglecting the legal requirement that it keep navigation lights on the bridge.

Then in November 1973, Penn Central, as part of its cost cutting, let go all of its bridge maintenance crew, among them the old-timers Ferris Davis and Tom Houston Sr. To take the place of the maintenance men, Penn Central only occasionally sent in a floating work crew from Danbury. The Brotherhood of Locomotive Engineers protested that Penn Central was disregarding its responsibility both to its employees and to the public. During the winter of 1973-74, no one drained the bridge's fire-fighting steel water pipe line, and consequently it froze and cracked. Penn Central was neglecting its fire-fighting equipment at the same time that it seldom kept any men on the bridge to spot fires.[6]

> **WHAT IF?**
> Long-term questions
>
> What if Americans had chosen to treat highways and railroads equally? What if our governments from the 1950s had created and then continued to support not only an interstate highway program but also a similar interstate railroad program? What if our governments owned railroad rights-of-way much as they own highways, if they built and maintained tracks on them as they also build and maintain highways, and if they allowed railroads to use the tracks basically for free as they allow autos and trucks to use highways for free?
>
> Would we then have a better balanced transportation system? Would the air be less polluted? Would our patterns of housing and commercial development be more concentrated, less sprawling, allowing more land to stay open? Would our downtowns be more lively, our tempers more benign?

FIRE ON THE POUGHKEEPSIE END OF THE BRIDGE, MAY 8, 1974. Rail fan Austin McEntee happened to be at the Hudson Line's Poughkeepsie station to take a train when he saw the fire and took this picture (McEN). Chunks of burning ties fell from the bridge down onto the Hudson Line tracks, forcing them to close. Visible at center right (about 1 1/4 inches to the left of the spire of Mt. Carmel Church, formerly St. Peters), is a white streak which rises at an angle to the top of the bridge. It is a stream of water, coming from a fire truck on a rise in the ground, which is playing on the fire, producing a cloud of steam.

44. THE BRIDGE BURNS (1974)[1]

Ties smoldered. Tracks twisted. Spikes popped loose and clanged down through the steel girders.

SHORTLY BEFORE 1:00 PM on May 8, 1974, a soft spring day, a train crept from the Highland shore, out over the Poughkeepsie Bridge. The train, the only regular eastbound train of the day, was as usual a long one, with over 100 cars. It was a train operated by Penn Central, in agreement with Erie Lackawanna. Originally made up in the Erie Lackawanna yard at Port Jervis, including cars which had arrived from Chicago, it picked up more cars in Maybrook. Its diesel units were Erie Lackawanna's, and it was heading for New Haven, Providence, and Boston.

About an hour later, a cloud of black smoke hung over the Poughkeepsie end of the bridge. On the bridge deck, wooden railroad ties were smoldering, and next to them, wooden walkways were burning, fanned by a moderate breeze.

Because Penn Central had no guards or maintenance men on duty on the bridge at the time, the fire was not quickly reported. When firemen arrived at the site, first firemen from Poughkeepsie and later from surrounding towns, they found that their equipment could not easily pump water up to the top of such a high bridge. When they tried turning on the water to flow into the steel pipe line which ran the length of the bridge, a line meant to help fight fires, they found that because it had not been drained the previous winter, it had burst at several points—Penn Central had known it but not repaired it.

On top of the bridge, firemen found that the walkways had already burned enough to make it difficult for them to know where they could safely walk on them, which was disconcerting on such a high bridge. When they found it necessary to pull up portions of the walkway to block the spread of the fire, they found it still more difficult to know where to walk.

On the bridge deck, the fire's heat became so intense that it expanded steel railroad tracks, twisting them. It also warped steel girders. Firemen felt the bridge move under them, making them wonder if the bridge would collapse. Poughkeepsie Fire Captain Thomas Ringwood considered that while the fire was not Poughkeepsie's biggest, it was its "most difficult."[2]

As the walkways and railroad ties burned, pieces of them dropped down through the bridge. They set grass on fire. They set buildings on fire. They threatened huge gas tanks. As the tracks buckled and the girders warped, they popped loose spikes, plates, and rivets, sometimes sending them clanging down through the bridge girders. By 3:20 PM, enough hot, burning debris had fallen down onto the Hudson Line's railroad tracks, where they passed under the bridge, to force Penn Central to close the tracks. By 3:49, similar debris falling onto Route 9, a major highway passing under the bridge, had forced its closing also. At 5:40, a burning tie hit the Hudson Line's signal wires, knocking out its signal system from Poughkeepsie all the way north to Rensselaer, near Albany. By 6:51, however, firemen considered the fire to be out.

The usual cause of fires on the bridge at about this time was the usual cause of track fires wherever diesel locomotives ran: sparks produced either by brakes or engine exhaust. The train which had crossed the bridge shortly before the fire was discovered could have produced such

247

sparks, and the sparks could have ignited the oil-soaked, creosoted railroad ties or the long-unpainted walkways.

While there was no official investigation of the fire, many observers believed that Penn Central, by not maintaining the bridge adequately, by not keeping guards on the bridge regularly to watch it, and by not caring for the bridge's fire fighting equipment, had invited the fire to happen. More provocatively, a few observers believed that the fire was deliberate. The *Hudson Valley Hornet* declared that while they lacked proof, they believed that the railroad "hired an arsonist to do the job."

Certainly it was plausible that some Penn Central's officials, as they understood its interests, welcomed the fire. Penn Central, weak as it was, had not wanted to take over the ailing New Haven Railroad along with its Poughkeepsie Bridge, but federal officials had forced it to. It was costly for Penn Central to operate both the Poughkeepsie and Selkirk Bridges and their accompanying rail yards. Even if trains heading for New England had to travel seventy miles north out of their way

FIREMEN ON THE BRIDGE DECK, WITH TRACKS TWISTED FROM THE FIERCE HEAT OF THE FIRE, THE DEC

to cross the Hudson at Selkirk, it was possible to believe that Penn Central, if it took a narrow view of its interest, would find it to its advantage to concentrate on using the modern Selkirk yard and its related Selkirk-Castleton Bridge, and welcome an excuse to abandon the outmoded Maybrook yard and its related Poughkeepsie Bridge.

With the 1974 fire, the Poughkeepsie Bridge, after being open to trains for eighty-five years, closed. Penn Central's dispatchers quickly rerouted the little traffic which had been going over it. They rerouted most of it, as expected, to pass through the Selkirk yard and over the nearby Selkirk-Castleton Bridge.

The fire had burned about 700 feet of the Poughkeepsie Bridge, only a small part of its total length. Within a few days, Penn Central reported that to reopen the bridge, it would have to replace 700 feet of railroad track, 600 railroad ties, 30 steel stringers, as well as accompanying walkways, hand rails, steel water line, and signal cable. That hardly seemed an impossible task in itself.[3]

ULGED OUT TO THE LEFT, AND BROKEN RAILING TO THE RIGHT (photo by an unidentified fireman, McEN)

THE ARMS OF CENTRAL HUDSON'S ELECTRIC TRANSMISSION LINES SHOW REACHING OUT SOUTHWARD FROM THE BRIDGE DECK, Nov. 10, 1978 (photo by Dennis O'Keefe, O'KEEF). Conrail, to divest itself of the onerous responsibility of owning a deteriorating bridge, offered to sell it for $1 to Central Hudson. Even though buying the bridge would facilitate Central Hudson's keeping its lines on the bridge, it declined, fearing the cost of maintaining it.

45. RESTORE OR DEMOLISH? (1974-1984)

Classic conflict.

IMMEDIATELY AFTER THE FIRE, Penn Central engineers estimated the cost to restore the bridge at about $650,000. They began to locate the necessary supplies. They called for bids. However, Penn Central being bankrupt, its management doubted that it could pay for the restoration by itself, and began negotiating with government agencies to share the cost.

Independent rail lines like the L&HR and Erie Lackawanna, whose traffic had been cut by the bridge's closing, wanted to see it restored to train service. A Pennsylvania state transportation official argued that the closing meant that freight on its way from Pennsylvania to New York City, Long Island, and Connecticut was being wastefully detoured 110 miles north to cross the Hudson at Selkirk. New York's Governor Malcolm Wilson claimed that the bridge was needed to reduce "congestion in the Northeast rail corridor." Connecticut's Governor Ella Grasso insisted that her state, having many defense-related industries, "has every right to be concerned that its proven, reliable and traditional rail route to the southern and western states be restored."[1]

Among broader arguments for restoring the bridge for train service, one was that transportation diversification is healthy. Another was that compared to truck traffic, rail traffic is both more energy efficient and less polluting. Still another was that it is risky to national security to depend heavily on any one Hudson River rail crossing, such as the one at Selkirk.

It was a boost to the bridge restoration cause that a prosperous private freight-only railroad, the Providence & Worcester (P&W), offered to assist in reviving "the Poughkeepsie Bridge route as a major gateway" into New England. The P&W even offered to buy the Poughkeepsie Bridge if governments would restore it first.[2]

Broad factors working against restoring the bridge included the long-term migration of heavy "smoke stack" industry out of the Northeast, and public preference for spending public funds on highways rather than railroads. A more specific deterrent to restoration was that, while earlier the Selkirk and Poughkeepsie gateways to New England had been owned by competing railroads, by the time of the fire and soon thereafter they were both owned by the same railroad, first Penn Central, later Conrail, so the owner was likely to have little interest in keeping them both open. Moreover, before the 1974 fire, Poughkeepsie Bridge traffic had fallen so low that, even if the fire had never occurred, any railroad that owned the bridge would quite possibly have closed it within a few years anyway.

NEGOTIATIONS

At the end of 1974, seven months after the fire had stopped train traffic on the bridge, three entities reached a tentative agreement to share the cost to repair it. The three were New York State, the U.S. Department of Transportation, and Penn Central. However, this agreement required approvals by the three parties, and the cost was going up.

By June 1975, the New York State legislature had voted $700,000 for its share of the repairs. But then when the state asked Penn Central to pay its share, the railroad admitted publicly that it regarded repairing the bridge as "against our best interests." As a Penn Central spokesman explained, it makes more sense for us to route freight through Selkirk because then we can use our $35 million modern Selkirk yard rather than have to repair the "outmoded" Maybrook yard."

In April 1976, two years after the fire, Conrail, a quasi-public corporation newly created by Congress to take over several failing northeastern railroads, took over the bankrupt Penn Central, becoming the new owner of the Poughkeepsie Bridge. At once, all four U.S. senators from New York and Connecticut urged Conrail to restore the bridge to service. But Conrail, which was under pressure from Congress to survive without subsidies, seemed determined to cut services which interfered with its making an immediate profit. In keeping with this attitude, Conrail, according to a skeptical congressional staffer, hoped the bridge would "fall quietly into the Hudson." One of Conrail's vice presidents, John L. Sweeney, explained more euphemistically that the conflict over restoring the bridge was a "classic" conflict between railroad and political forces, both of them legitimate. "Everybody's right", he said, but Conrail has to make the choice that "serves the most people."

In 1978, four years after the fire, when Connecticut's Senator Abraham Ribicoff proposed a bill for Congress to authorize $9 million to repair not only the bridge but also its connecting rail lines and even the Maybrook yard, both Conrail

AFTER THE FIRE CLOSED THE BRIDGE, PENN CENTRAL NEVERTHELESS CONTINUED TO RUN LOCAL FREIGHTS ON THE BRIDGE'S MAJOR CONNECTING LINES ON BOTH SIDES OF THE HUDSON. This Penn Central engine, an ex-New Haven ALCO RS-11 built in Schenectady, was formerly based in Maybrook. Here it is shown, Nov. 9, 1974, at the old CNE freight station, on Poughkeepsie's Parker Ave. which had become its new base (photo by Heyward Cohen, COHEN). From this location, the engine hauled freight cars along the Hospital Branch, as to switch them onto the Hudson Line, and it also hauled freight cars on regular runs to Hopewell Jct. As shown, the engine has behind it a flatcar loaded with a large Caterpillar machine for delivery in Manchester Bridge.

ALTHOUGH PENN CENTRAL WAS ALLOWING THE MAYBROOK YARD TO DETERIORATE, PAYDAYS CONTINUED THERE, ON A REDUCED SCALE. Seated is E. M. Lucas who had come from Danbury to issue the pay. Standing are, left, engineer William Trepton, and, center, brakeman Sam Christian (*Middletown Record*, Jan. 11, 1976).

and the federal Department of Transportation opposed it. But Congress, at a time when it was facing elections, passed it anyway. However, the money could not be spent because Congress balked at taking the additional necessary step of appropriating the money. By early 1980, at the request of the federal Department of Transportation, the engineering firm of DeLeuw, Cather had studied the bridge. They reported it could carry moderate rail traffic provided not only that the fire damage was repaired but also that other needed bridge rehabilitation work was done, which would cost altogether $19.8 million. By request, they also studied the cost of renovating the Maybrook yard and the whole 127 miles of track from Hopewell Junction, NY, across the bridge through Maybrook and on via the L&HR to Phillipsburg NJ, on the Delaware River, which they estimated at an additional $10.8 million. Alarmed by these high estimates, the friends of the bridge cut their request to Congress to $1 million, with the hope that the reduction would help Congress at least get reconstruction started. New York and New England congressmen overwhelmingly supported this scaled down proposal, but in the summer of 1980, with the federal Railroad Administration warning that the cost might go sky high, the House rejected it. Poughkeepsie's Mayor Thomas Aposporos called the rejection a "tragedy" for Poughkeepsie and the whole "beleaguered Northeast."[3]

During 1980, President Carter led in the drastic reduction of federal regulation of railroads. This was followed in 1981, when Ronald Reagan

CRANE SEEN THROUGH THE FENCE AT THE ENTRANCE TO THE BRIDGE, near Washington St., Poughkeepsie, looking west, July 3, 1983 (photo by Austin McEntee, McEN). Conrail had been using the crane to carry out a court order to strip the Poughkeepsie end of the bridge of anything likely to fall down through it.

became president, with a new emphasis on less government spending. These changes offered less chance for governments to lead in restoring the bridge, but more encouragement for private interests to do so. At the time, several private, independent railroad interests considered buying the bridge from Conrail. Among these interests were Arthur G. Adams, of Mahwah, NJ, and his associates, whose holding company, the Lehigh, Erie, & Wallkill Transportation Co., offered to buy from Conrail both the bridge and related track. While the holding company's plans varied, at one time it wanted to buy much of the former Erie Lackawanna track to create a major connecting line for the bridge across New York State's southern tier through Port Jervis and Binghamton into the Midwest. Adams' holding company also negotiated with the Providence & Worcester Railroad to provide a matching privately-owned rail route from the bridge into New England. By 1983, however, it was clear that Conrail, taking a narrow view of its own interest, was determined to avoid selling enough track, whether ex-Erie Lackawanna track or any other track, to allow any significant competition with Conrail to develop. As the years stretched out after the fire, it seemed to Adams that much of what happened to the bridge was determined not so much by either the public interest or the open operation of the market as by greedy power play.[4]

CLEAN-UP

After fire closed the bridge, its owners did not maintain it. Spikes and chunks of charred wood occasionally fell from it, some of them close enough to Poughkeepsie houses to alarm householders. When a piece of a railroad tie fell from the bridge and hit the windshield of an automobile passing under it, drivers also became alarmed.

Responding to the alarm, in 1979, the

Poughkeepsie Common Council asked Conrail either to repair the bridge or demolish it. When Conrail did neither, in 1981 the city sued Conrail to force it to act. By 1983, a court required Conrail to reduce the likelihood of debris continuing to drop by stripping the bridge approach in Poughkeepsie down to its steel frame, removing all tracks, ties, walkways, and railings.

ABANDONING RELATED RAIL LINES

By 1979, with the Maybrook yard having become what Conrail called "a desolate shell," Conrail had sold off forty-three acres of it to Yellow Freight, one of the largest trucking firms in the nation. Maybrook is a magnificent site for a truck terminal, as it had been for a rail terminal. It is close to Stewart Airport. It is also close to the intersection of two interstate highways, the north-south Thruway, I-87, which runs from Albany to New York, and the west-east I-84 which runs from Scranton, PA, across the Hudson at Newburgh and on into Connecticut through Danbury. It was symbolic of the national shift from railroads to trucks that by 1980 much of the Maybrook rail yard had been transformed into an efficient truck terminal.

Up to 1981, Conrail's two major rail lines reaching out from the bridge, both the one westward to Maybrook and the one eastward to Hopewell Junction, were still operating, although because the bridge was closed, they were not profitable. By 1982, Conrail had received permission from the ICC to abandon both of these lines. In March 1982, Conrail ran its last trains over them, and by 1983 was taking up their tracks. Workmen loosened rails, prying them up. "Too bad this old

SECTION OF THE BRIDGE OVER TALMADGE ST., POUGHKEEPSIE, July 3, 1983. CONRAIL, BY COURT ORDER, HAD JUST STRIPPED IT DOWN TO ITS STEEL WORK, removing railroad ties, board walk, railings, anything likely to fall (photo by Austin McEntee, McEN)

In 1982, CONRAIL STOPPED TRAIN SERVICE ON THE LINES CLOSELY CONNECTED TO THE BRIDGE. BY THE NEXT YEAR, CONRAIL WAS PULLING UP THEIR RAILS. Above, PULLING UP RAILS WEST OF THE BRIDGE, near Little Italy Rd., Highland, April 5, 1983. Right top, RACKS ON TRAIN CARS READY TO RECEIVE RAILS BEING REMOVED, east of the bridge, near Creek Rd., Poughkeepsie, March 5, 1984 (both photos by Austin McEntee, McEN). Right bottom, RAILS LOADED ONTO RACKS ON TRAIN CARS, READY TO BE CARRIED AWAY, west of the bridge in Modena, May, 1983 (photo by Shirley Anson, PLATT).

257

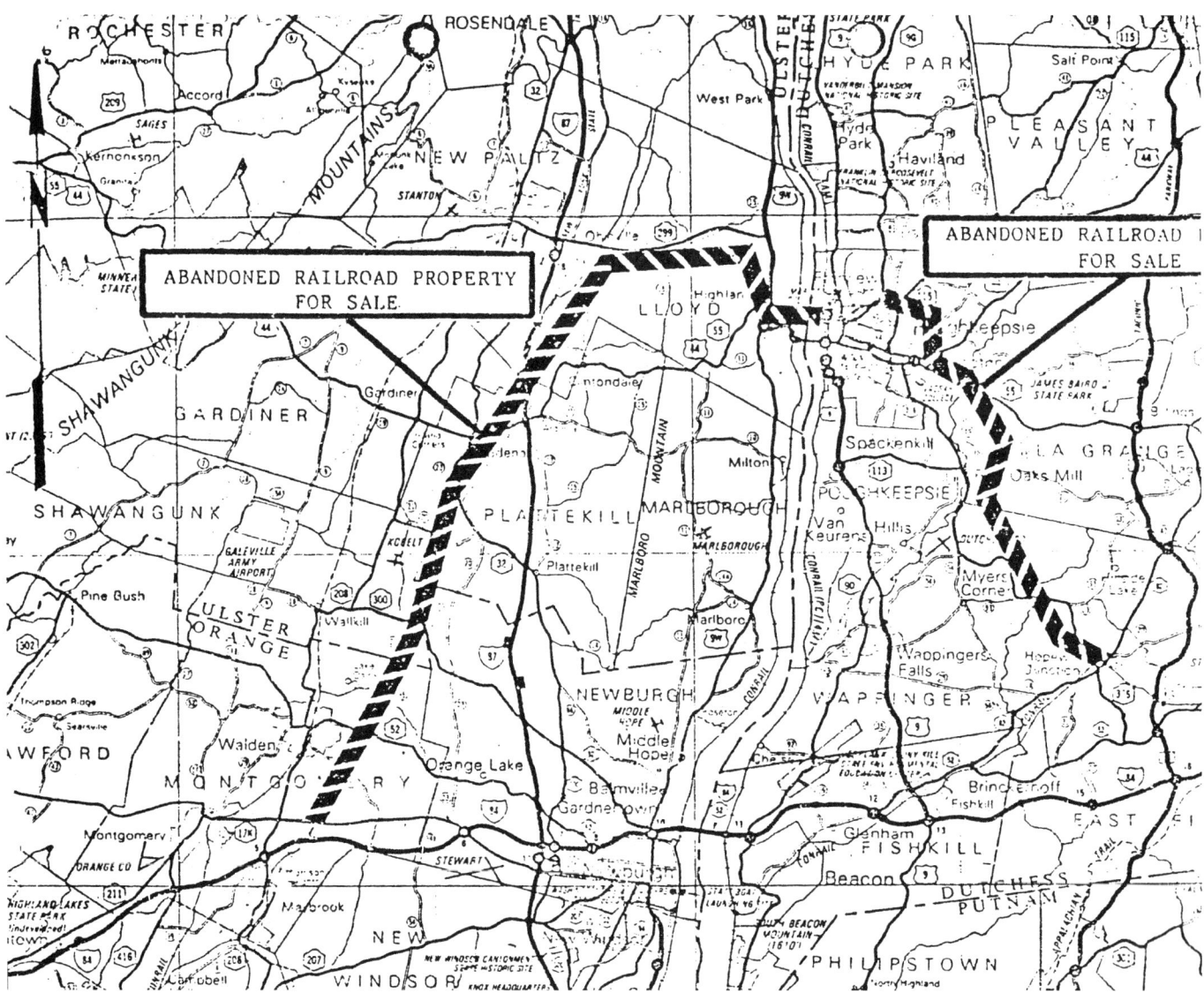

CONRAIL'S ABANDONED RIGHT OF WAY ANNOUNCED BY THE NY STATE DEPT. OF TRANSPORTATION AS FOR SALE, MARCH 1, 1984 (PLATT) This right of way, 35.1 miles long, reached altogether from near Maybrook to Hopewell Jct. (except for a short stretch in Poughkeepsie, with a connection to the Hudson Line, on which Conrail continued to run trains). As the law required, the state gave preference for buying it to various governmental units. Dutchess County took advantage of the sale, buying its portion. Since at first neither Ulster County nor any other governmental unit bought the Ulster County portion, Conrail was free to sell it to private parties.

road has to go," a workman reflected. "The rails are good, the ties aren't rotted." A salvage train pulled in the rails with a winch, loaded them onto its cars, and hauled them away.[5]

TEAR DOWN THE BRIDGE?

In 1977-78, the New York State Bridge Authority, watching the increasing glut of auto traffic on its Mid-Hudson Bridge, was considering building a new Hudson River highway bridge, but raised the question whether instead it could adapt the abandoned railroad bridge to highway use. When the Authority asked the Modjeski engineering firm for advice on such a proposal, the firm advised, much as it had advised about 1920 when a similar question had arisen, that, considering the extensive modifications in the bridge that would be necessary, it would not be wise. While as late as 1983 the Authority was continuing to speculate about buying the old bridge, and both Poughkeesie's Mayor Aposporos and Conrail were encouraging it to do so, the Authority did not seriously move toward buying it.[6]

Conrail, however, was anxious to relieve itself of the responsibility of owning the deteriorating bridge, especially since certain public entities were insisting it would have to be torn down at the owner's expense as a public nuisance. With the state seeming to be reluctant to purchase the bridge, Conrail made an offer to sell it for one dollar to Central Hudson, the electric utility which had long kept high voltage power lines on the bridge. When Central Hudson balked at buying the bridge, Conrail asked Central Hudson a drastically increased rent for keeping its power lines on it, and threatened that if Central Hudson chose not to pay the increase, it would demolish the bridge, depriving the power company of a convenient means for its power lines to cross the river.

In 1982, preparing to carry out its threat, Conrail asked Central Hudson to remove its transmission lines from the bridge. Central Hudson responded by asking the DeLeuw, Cather engineering firm to study whether it should buy the bridge for the sake of keeping its transmission lines on it, but the firm advised that maintaining the bridge for such a purpose would cost what Central Hudson considered to be too much. By 1984, Central Hudson had arranged to place new transmission lines under the river, and to abandon its lines on the bridge. Conrail then prepared to demolish the bridge, at the same time that it continued to hope to find a buyer.[7]

In early 1983, Conrail had already called for bids to tear the bridge down, with contractors allowed to keep the scrap metal for salvage. While the bids ranged up to $19.5 million, the low bid, submitted by a Texas demolition firm, Jet Research, was for $1.3 million. Conrail tentatively accepted Jet's bid, contingent on Jet's securing the approximately dozen permits that were necessary. The permits were likely to be difficult to obtain both because by 1979, on the initiative of local preservationists, the bridge had been entered on the National Park Service's Register of Historic Places, and because of the questionable methods the demolition firm proposed to use. The firm proposed to cut the bridge in pieces, using 114 separate explosive charges for each span, drop the pieces into the river, and then pick them up by crane. Such a process would be likely to disrupt river traffic as well as disturb the river water which both Poughkeepsie and Highland drew on for drinking; especially it might stir up the cancer-threatening PCBs that over many years upriver General Electric factories had dumped into the river. However, in May 1984, Conrail, apparently pushing hard to proceed with demolition, flew a representative from its headquarters in Philadelphia to Texas, and ordered him not to return until he had negotiated the terms of a final contract with Jet to do the demolition. By negotiating from 10:00 AM one morning to 5:00 AM the next, he secured a contract, the amount said to have risen by this time to about $7 million. In the summer of 1984, the Jet firm was still pursuing the necessary permits.[8]

By this time, it had been ten years since fire had stopped trains from running over the bridge. Conrail was not only proposing to demolish the bridge, but also had already hauled away significant stretches of the track that would be needed to enable trains to resume crossing the bridge. Under these circumstances, friends of the bridge were likely to have a better chance to preserve it if proposals to use the bridge for non-railroad purposes came to the fore, and the sooner the better.

PROPOSAL TO REMODEL THE BRIDGE, BY ARCHITECT EDMOND LOEDY, 1978 (LOEDY). According to Loedy's design, the entire bridge was to be strengthened, widened, and enclosed so that its steel structure would no longer be exposed to the weather. Shops, hotels, and restaurants would be built along the length of the bridge. Each of the bridge's four piers would have sloping condominiums built against them, with their windows facing south, for sunlight. All the piers would have observation areas, with elevators available. The tower over the second pier in from the Poughkeepsie shore was specially planned as a high observation tower. Since at the time of Loedy's drawing railroad tracks still crossed the bridge, he planned for trains to run across it (a streamlined train shows on top of the bridge at the right). Loedy believed, however, that since ordinary trains crossing the bridge could shake it enough to endanger any buildings on it, the trains would have to be of advanced reduced-shaking design.

46. ALTERNATIVE PROPOSALS: FOR SHOPPERS, TOURISTS, JUMPERS? (1977-1990s)[1]

"Not as kooky" as the *"press made it seem."*

WHEN the Hungarian-born Edmond Loedy first came to live in Poughkeepsie as a youth of fifteen, he did not pay much attention to the bridge. But a few years later when an engineer uncle of his, visiting from Hungary, told him that the bridge was a renowned engineering monument, he became more aware of it. One day after Loedy had become an architect, he was looking at the abandoned bridge when he suddenly saw it as a structural skeleton which could be rebuilt into whatever he might imagine. He began to play with redesigning it, as by constructing on different levels of it walkways, shops, cafes, a hotel, and a bungee jumping platform. He proposed building museums too, if not on the bridge itself, then on the shore below the bridge. While he believed that such a project would require both private and tax support, it would be worth it, he argued, because it could become a striking attraction like the Eiffel Tower, which would draw visitors from all over the world.

Meantime, Donald L. Pevsner, a Florida transportation lawyer who believed in historic preservation, approached Conrail with a plan somewhat similar to Loedy's, if less elaborate. Pevsner, who Conrail's management considered to have substantial political and financial contacts, proposed to buy the bridge outright or simply to secure air rights to build a restaurant and other tourist amenities on the bridge, without interfering with any train traffic on it. When Pevsner had been growing up in New York City, he had once taken a rail-fan excursion train which stopped on the bridge to allow the fans to step out onto it and view the historic river. This experience "hooked" Pevsner on the bridge as a potential tourist attraction. After all, the bridge was not only unusually high for any bridge, but also remarkably free from any impediments to viewing from it—it had no heavy fences, no superstructure above the deck.

Some people giggled at Loedy's and Pevsner's proposals. They wondered if buildings constructed on the bridge would survive when the bridge vibrated in the wind. They wondered how hotels or restaurants built on the bridge would dispose of their waste water. Preservationists asked, would commercial development of the bridge destroy its historic nature?

However, as Loedy's wife, Ann Loedy, who was a member of the Dutchess County legislature, pointed out, such commercial proposals were in line with the medieval custom of building shops and houses on bridges, as exemplified by the Ponte Vecchio and the London Bridge. They were also in line with a recent international competition—sponsored by London's Royal Academy of Arts—for the design of habitable, tourist-friendly bridges, having on them such amenities as cafes, boutiques, and gardens. The radio commentator Lowell Thomas, from his home in nearby Pawling, NY, wrote Pevsner praising his proposal as imaginative. When Pevsner took the New York columnist William F. Buckley Jr. for a walk on the bridge, Buckley found Pevsner's proposal appealed to his sense of adventure, and while he was not willing to give money to the project, he decided he was willing to give it time, if it was clear from the engineering point of view that it was not totally mad. The engineering firm with which Pevsner was working, Rowell & Associates of Syracuse, after examining the bridge, considered Pevsner's project to be questionable considering the cost necessary to strengthen the bridge for it, but felt certain that if Pevsner carried through with it anyway, the spectacular view that a restaurant on the bridge would have, would make it successful, and therrefore, Rowell & Associates was willing to work with Pevsner on it. In 1978, the Rowell firm, on Pevsner's behalf, secured from Conrail an informal first-refusal option to buy the bridge.[2] However, in the next few years neither Pevsner, the Rowell firm, or any of their supporters located developers ready to carry the project into effect.

In early 1984, when Conrail was driving toward tearing down the bridge, Pevsner, in a renewed push, secured an option from Conrail to buy both the bridge and certain related rights of way for one dollar. By this time, in a somewhat modified proposal, Pevsner planned to build on the bridge a hotel, shops, and the like, which would require, the Rowell engineers estimated, $5 million to rehabilitate the bridge as well as more to build the new structures on it. Pevsner proposed building boat docks beside one of the bridge piers and providing glass-walled elevators to rise from the docks up to the top of the bridge. He also proposed building a light rail line from the State Thruway exit at New Paltz to the west end of the bridge, and a moving sidewalk from there out onto the bridge. Pevsner admitted that nothing like his proposal had ever been tried, but claimed it might turn out to make the Ponte Vecchio look second rate. A columnist regarded Pevsner's proposal as "outrageous" but was thankful that there are dreamers in the world like Pevsner. Meantime, Pevsner and his associates approached several developers to undertake the project, including James Rouse, of Columbia, MD, and Donald Trump, of New York City, without success.

By fall 1984, local citizens had formed a "Save the Railroad Bridge Committee," with Ann Loedy as co-chair, which was open to any proposals to prevent the bridge's demolition, including Loedy's and Pevsner's. Moreover, Pevsner's proposal, as Pevsner came to modify it, seemed to railroad promoter Arthur G. Adams "not as kooky" as the "press made it seem." To reduce costs, Pevsner conceded that the hotel might be built on the shore at the west end of the bridge rather than on the bridge itself.[3]

SELLING THE BRIDGE:

On November 1, 1984, Pevsner, not having found the financial support necessary to carry out his bridge plans, and believing it would be irresponsible to buy the bridge without it, let his option to buy the bridge expire. The very next day Conrail sold the bridge, along with certain related rights of way, to someone else.

Conrail was delighted to sell off the bridge. By selling it, Conrail avoided upkeep charges. It avoided taxes. If any government agency ever required its owner to tear the bridge down, Conrail would avoid the cost of doing so.

Conrail sold the bridge to Gordon Schreiber Miller, an agent for Railway Management Associates who called themselves "Consultants to the Railroad Industry." Miller, an agreeable gentleman of about sixty years who drove a Cadillac limousine and claimed he knew important railroad people, had told Conrail he was interested in purchasing the bridge as early as 1982, but Conrail put him off while threatening to tear down the bridge or sell it to more plausible purchasers like Central Hudson or Pevsner. Once it seemed clear that neither Central Hudson or Pevsner would buy it, Conrail's President L. Stanley Crane seemed to abandon his sense of responsibility for

DONALD PEVSNER HELD AN OPTION TO BUY THE BRIDGE (PEVN)

the bridge. To the embarrassment of other Conrail officials, Crane arranged for Conrail to sell the bridge to Miller without considering who he was.

When Miller bought the bridge, he seemed uncertain what he wished to do with it, but said he was open to suggestions, including how to return it to railroad use or remake it into a tourist attraction. Both Loedy and Pevsner offered to work with him.

However, Miller proved to be evasive. He refused to identify who his associates were or where they had their offices. He said he could only be reached by phone at a taxicab office in St. Davids, PA, a Philadelphia suburb. About two weeks after Miller's purchase of the bridge, it became known that Miller had been convicted of fraud as a bank official, and had served time in federal prison. After learning this, Dutchess County Executive Lucille Pattison charged that Conrail was so anxious to unload its bridge "albatross," that it walked away without taking into account who Miller was. The *Hudson Valley Hornet* added sardonically that Miller was probably someone like the brother-in-law of the arsonist who had set the 1974 bridge-closing fire.[4]

At one time Miller considered developing the bridge as a toll road for autos. However, the State Bridge Authority kept reporting that engineering studies showed the bridge superstructure was not sound enough to be adapted for highway use. At another time, Miller retained consultants, including Arthur G. Adams, to develop plans for a light rail line to run across the bridge. Adams, with the help of Engineer Edson L. Tennyson, who had figured in devising San Diego's fabulous new electric street car system, proposed a light commuter rail line to run over the bridge. This commuter line, like the similar joint trolley-train line which had run street cars over the bridge in about 1900, would reach west from the bridge as far as New Paltz, to the state college campus there. From the east end of the bridge, it would reach to Poughkeepsie's city center, and also to the huge IBM plant to the south of the city, Vassar College to the east, and Hyde Park to the north, including the F. D. Roosevelt Historic Site. Adams expected this commuter system would reduce the traffic that often clogged the Mid-Hudson bridge, and certainly would be far cheaper to build than a new bridge. When Adams, for Miller, outlined such a plan to the Save the Bridge Committee, it endorsed the plan in principle.

Meantime, the architect Loedy wanted to find out whether he could raise the funds necessary for his bridge plan. To do this, he needed authority from Miller. He called Miller repeatedly, at first without getting any response, but kept trying, and finally got his consent to meet with him. They met several times in Miller's neighborhood in suburban Philadelphia, over long lunches. They talked about the possibility of Loedy's becoming Miller's agent, or co-owner with him of the bridge, or obtaining an option from him to buy it. It often seemed to Loedy that Miller was on the verge of taking some such action, but he never quite did it—why Loedy never understood. Without authority from Miller, Loedy never had the chance to find out if he could raise the necessary capital. Pevsner too tried to work out a deal with Miller which would enable Pevsner to move ahead with his bridge development plan, but he also found Miller evasive.

Year after year Miller neglected to keep navigation lights burning on the bridge even though the Coast Guard kept fining him for it. Year after year, he neglected to keep adequate liability insurance on the bridge, or do basic maintenance on it, or pay either the taxes or fines he owed for it. Miller's neglect made it difficult for him to build on the community support for his proposals which originally seemed available to him. As for Adams' light commuter line proposal, in the next few years as IBM savagely downsized its operations in the Mid-Hudson region, traffic using the Mid-Hudson Bridge declined, rendering the proposal less viable.

During these years, the Save the Bridge Committee, with considerable political support, approached New York City developers for help in redeveloping the bridge, in cooperation with Miller or otherwise. As the years passed and the committee seemed unable to secure the requisite capital, it met less often, and co-chair Loedy admitted sadly that it might take as long to redevelop the bridge as it took its original promoters to build it.[5]

In 1990, the phantom Miller, having personally gone bankrupt, and feeling harassed by those who pushed him to maintain the bridge and pay taxes on it, sold the bridge along with certain related abandoned railroad right of way. He sold it for one dollar to Vito Moreno, who was understood to be an electronics engineer who had a postbox in King of Prussia, PA, another Philadelphia suburb.

Miller later explained to the Coast Guard that he had sold the bridge to Moreno only under certain conditions which might not yet have been met, and that muddied the situation. Friends of the bridge, however, scarcely understood this. Nor did they easily discover that Moreno and Miller had long been buddies, and that Miller had apparently sold the bridge to Moreno with the expectation that he would work closely with him, if not outright control him. Architect Loedy, persisting in his dream of remaking the bridge, tried to work with Moreno, but like Miller, Moreno, while friendly, declined to give him firm enough authority over the bridge to allow him to proceed to raise capital for his plans. Like Miller, Moreno made himself hard to find, and did not maintain the bridge or pay taxes on it or on its related track. The bridge seemed to be a grand old lady lurching from one shady suitor to another.[6]

RIGHTS-OF-WAY BEING TRANSFORMED

Miller's and Moreno's failure to pay their taxes for the bridge and its related railroad rights of way meant that their title to them was doubtful. Of course, the governments concerned knew they could seize the bridge for non-payment of taxes. But in fact, they were reluctant to seize it because doing so would mean taking on the onerous responsibility for it, which they feared as much as Conrail had. There was no such responsibility, however, if they merely seized the related rights of way. On the west side of the bridge, Conrail had already sold off its bridge-related right-of-way in the town of Plattekill to private persons, but by 1991 Ulster County seized the right of way closer to the bridge, in the town of Lloyd, which Miller

GORDON S. MILLER, THE ELUSIVE BUYER OF THE BRIDGE (*Middletown Record*, Nov. 22, 1987)

had bought from Conrail and on which neither he nor later Moreno had paid taxes. The county seized this right of way and turned over nearly five miles of it to the town of Lloyd, which eventually, with the help of volunteers and government grants, began to develop it as a rail trail, for walking and bicycling. Earlier, on the east side of the bridge, Dutchess County had bought from Conrail the twelve mile railroad corridor reaching from the edge of Poughkeepsie to Hopewell Junction. At first it had talked of turning the corridor into a highway, but later, like the town of Lloyd, it also decided, also with the help of volunteers and government grants, to turn it into a rail trail for walking and bicycling.[7]

BUNGEE JUMPERS

In the fall of 1991, two bungee jumpers, Michael Magee and Jeff Venier, both 29, arrived from Colorado, proposing to start a bungee jumping business on the bridge. It had been 103 years since Steve Brodie had made the bridge famous as a jumping site.

The two jumpers secured permission from bridge owner Moreno to jump from the bridge, and one day walked up on the bridge, prepared to jump. Magee, with a bungee cord attached, jumped first. As he intended, he fell into the water and the bungee cord yanked him back out of it.

Venier lowered a rope to him on which he climbed back up onto the bridge. Then Venier also jumped and climbed back up.

However, an owner of land near the bridge had seen the two jumpers cross his land to get to the bridge, and complained to the police. The police arrested the two men for trespassing. Moreover, the police told them the bridge was not safe for anyone to walk on, with or without permission, much less to jump from.

The next year, two other bungee jumping enterprises, one based in Colorado, the other in France, also sought to establish bungee jumping on the bridge. While their proposals were pending, the *Poughkeepsie Journal* asked in an informal poll if bungee jumping should be allowed on the bridge; most of those responding said no, giving such reasons as that it was dangerous and would detract from the bridge's dignity. The town bodies concerned balked at giving the necessary permission.[8]

RESTORE FOR HEAVY RAIL USE AFTER ALL?

By the late 1980s, contrary to expectations, railroad freight traffic in the nation at large was beginning to grow. Railroads increased their share of the nation's freight tonnage at the expense of trucking from 36% in 1985 to 41% in 1995, reversing a longtime trend.[9]

Several factors contributed to this surprising turn around. The Interstate Commerce Commission was shorn of much of its power in 1980, and

A MODEL FOR RENEWING THE BRIDGE, BY ROBERT TURNER, ARCHITECTURE STUDENT. On top or inside the structure, Turner proposed not only a light rail line but also paths for sightseeing, jogging, and bicycling. At water level, he proposed docks. At each end, he proposed towers for offices, restaurants, and museums (Photo by Robert V. Niles, *Poughkeepsie Journal*, July 13, 1986)

abolished in 1995, allowing railroads to become more flexible. In place of cars built of steel, railroads increasingly acquired cars built of lighter metals like aluminum. Railroads installed along the tracks scanners which reported to centralized computers where cars were. Railroads learned, if belatedly, not only to encourage truckers to load their containers onto rail cars for long trips and then pick them up again for local delivery, but also, when rail cars were carrying containers, to stack containers on top of each other. Such changes made it possible to lengthen trains, haul heavier freight, eliminate cabooses, and reduce train crews from the traditional five men to two. By the 1990s, the railroad freight boom renewed proposals to restore the bridge for rail freight use.

For many years the federal government had pushed Conrail, partially government owned as it was, to sell itself to private owners, and to do so in such a way as to increase competition among railroads. By 1997, with government prodding, Norfolk Southern and CSX, both prosperous private railroads based in Virginia, arranged to buy Conrail and split it between them, to take effect in 1999. By this arrangement, CSX acquired much of the ex-New York Central tracks, including those running along both sides of the Hudson River, and Norfolk Southern acquired much of both the ex-Pennsylvania and ex-Erie Lackawanna tracks, including those which reached to Maybrook.

Most of those who know railroads believe that there is no longer the slightest possibility railroads will ever restore the bridge to regular freight use. However, Arthur G. Adams insists that New York City needs to secure better rail freight access. It is served by rail freight less than other cities and is choking on trucks. The old proposal to build a rail freight tunnel under New York Harbor, though recently revived, seems unlikely to succeed any time soon. Adams believes that rehabilitating the Poughkeepsie Bridge for freight trains would cost much less than constructing the proposed rail tunnel, and would be a significant step toward keeping the New York port competitive. Also he believes, as many informed observers do not, that it is still possible that Norfolk Southern might eventually decide to rehabilitate the bridge. More likely, however, is that Norfolk Southern will ignore the bridge, but redevelop part of the Maybrook yard into a rail-truck terminal which would transfer freight brought from the South and West by train into trucks for delivery in New England and metropolitan New York.

In the 1990s, restoring the bridge for either light or heavy rail use continued to seem unlikely. Moreover, rebuilding it for highway use seemed impossibly expensive, and Pevsner's and Loedy's plans for commercial development had failed to attract the necessary capital. Was there any other way to make use of it and thus preserve it?

ARTHUR G. ADAMS, persistent advocate of restoring the bridge to rail use, whether for commuters or freight (photo by Gretchen McHugh, ADAMS)

47. WALKWAY (1991-   )[1]

A chance to "hear the quiet," bask in a vast panorama, and let your imagination soar.

WHEN THE BRIDGE first opened, some people expected it would immediately open to public walking, as the bridge charter originally provided that it could, but they were disappointed. In the 1890s and again about 1920, citizens campaigned to open the bridge to walking. Despite some encouragement, the campaigns did not succeed.

Then in 1991, another such campaign began to surface. It was sparked by the Coast Guard. When the bridge's navigation lights were out and no one was taking responsibility for them, the Coast Guard called for the bridge to be torn down, to make navigation safer. Hearing this, Jonas Lenktaitis, of Fishkill, NY, proposed to demolish the bridge to use its steel for scrap. He asked the Coast Guard for permission. In response, the Coast Guard, "authorized " him to demolish the bridge, provided he met certain safety stipulations.

This threat of demolition stirred a few citizens into struggling to preserve the bridge by remaking it into a public walkway. The group came to include former Poughkeepsie Mayor Jack Economou, Theodore Frank of the Hudson Valley Railroad Society, Hyde Park, and a man who lived near the bridge in Poughkeepsie, Bill Sepe, a large man with a deep voice who emerged as a leader. They suspected that the cost for rehabilitating the bridge as a walkway could be less than the cost for demolishing it safely, including appropriate measures to avoid contaminating the river.

WALKWAY VOLUNTEERS IMPROVING THE BRIDGE FOR PUBLIC WALKING, July 15, 1995, looking toward Poughkeepsie (WALK). Tom Lepre (left) and Bill Sepe hauling a section of metal grating -- similar to the grating which shows at the extreme right -- to add to the Walkway. To the left the bridge railings have been removed temporarily as part of the rehabilitation process.

At the time, New York State had recently created the new Hudson River Valley Greenway, a tax-supported agency intended to encourage the creation of public parks along the Hudson River between New York City and Albany. While the walkway promoters were later to oppose seeking tax funds for making the bridge into a walkway, at this stage they believed that if such a walkway were to become part of the Greenway park system, the Greenway could help pay for it.

The walkway proposal won encouragement from the Poughkeepsie Common Council, and the legislatures of both Ulster and Dutchess counties. When the *Poughkeepsie Journal* took an informal poll of residents on whether they favored a walkway, 63% said yes. By May 1992, Bill Sepe was leading in organizing the walkway promoters into a group which eventually adopted the name Walkway Over the Hudson. The Walkway became strong enough to discourage Lenktaitis from pursuing his plan to demolish the bridge. With the assistance of John Mylod, who by this time was executive director of the Hudson River Sloop Clearwater, a river preservation project, the Walkway secured funding from the Greenway to study whether the bridge was structurally sound enough to serve as a path for walking and bicycling.[2]

COSTS

The Lichtenstein engineering firm, which did the study, reported in December 1992, that the bridge, a "magnificent structure," could bear not only the weight of pedestrian and bicycle use but also of a "light rail" system if that became desirable. Necessary work to prepare the bridge included, in Lichtenstein's view, structural rehabilitation, cleaning and painting to prevent further deterioration (including the removal of toxic lead-based paint, so that it did not peel and fall into the river), new railings, and a new navigation lighting system, at an estimated cost of $21.5 million.

The Lichtenstein estimate was much higher than the $5.9 million estimate which DeLeuw, Cather & Co. had made in 1983 for Central Hudson to rehabilitate the bridge for the similarly limited purpose of carrying high voltage transmission wires. While some friends of the walkway took the new estimate as a warning that walkway costs might be prohibitive, Bill Sepe and other walkway advocates believed that engineering estimates are often conceived in a world of bloated budgets, and that if walkway volunteers operated simply, providing much of the labor and tools themselves, they could raise the necessary funds.

By temperament, Bill Sepe is a seeker and student. Of Italian-Scandinavian descent, he has been at various times a carpenter, school bus driver, and landlord. He is articulate, forceful, and uncompromising. He is not sure he is good at organizational matters. He doubts he would have gotten into walkway promotion if he had seen where it would lead. He emphasizes that transforming the bridge into a walkway—with lanes reserved for bicycling and jogging—would offer "awe inspiring panoramas." It would offer a chance, he says, to "hear the quiet." Sepe cites Frederick Law Olmsted, the creator of parks in Manhattan and Niagara Falls, as insisting that walking within sight of natural vistas, free from commercialism, can help people develop inner balance.

Despite the Lichtenstein warning of high costs, Sepe and other Walkway directors, shied away from encouraging the Walkway to cooperate in commercial development of the bridge. Sepe is skeptical that shopping malls add to the quality of life, whether on top of a bridge or elsewhere. In September 1992, when Poughkeepsie city officials arranged for Loedy, who was still advocating placing shops and housing on the bridge, to meet with the Walkway directors to see if they could work together, Loedy believed they could, but the Walkway directors decided their proposals were too much in conflict.

At about the same time, the Walkway tried to persuade bridge owner Moreno to give the bridge to them. The Walkway hoped that because the Coast Guard was continuing to push Moreno either to restore the bridge or tear it down, neither of which Moreno seemed able to do, Moreno would see the advantage for him of giving the bridge away.[3]

TAKING POSSESSION

In the fall of 1993, the Walkway leaders decided that since Moreno, the owner of record, had not paid his taxes or Coast Guard fines, he had in effect abandoned the bridge. Convinced that Moreno would not give them the bridge any time soon, they decided to take possession of it anyway. They took it by squatting on it, or what lawyers call "adverse possession." If the legal owner still was Moreno, they doubted he would proceed against them for trespassing because to do so he would have to prove that he owned the bridge, which would open him to prosecution for not having paid taxes and Coast Guard fines. In fact, Moreno merely offered to join the Town of Lloyd in taking legal action against the Walkway's

CHANGING OWNERSHIP OF THE POUGHKEEPSIE BRIDGE: THE LATER YEARS	
1968	Penn Central Company
1976 (after the 1974 fire had closed the bridge to trains)	Consolidated Rail Corporation (Conrail)
1984	Gordon Schreiber Miller for Railway Management Assosiates
1990	Vito Moreno
1993	Walkway claims ownership by squatting
1998	Walkway takes full legal ownership

WALKWAY VOLUNTEERS USING ROPE AND A LEVER TO LIFT LONG-COLLAPSED RAILINGS INTO A STANDING POSITION, on the bridge approach, in Highland, May, 1994 (WALK). Left to right are Lee Crowell, Klaus Jonietz, and Rick Kjeldsen.

WALKWAY OVER THE HUDSON: BOARD OF DIRECTORS (selected), 1994-1995		
Sepe, William, Chair	Carpenter, school bus driver, landlord	Poughkeepsie
Anson, Robert Sr.	Building contractor	Clintondale
Coller, Richard	IBM engineer, retired	Staatsburg
Crowell, R. Lee	Electrician	Poughkeepsie
Decker, Jon	Former Lloyd town supervisor	Highland
Economou, Jack	Attorney, former Poughkeepsie mayor	Poughkeepsie
Hansut, Robert	Former Lloyd town board member	Highland
Jonietz, Klaus	Insurance agent	Highland
Schaeffer, Fred W.	Attorney	Pleasant Valley
Valentino, Joseph	Retired from IBM, former Highland fire chief	Highland
Wadlin, Vivian	Graphics designer	Highland

pride in what volunteers could do at little expense, with their own arms and hammers. Feeling confident, they formally adopted the policy that they would raise all their funds privately, none at all from government sources. At the center of the bridge, they flew an American flag, with a nameplate claiming ownership for the Walkway. They secured liability insurance for use of the bridge, and when it was questioned whether the coverage was enough, they increased it.

To demonstrate the feasibility of the walkway, in 1994 the volunteers opened much of the western half of the bridge to public walking, the first time in the history of the bridge that it had been easily open to such walking. In some places the railings were still missing or collapsed. The new metal grating walkway, as far as it reached, was sturdy, but it was narrow. Moreover, uncovered

"criminal trespass" on the bridge, if the town were to take such action, but it did not.

While the Walkway did not claim that it took possession of the bridge with the permission of local governments, nevertheless soon afterward the legislatures of the two counties concerned, Ulster and Dutchess, formally urged citizens to assist the Walkway in constructing a "pedestrian/bicycle pathway" on the bridge. In effect the two counties seemed to be recognizing that the Walkway now had a claim to own the bridge, if only a tentative one, and were cheering.[4]

OPENING THE BRIDGE TO WALKERS—TEMPORARILY

By 1994, Walkway volunteers—the Walkway was an all-volunteer organization—were removing some of the bridge's rotting wooden walkway. Working from the Highland end, they began to lay a new metal-grating walkway and otherwise make the bridge safer for walking. They took spaces between the railroad ties showed pedestrians the view straight down to the water. However, during 1994 and 1995, with the help of Walkway guides, over 5,000 people, including the author, walked from the Highland end of the bridge well out on it—they were not allowed to walk out from the Poughkeepsie end as the fire and the subsequent court-ordered stripping had rendered walking there impossible.

Many of the walkers were awed by how high over the water they were, and if sometimes a little queasy, felt exhilarated by the view. Others relished the quiet, or became fascinated to see the historic bridge structure close up and tried to figure out how it was put together. According to the Walkway, almost all of the walkers thereafter became supporters of the Walkway.[5]

In the spring of 1998, Vito Moreno, discouraged that he would never be able to do anything with the bridge, arranged to give his claim for it to the Walkway. Miller, when he heard this, called Loedy, asking him to take legal steps to stop Moreno from giving his claim away. But Loedy refused. On June 5, 1998, the Walkway filed the new deeds in the Ulster and Dutchess County courthouses, becoming no longer a squatter, but the owner of record of the entire bridge.

OBSTACLES

Meantime, the Walkway ran into obstacles, as proposals to use the bridge as a walkway also had in the 1890s and 1920s. The Walkway became mired in negotiations with Central Hudson, the electric utility. Central Hudson had stopped using its high voltage electric transmission lines on the bridge in 1985, and argued it was not its responsi-

Opposite: VOLUNTEER JACK BARNARD, a retired IBM engineer, STRENGTHENING THE WALKWAY RAILING, Sept. 2, 1995 (WALK). He is high on the bridge's Highland approach, wearing a safety belt. He is sitting on a cross beam to which he is attaching a metal brace for the railing (which itself is out of sight to the left).

Right: VOLUNTEER MIKE BALDLWIN, a heavy equipment mechanic, OFTEN CLIMBS ON THE BRIDGE FOR THE WALKWAY. In this case, he has climbed down from the top of the bridge to bring back up the electric cable-powered navigation light which he holds. Spring, 1999 (WALK). By 1901, the Walkway had gradually replaced nearly all such lights with solar-powered lights. The Mid-Hudson Bridge shows in the background.

WALKWAY FUND-RAISING TABLES, AT A YARD SALE OF THE UNITED METHODIST CHURCH, MODENA, NY, Spring, 1996 (WALK). One of many Walkway fund-raising efforts, held at various Hudson Valley locations. Judy Gridley, Walkway volunteer attending the tables, is the second person from the right.

REMOVING BRUSH AND BOTTLES FROM UNDER THE BRIDGE IN POUGHKEEPSIE, near Washington St., Fall, 1998, are Walkway volunteers David Black and Ingrid Sanchez (WALK). The back hoe was on loan to the Walkway to help in clean-up.

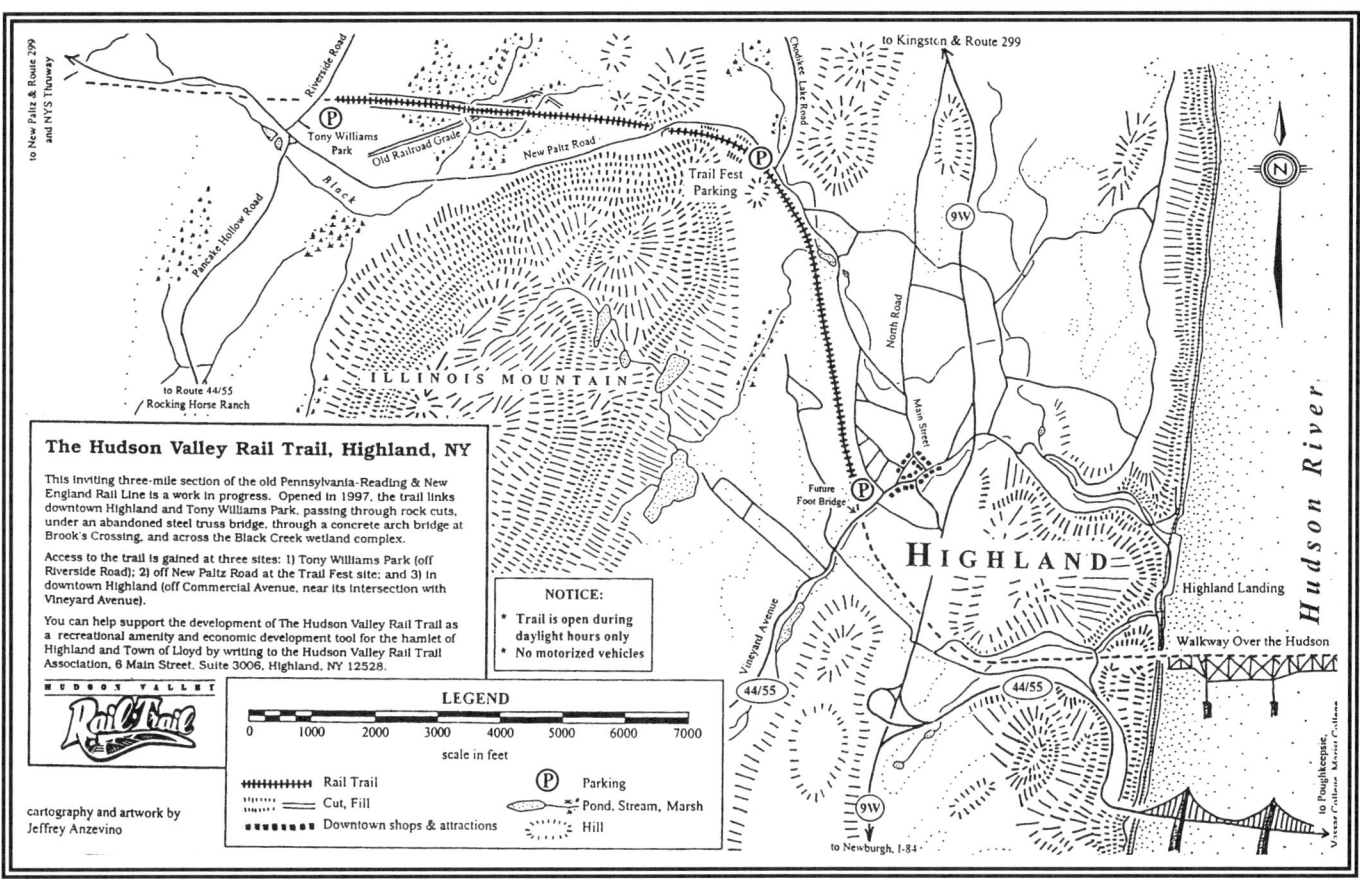

MAP SHOWING THE EXISTING HUDSON VALLEY RAIL TRAIL, ALONG THE BRIDGE ROUTE'S ABANDONED RIGHT OF WAY IN THE TOWN OF LLOYD. ALSO SHOWING THE TRAIL'S PROPOSED EXTENSION EASTWARD (broken line), ALONG MORE OF THE BRIDGE ROUTE'S RIGHT OF WAY, TO CONNECT TO THE "WALKWAY OVER THE HUDSON" ON THE POUGHKEEPSIE BRIDGE, 1998 (By permission of Jeffrey Anzevino, HVRTA). Completing this connection would require such work as building a "Foot Bridge" over Vineyard Avenue, as the map indicates, because the railroad bridge there was removed, and also opening up a clogged railroad right of way under a Rt. 9W bridge. Connections for this rail trail have also been proposed westward (broken line) to the Wallkill Valley Rail Trail in New Paltz.

bility to remove the lines. But the Walkway contended it was Central Hudson's responsibility to remove them, pointing out not only that the wires, as they deteriorated with age, were increasingly likely to become a hazard, but also that the wires interfered with the view from the bridge. Moreover, at the western end of the bridge, Central Hudson still used a major power line, on what the Walkway claimed had become its property, for which the Walkway asked them rent, more rent than Central Hudson wanted to pay. Also Central Hudson owned other land nearby which provided access from Haviland Road onto the bridge, the best access on the western side of the river—it was level, while other potential access routes were steep. Central Hudson at first allowed the Walkway to use this level access by verbal agreement—and in 1994 and 1995, thousands of people were able to walk out onto the bridge by this route. But when Central Hudson asked the Walkway to sign a written agreement limiting its use of this access to walkers only, the Walkway balked, saying it also needed use of this access for maintenance and emergency vehicles. By October 30, 1995, Central Hudson then closed off this access to the Walkway, and the two parties since then have been unable to reach an agreement about it. In 1999, the Walkway took Central Hudson to court over several of these issues. A judge ruled that Central Hudson remained liable for the wires that it had left on the bridge, but Central Hudson said it would appeal.[6]

The Walkway also became mired in negotiating with the town of Lloyd. The town had long supported the Walkway in principle. It had encouraged plans to develop the abandoned railroad right of way that stretches westward from the bridge into a walking and bicycling path which would connect the bridge Walkway all the way westward to the existing Wallkill Valley Rail Trail in New Paltz, thus holding out the prospect of a magnificent walking and bicycling network for the region. But the town of Lloyd, claiming it has jurisdiction over the western half of the bridge, required the Walkway to secure zoning and building permits, including certificates that the bridge was safe for walking. Since the Walkway did not yet have those permits, in early 1996 the town went to court to enjoin the Walkway from continuing to open the bridge to public walking. In response, the Walkway argued that the town lacked jurisdiction over the bridge because the river, being navigable, was under state and federal authority. The court, in its decision in spring 1997, said it had "the utmost respect" for volunteers like the Walkway members "who strive to preserve, protect and make available for public enjoyment the historical sites of the Hudson Valley." It accepted the Walkway, not Moreno, as the "undisputed" owner of the half of the bridge in the town of Lloyd. Nevertheless it decided that while the federal government had jurisdiction over the bridge with respect to its effect on navigation, in other respects the town had jurisdiction over the half of the bridge located in the town, and "the risks implicit in permitting visitors on a very old bridge some 200 feet above the water without any prior safety inspection by the town presently outweigh public enjoyment of the bridge." Thereafter the Walkway applied to the town of Lloyd for the necessary permits. The permits are still pending, and the town doubts that the Walkway has the resources to do what is necessary to secure them. Meanwhile the Walkway continued to work at improving the bridge, improving it for walking, and improving its navigation lights.

But eventually the town, insisting on its prerogatives, asked a court for an injunction to prevent the Walkway from doing any work on the bridge whatsoever until the town's building department gave permission, and in 2001 the court granted the injunction.[7]

Recently friends of the Walkway differ on how to overcome the obstacles to its opening. Some believe that if the parties concerned made a fresh start, approaching each other without the baggage of accumulated resentments, willing to compromise, they could find a way to open the Walkway permanently. Some believe the Walkway would be stronger if it would propose to assist those visitors for whom walking might be too far or too scary by offering some form of supplementary transportation, such as motorized carts or light rail. Others believe the Walkway cause would be stronger if it were willing to accept government as well as private funding, much as the related rail trails on each side of the Hudson accept both. Still others believe that some form of direct government participation in the Walkway may be necessary. While this would be a cost for taxpayers, it might not be as costly as allowing the Walkway dream to fail, because that might force taxpayers to pay for tearing down the bridge.

While other options for use of the bridge are still conceivable, many observers see a walkway as the option which is most likely to succeed. After visiting the bridge, a representative of Rails to Trails Conservancy, based in Washington, DC, welcomed the walkway proposal as part of the burgeoning nationwide effort to convert abandoned railroad corridors into walking trails, and declared the bridge to be for this purpose "one of the most outstanding structures of its kind anywhere in the nation."[8] Many promoters of the Hudson River region believe that walkers, bicyclists, and all kinds of tourists would be drawn to a completed walkway. They would be drawn to it for its remarkable height, its stunning views, and the rich history of both the bridge and the river. They would be drawn to it as one of the longest pedestrian bridges in the world.

If the walkway opens to the public, the curious

In 1994 and 1995 while Walkway volunteers were improving the bridge for safer walking, THE WALKWAY OPENED THE BRIDGE TO REGULAR PUBLIC WALKING FOR THE FIRST TIME IN ITS MORE THAN A CENTURY OF HISTORY, and thousands of people took advantage of the opportunity. Here a group of walkers has proceeded from the bridge's Highland end about half way across the bridge, as far as they were allowed to walk, to a viewing platform. They are looking toward the Poughkeepsie shore. They are being led by two volunteer Walkway guides: at the left, Lee Crowell (beard), a retired electrician, and next to him, Richard Coller (black cap), a retired IBM engineer (WALK). Unfortunately, by early 1996, legal pressure from Central Hudson and the town of Lloyd had effectively closed the bridge to public walking.

will have a chance to puzzle out how such engineers as Clarke and O'Rourke could have built the bridge so well more than a century ago, that even though it has been neglected for more than twenty-five years, it still stands tall over the mighty Hudson. If the walkway, as proposed, sets aside lanes for bicycles, creates parking and picnic areas at each end, creates a museum featuring the history of the river and the bridge, and connects with other bicycle and walking paths on both sides of the river, it might make the bridge more appealing to the public than it has ever been before. If visitors savor the rich past of the river and bridge, if they exult in rising high over the river, basking quietly in the vast panorama of boats, trains, and distant misty mountains, the bridge might rouse the public's imagination even more than it did when trains passed over it night and day, carrying the wealth of the continent.

APPENDIX

LOCOMOTIVES OF THE HARTFORD & CONNECTICUT WESTERN RAILROAD (H&CW), 1881-1888

A SAMPLE ROSTER OF LOCOMOTIVES WHICH MAY HAVE RUN ACROSS THE BRIDGE

Compiled by Leroy Y. Beaujon

THIS ROSTER lists the H&CW's locomotives beginning in 1881 when the H&CW took over the Connecticut Western Railroad (CW). At that time the promoters of the Poughkeepsie Bridge planned that the H&CW was to be the principal road by which Poughkeepsie Bridge traffic would enter New England. The roster continues to 1889 when the bridge opened, and the H&CW was absorbed into the newly created Bridge Route line, the Central New England & Western.

However, the roster follows these locomotives longer than the 1881-1889 dates for the H&CW's holding them might suggest. The roster reaches from 1871 when, as the roster table indicates, the earliest locomotives in the list were built, to 1926 when, as the Notes indicate, locomotive #18 was condemned.

To relate this roster especially to the Poughkeepsie Bridge, we have included here only those locomotives which Beaujon believes may have run across the bridge. All the included locomotives were built by Rogers, in Paterson, NJ. Some were considered to be primarily for passenger service, others for freight service.

NUMBER ASSIGNED by the Railroad (originally the CW, but from 1881 the H&CW)	NAME ASSIGNED BY THE CW (the H&CW did not name locomotives)	NUMBER ASSIGNED BY THE BUILDER, Rogers Locomotive Works	DATE BUILDING FINISHED	DRIVERS (diameter, in inches)	WHEELS (pattern of their number, front to back, with drivers in middle)
3	Winsted	1955	10-3-71	60"	4-4-0
4	Salisbury	1968	10-28-71	60"	4-4-0
5	Norfolk	1969	10-30-71	60"	4-4-0
6	Bloomfield	1838	1-09-71	66"	4-4-0
7	Canton	2079	6-07-72	54"	4-4-0
8	Canaan	2080	6-10-72	54"	4-4-0
9	Simsbury	2363	11-17-73	54"	4-4-0
10	Tariffville	2429	12-07-75	60"	4-4-0
15	None	3052	8-07-82	60"	4-4-0
16	None	3096	9-30-82	62"	4-4-0
17	None	3299	7-06-83	62"	4-4-0
18	None	3346	9-10-83	62"	4-4-0
19	None	4108	2-11-89	56"	2-6-0

ROSTER NOTES
By the Numbers Assigned by the Railroad

3. Sent for repairs to CNE&W shops at Hartford, CT, Oct., 1890. Taken to the shops of the P&R (the PR&NE's parent) in Reading, PA, Fall, 1893, for repairs that were never made. Scrapped there in late 1894.

4. Removed by CNE&W from road service in late 1892 due to poor performance, and used at Hartford as a stationary boiler for passenger cars. Shipped March 18, 1899, for scrap to the Penn. Bolt and Nut Co., Lebanon, PA.

5. Wrecked in PR&NE service at Modena, NY, Nov. 13, 1892, but repaired. In March, 1899, when the regular CNE locomotive that hauled trolley cars from Highland over the Poughkeepsie Bridge was being repaired, this locomotive took its place. Shipped for scrap, Jan. 16, 1901, to American Iron & Steel Mfg. Co., Lebanon, PA.

6. Knocked by a 5-ton boulder off the track and down a bank at Stoney Lonesome, near West Norfolk, CT., March 8, 1882. It took ten days to bring it back up to track level. On March 9, 1899, the CNE shipped it for scrap to the Penn. Bolt & Nut Co., Lebanon, PA.

7. Served the CNE&W, and then the PR&NE, but was scrapped by the P&R at its shops in Reading, PA, in late 1893.

8. Scrapped by the P&R at its Reading shops, early 1893.

9. Taken to the P&R's shops at Reading, Fall, 1893, for repairs that were never made. Scrapped there late 1894.

10. Wrecked in the collapse of the Tariffville Bridge in 1878 but rebuilt. Was the first locomotive to cross the Poughkeepsie Bridge when it was tested in late Dec., 1888, and formally opened Jan.1, 1889. Rebuilt again at the Hartford, CT, shops, 1890, 1898, and 1902. Sold to a Hartford scrap dealer, I. U. Harris, Sept., 1906, for $625.

15. Cost new $13,224. Used primarily on the Rhinecliff Branch because of its light weight. Rebuilt by Rogers, in Paterson, NJ, 1892. Sold as scrap, April, 1908, for $200.

16. Cost new $12,500. Rebuilt in July, 1900, at the CNE's Hartford shops, with a new Baldwin boiler, for $5231. Condemned 1919.

17. Rebuilt with an extended front end, 1890. Reboilered, 1901, for $5651. Condemned 1925.

18. Cost new $9572. Reboilered and fitted with an extended saddle, 1901, for $5061. Condemned 1926.

19. Cost new $9825. Sold in 1911 for scrap for $300.

CHAPTER NOTES

CH. 1. DREAMING OF A BRIDGE

1. For chapter: Poughkeepsie Bridge Co., *Bridging the Hudson*, 1871; Poughkeepsie Bridge Co., *Charter...and Amendments*, 1887
2. *Poks. Eagle*, Oct. 27, 1855
3. *Poks. Daily Eagle*, Jan. 22, 1868
4. *Poks. Daily Eagle*, Jan. 25, 26, Feb. 15, 1871
5. Poks. Bridge Co., *Charter...and Amendments*, 4, 9
6. *Poks. Daily Eagle*, May 15, 1871
7. *Poks. Daily Eagle*, July 20, 1871; Poks. Bridge. Co., *Bridging the Hudson*, 1871, 29; Poughkeepsie, *The Bridge*, 1889, 6-7

CH. 2. LAYING A CORNERSTONE

1. For chapter: *Laying the Corner-stone of the Great Bridge to Span the Hudson at Poughkeepsie*, Dec. 17, 1873; Poughkeepsie, *The Bridge*, 1889; James A. Ward, *J. Edgar Thomson*, 1980; Joseph F. Wall, *Andrew Carnegie*, 1989
2. "Corrupt": *NY Times*, May 7, 1887
3. "King": *Poks. Daily Eagle*, Dec. 18, 1873
4. "Enormous": Ward, 183. The next highest amounts subscribed were Eastman's $25,000 and Engineer Dickinson's $10,000 (Poks. Bridge Co., Journal, Oct. 18, 1873, UCONN)
5. Keystone Bridge Co., *Descriptive Catalogue of Wrought Iron Bridges*, 1874, 12-13
6. *Poks. Daily Eagle*, Dec. 18, 1873; *Poks. Telegraph*, Dec. 6, 1873
7. *Laying the Corner-Stone*, 9-11; *NY Herald*, Dec. 18, 1873. In 1887 when the push was underway to complete the construction of the bridge, because the pier foundation in which the cornerstone had been laid was moved slightly, the cornerstone itself was moved. It was moved without formal ceremony, but in the presence of Chief Engineer O'Rourke, into the nearby southwest corner of the anchorage pier (*Poks. Daily Eagle*, Feb. 15, 1887).
8. *Laying the Corner-Stone*, 6-7, 13-15; *Kingston Daily Freeman*, Dec. 19, 1873; *Albany Argus*, Dec. 18, 1873; *NY Herald*, Dec. 18, 1873; Lossing: *Poks. Dutchess Farmer*, Dec. 30, 1873; *NY Times*, Dec. 18, 1873
9. Poks. Bridge Co., Bd. Minutes, March 30, 1875, UCONN

CH. 3. BOSTONIANS INVESTIGATE

1. For chapter: Poughkeepsie Bridge Co., *Bridging the Hudson at Poughkeepsie*, 1875; J. P. Gould, *A Review of a Project for Bridging the Hudson River at Poughkeepsie*, 1876
2. *Boston Herald*, Nov. 24, 1874; quotes: *NY Times*, Feb. 27, 1875
3. "Practical": *Hartford Daily Times*, Oct. 6, 1874; "disagreeable": *Boston Daily Evening Traveller*, Dec. 2, 1874
4. "Wolf": *Winsted, CT, Herald*, Jan. 29, 1875; "Almighty:" *Poks. Daily Eagle*, Jan. 21, 1875; on ice: *Poks. Daily News*, Jan. 22, 1875. It was not until about 1912 that ice breakers began to keep the Hudson open for winter navigation.
5. "Pullman," "crazy": *Poks. Daily Eagle*, Jan. 22, 1875; "palatial": *Boston Daily Globe*, Jan. 27, 1875; eels: *Poks. Daily Eagle*, Jan. 25, 1875
6. *Poks. Daily Eagle*, Feb. 27, 1875
7. "Princely": *NY Times*, Oct. 7, 1875; "capital": Feb. 27, 1875; *Hartford Daily Courant*, Dec. 24, 1874; *Boston American Union*, in *Poks. Daily Eagle*, Nov. 4, 1875
8. *Boston Journal*, Dec. 29, 1874; *Poks. Daily Eagle*, Feb. 24, 1875; Gould, *A Review*, 18, claimed his "lucubrations have been neither suggested, inspired, nor paid for by rival interests."
9. Contract, Jan. 21, 1876: Poks. Bridge Co., Bd. Minutes, pp. 85-93, UCONN; *Poks., The Bridge*, 7-8; *New Paltz Independent*, June 1, 22, Aug. 10, 1876; *NY Times*, Sept. 1, 1876

CH. 4. PLACING PIERS IN THE RIVER

1. For chapter: *Poughkeepsie, The Bridge*, 1889
2. "Largest": *Poks. Daily Eagle*, Apr. 23, 1878; "expensive": July 13, 1877; "colossal": Nov. 15, 1876; "longest": March 21, 1877
3. *Poks. Daily Eagle*, Sept. 18, Oct. 12; "air lock": Nov. 16, 1877
4. Poks. Bridge Co., Minutes, Nov. 7-Dec. 29, 1877, UCONN
5. Eastman obituary: *NY Times*, July 14, 1878; monument: Platt, *The Eagle's History of Poughkeepsie*, 1905, 228

CH. 5. DESIGN, FINAL PHASE

1. For chapter: Thomas C. Clarke, "Railway-Engineering in the US," *Atlantic*, Nov. 1858, 2: 641-656; Clarke, "Hudson River Bridge at Poughkeepsie," *Scientific American Supplement*, May 19, 1888, 25:10311-10312; Clarke, "Building of a Railway," *Scribner's Magazine*, June, 1888, 3:643-670; Clarke, "Engineering," in Alfred R. Wallace, *Progress of the Century*, 1901, 421-452; Charles Macdonald, "The Six-Hundred Ton Testing Machine at the Works of the Union Bridge Company at Athens, Pa.," American Society of Civil Engineers (ASCE), *Transactions*, Jan. 1887, 16:1-29; Macdonald, "Address," ASCE, *Transactions*, Dec. 1908, 61: 544-552
2. Poks. Bridge Co., Minutes, UCONN. *Poks., The Bridge*, 8, says bridge construction work "lay untouched" for "nearly nine years." But according to Poughkeepsie newspapers (and confirmed in *Engineering News*, Oct. 29, 1887, 18:306), bridge work ceased in Sept. 1878, and began again by Sept. 1886, which is eight years.
3. "Led": *NY Times*, Nov. 5, 1886; "houses": *NY Sun* in *Poks. News-Telegraph*, Jan. 15, 1887
4. "Leading": Poks. Bridge Co., *The Poks. Bridge: A New Route*, Feb. 1887, 6; *Towanda, PA, Reporter-Journal*, Dec. 4, 1884

5. Field lobbied on behalf of the bridge cause in Albany. Maurice, who graduated from Williams College in 1861 and then attended RPI, wrote in 1885 in a Williams class letter (WILLM) that his Union Bridge Co. duties were "confined" to its Athens plant. The *NY Herald*, Jan. 6, 1887, claimed that Clarke and Macdonald "designed" the bridge. Clarke ("Building of a Railway," 663-664) called himself "one of the designing and executive engineers" of the bridge. Macdonald's obituary in the Gananoque, Ont., *Reporter*, July 12, 1928, said he was "specially connected" to the bridge. *Biographical Dict of Am. Civil Engrs.*, v.2, 1991, reported that Macdonald "supervised design" of the bridge.
6. "Best-known": *NY Times*, June 17, 1901; "great:" Clarke, "Hudson River Bridge," 10311-10312; "ignorant": Clarke, "Engineering," 451
7. Clarke, "Architects and Architecture," *Christian Examiner*, Sept. 1850, 49:285-286; Clarke, "Vincent Bourne and the Modern Latinists," *Christian Examiner*, Sept. 1848, 244-245, 257
8. Macdonald, "Address," 546; "contempt": Clarke, "Building of a Railway," 644; 'courage": Clarke, "European and American Bridge-Building Practice," *Engineering Mag.*, Apr. 1901, 21:58
9. Longest: *NY Herald*, Jan. 6, 1887; "notable": Clarke, "Building of a Railway," 664; Waddell, *Bridge Engineering*, 1916, 568-569, 603
10. "Shuddered": Memoir of Thomas Curtis Clarke, ASCE, *Transactions*, June, 1903, 50:498; Lindenthal: *Poks. Eagle-News*, Jan. 27, 1921; "uniform": Clarke, "Building of a Railway," 648-650
11. *NY Times*, Nov. 8, 1887; Clarke, "Hudson River Bridge," 10312. Despite Clarke's intention, according to engineer Ralph Modjeski (to E. J. Pearson, Oct. 21, 1916, box 2, file 2, PBEN) who was an inspector for another project in the Athens shops in 1887-88 when Poughkeepsie Bridge steel was being fabricated there, some of it was "brittle" Bessemer steel.
12. "Heaviest" and "blow": Clarke, "Hudson River Bridge," 10312; "links": Clarke, "Building of a Railway," 662, 665
13. Brooklyn Bridge Trustees, minutes, reports, clippings, ROEBL. Both Clarke and Macdonald were appointed Brooklyn Bridge trustees by New York's Mayor W. R. Grace. At a trustees meeting in 1882, because Brooklyn Bridge chief engineer W. A. Roebling had become a long-term invalid, both Macdonald and Clarke voted, with Brooklyn's Mayor Seth Low and New York's Mayor Grace, to remove him from his position, but the majority of the board voted to retain him (*NY Times*, Sept. 12, 1882).
14. "American": Clarke, "Building of a Railway," 663

CH. 6. BUILDING, FINAL PHASE

1. For chapter: *Poughkeepsie, The Bridge*, 1889; Thomas C. Clarke, "Hudson River Bridge," *Scientific American Supplement*, May 19, 1888, 25: 10311-10312; John F. O'Rourke, "Construction of the Poughkeepsie Bridge," American Soc. of Civil Engineers, *Transactions*, June, 1888, 18:199-215
2. *Poks. News-Telegraph*, Sept. 8, 1888
3. Contrary to *Poks., The Bridge*, 1889, 13, O'Rourke was born in Ireland (his obituary, *NY Times*, July 30, 1934; legal documents in the possession of his great grandson, O'ROUR). Mosquitoes: *Poks. News-Telegraph*, Jan. 5, 1889. "Close": *Engineering News*, Oct. 29, 1887, 18:306.
4. Cement: Union Akron Cement Co., Buffalo, advertisement, in *Poks., The Bridge*, 1889, 42. The F. O. Norton Cement Co. of Rosendale, NY, claimed in an undated flyer that its natural cement was used in building the Poughkeepsie Bridge (Century House Historical Society, Rosendale, *Newsletter*, Fall, 1996, [21]), which seems reasonable considering the company's location near the bridge site, but unambiguous evidence to support this claim is not available. O'Rourke, "Construction," 205, reported, "Domestic cement is used in the concrete for the cribs, Portland cement in the masonry." Ralph Modjeski wrote New Haven's Vice President E. J. Pearson, Oct. 21, 1916 (file 2, box 2, PBEN) that both "Portland and Rosendale Cement" were deposited in the cribs. But it is uncertain whether Modjeski and other engineers who commented similarly (PBEN) meant cement from the Rosendale region or generic Rosendale cement which could come from anywhere.
5. *Poks. News-Telegraph*, ("hard") May 21, ("Teutons") Apr. 2, ("clink") March 19, 1887
6. "Nothing": *Poks. Daily Eagle*, Feb. 25; "Gypsey": *Poks. News-Telegraph*, May 21, 1887
7. "Coat": *NY Herald*, Jan. 6, 1887
8. "Freshet," "unexampled," "worst," "wind," " feat": O'Rourke, "Construction," 210-212; other quotes: *Poks. News-Telegraph*, Apr. 9, 16, 1887; *Engineering News*, Sept. 8, 1888, 20:179
9. *Poks. News-Telegraph*, July 9, 1887
10. *Scientific American*, Feb. 5, 1887, 56:79. The machine was designed by former Union Bridge Co. partner Charles Kellogg. "Admirable": *Engineering News*, Jan. 7, 1888, 19:11
11. "Gentlemanly": *New Paltz Times*, Dec. 12, 1888; Poks. Bridge Co., Minutes, Nov. 22, 26, 1886, UCONN; Poks. Bridge Co., *Charter. . .and Amendments*, 1887, 4, 9-10, 21-22
12. "Mazy": *Poks. News-Telegraph*, May 14, 1887; "largest": Clarke, "Hudson River Bridge," 10312
13. *Engineering News*, Jan. 7, 1888, 19:11; Oct. 22, 1887, 18:287; *Poks. News-Telegraph*, July 16, 1887; May 21, 1887
14. *Engineering News*, May 26, 1888, 19:436; June 25, 1887, 17:405; Clarke, "Hudson River Bridge," 10311-10312
15. *NY Times*, July 9 (pp. 2, 4), 1887. The governor signed the charter amendment by June 25, 1887. By Aug. 1887, Husted was pres. of the H&CW, and by 1889 pres. of the PP&B
16. *Scientific American*, Feb. 4, 1888, 58:70; *Poks. News-Telegraph*, Apr. 9, 1887; Skinner, "Great Achievements in Modern Bridge Building," *McClures*, Jan. 1901, 254-255
17. *Poks. News-Telegraph*, ("familiar") Aug. 18, July 14, Sept. 1, ("scream") 8, 1888

CH. 7. THE BRIDGE OPENS

1. *Poks. News-Telegraph*, Dec. 29, 1888; Jan. 5, 1889
2. *Poks. Daily Eagle*, Dec. 31, 1888
3. "Speed": *NY Times*, Jan. 1, 1889; *Poks. Daily Eagle*, Jan. 1, May 24, 1889; Poks. Bridge Co., Minutes, Jan. 7, Feb. 6, 1889, UCONN
4. Macdonald, "Address," ASCE, *Transactions*, Dec. 1908, 61:551-552; Clarke, "Hudson River Bridge," *Scientific American Supplement*, May 19, 1888, 25: 10311; *NY Herald*, Jan. 6, 1887; *Engineering News*, Oct. 22, 1887, 18:287; *Rhinebeck Gazette*, in *Poks. News-Telegraph*, Sept. 21, 1889
5. *Albany Argus*, May 20, 1889; promoters: PR&NE, *Summer Homes*, 1895, 32, 34; *Engineering News*, June 25, 1887, 17:405; Jan. 7, 1888, 19:11; Clarke, in Canadian Society of Civil Engineers, *Transactions*, 1889, 3:88; *Scientific American*, Feb. 4, 1888, 58:70

CH. 8. CONNECTING TO THE WORLD

1. *Engineering News*, July 28, 1888, 20:59; Dec. 8, 1888, 20:453
2. "Shrewd": *Poks. Sunday Courier*, July 10, 1887; *Poks. Daily Eagle*, ("hundreds") June 15, 19, 1889
3. "Thieves": C. L. Kimball to J. S. Schultze, Apr. 15, 1892, ND&C; decided: Hudson Connecting Railr'd, Minutes, Sept. 12, 1888, NYNH&H papers, v. 223, HARV; *New Paltz Times*, ("thousands") Nov. 7, Dec. 19, 1888; "gang": *Poks. Daily Eagle*, May 11, 1889

4. *Poks. Daily Eagle,* May 22, June 14, 17, 1889; *New Paltz Independent,* July 19, 1889
5. *Poks. Daily Eagle,* ("ready," "steadily,") June 19, Aug. 26, 27, 1889; "Route": *Poks. News-Telegraph,* Aug. 3, 1889
6. *Poks. Daily Eagle,* Jan. 8, 1890; *Poks. News-Telegraph,* Jan. 11, 1890. O'Rourke continued his connection with the bridge as a salaried consulting engineer for the PR&NE (O'Rourke, receipt from PR&NE, Dec. 31, 1897, Clifford B. Burr Collection, box 4, HARV).
7. "Open," "prosperous": *Poks. Daily Eagle,* in *New Paltz Times,* Nov. 26, 1890; "crooked": Clarke, "Building of a Railway," *Scribner's,* June, 1888, 3: 654; *New Paltz Independent,* ("mudhole") Aug. 30, 1889; Aug. 22, 1890
8. Tingley, "Bridging the Mighty Hudson," *Railroad Man's Magazine,* Aug. 1918, 36:727-730
9. Warwick, NY, *Valley Dispatch* ("L&H Centennial Supplement" issue), June 1, 1960
10. *NY Times,* Oct. 29, 1891; *Poks. News-Telegraph,* Jan. 30, Feb. 6, 1892
11. *Poks. Daily Eagle,* March 8, 1893; *NY Times,* Feb. 9, Sept. 26, 1893

CH. 9. IMMIGRANT LABORERS

1. For chapter: Federal Writers' Project, *Italians of New York,* 1938; Patricia S. Weibust, *Italians. . .in Connecticut,* 1976; Louis C. Zuccarello, "The Catholic Community in Poughkeepsie," Dutchess County Historical Society, *Year Book,* 1987, 112-115
2. *NY Times,* Oct. 19, 1886; *Poks. News-Telegraph,* May 14, 21, 1887
3. *New Paltz Independent,* Feb. 10, 1882; June 8, 1883; Roosevelts: Ward, *Before the Trumpet,* 1985, 138; Ward, *A First Class Temperament,* 1989, 448
4. *Poks. Courier,* July 3, Apr. 24, 1887; *Poks. News-Telegraph,* Aug. 18, July 14, 1888
5. *Poks. News-Telegraph,* July 27, 1889
6. *Poks. New Yorker,* Dec. 5, 1954; *New Paltz Independent,* Dec. 11, 1903; Anthony Canino, interv., Oct. 22, 1998
7. *Winsted Herald,* Sept. 23, 1892; Moffett, *Down to the River by Trolley,* 1993, 22, 64, 65, 73; G. Hunter Brown to Joseph Rico, Oct 10, 1898, ND&C; *Poks. New Yorker,* Dec. 5, 1954; *New Paltz Independent,* Oct. 21, 1904; Ward, *A First Class Temperament,* 116-117; Rinaldi et al., *History of the Italian Center. . .Poughkeepsie,* 1981, 22
8. Sam and Doris Christian, interv., 1997; Antonio Marano, interv., 1998

CH. 10. THROUGH PASSENGER TRAINS

1. For chapter: Boston & Maine Railroad Hist. Soc., *The Central Mass.,* 1975; Pennsylvania Railroad, *History: Passenger. . .Service,* 1974; Winfield W. Robinson, "Locomotives of the New Haven Railroad," Railway & Locom. Hist. Soc., *Bulletin,* Oct. 1939, 7-27. Summer travel guides: Philadelphia, Reading, & New England Railroad (PR&NE), *Summer Homes,* 1895-1898; Central New England Railway (CNE), *Summer Homes,* 1899-1901; CNE, *Poughkeepsie Bridge Route,* 1903 (a shorter version of the same guide under a different title) (quotations are from the 1899 guide except as noted)
2. "Coach": CNE&W's Gen. Supt. S. B. Opdyke, circular letter, May 10, 1890; "Faint hope": John S. Wilson to Jas. T. Furber, Apr. 12, 1890, both BEAUJ; Planned schedule for Harrisburg express: letters among Opdyke, Wilson, Arthur Brock, et al., Apr. 12-May 12, 1890, BEAUJ; *NY Times,* May 13, 1890; *Poks. Daily Eagle,* May 23, 1890
3. The Washington express was being planned according to the *Poks. Daily Eagle,* May 31, June 26, 1890, and was announced to begin running June 30, according to the *Poks. Daily Eagle,* June 27, 1890, and Boston & Maine, Time Table, June 30, 1890. The *Poks. Daily Eagle* inserted it into its daily CNE&W time tables on July 8, 1890.
4. "Parlor": CNE&W, Time Table, June 30, 1890; "polite": *Poks. Daily Eagle,* May 28, 1890; "magnificent": *Official Guide of the Railways,* Aug. 1890, 181; *NY Times,* Aug. 5, 1892; Howells, *The Sleeping-Car and Other Farces,* 1904
5. Day express revival, ca. June 26-Dec. 3, 1892: *Poks. Daily Eagle,* June 24, ("Quaker City Day Express") Nov. 28, Dec. 3, 5, 1892; Bob Adams to Bill Wintringham, March 23, 1976, BEAUJ; Inauguration: *Railroad Gazette,* Apr. 28, 1893, 25:317; Haight, *Pine Plains and the Railroads,* 1976, 46
6. Reading Railroad System, New England Div., Time Table, June 11, 1893; NY&NE, Time Table, Aug. 13, 1893; *Poks. News-Telegraph,* Apr. 22, 1893; *Poks. Daily Eagle,* Sept. 11, 1893
7. PR&NE, Time Table, Oct. 29, 1893 (in *Winsted Herald,* Nov. 3, and *Poks. Daily Eagle,* Nov. 6, 1893) shows the Washington Express still running, but Time Table, Nov. 19, 1893 (in *Poks. Daily Eagle,* Nov. 20, and *Winsted Herald,* Nov. 24, 1893), does not show it. Washington express revival: Bob Adams to Bill Wintringham, et al, March 23, 1976, BEAUJ
8. *NY Times,* July 10, 1892
9. "Undesirable": G. Hunter Brown to A. E. Dieterich, Brown to Henry F. Gillig, both May 1, 1902, ND&C
10. "Industry," "thundering": CNE, *Poughkeepsie Bridge Route,* 1903, 40, 42
11. *NY Times,* Jan. 17, 1915; March 11, 1917; CNE, Time Table, for employees, June 6, 1915; Pennsylvania Railroad, *History,* 15-16; locomotives: Kulp, *Railroads in the Lehigh River Valley,* 1962, 37-38; coaches: *Official Guide of the Railways,* June 1915, 150; *Engineering News,* March 15, 1917, 77:453

CH. 11. TROLLEY CARS ON THE BRIDGE

1. For chapter: Robert B. Adams, "Poughkeepsie's Unusual 'Rapid Transit'," National Model Railroad Association, *Bulletin,* Apr. 1975, 8-15; Glendon L. Moffett, *Down to the River by Trolley,* 1993; Charles Benjamin, intervs.
2. "Suburban": CNE, Time Table, July 3, 1899; "Rapid Transit": CNE, Time Table, June 3, 1900
3. "Luxurious," "greatest": PR&NE, *Summer Homes,* 1898, 148, 150; "slowly": New Paltz & Wallkill Valley RR, Time Table, [1898?], 4; *Poks. Daily Eagle,* Aug. 13, 1897
4. W. J. Martin to John W. Brock, Aug. 31, 1897; Oct. 28, 1898; July 7, 1899, BENJ; Moffett, *Down to the River,* 38; *New Paltz Independent,* Aug. 22, 1930

CH. 12. LOCAL PASSENGER TRAINS

1. For chapter: Deyo Diaries, 1905-1917, DEYO; Robert B. Adams, "Decline and Fall of the CNE," Lakeville, CT, *Journal,* Oct. 16, Nov. 6, 26, 1980; Robert Ashman et al., *Central New England Railroad,* 1972; Wm. P. McDermott, *Dutchess County Railroads,* 1996
2. Number of stations: PR&NE, Time Table, Nov. 17, 1898; CNE, Time Table, Sept. 24, 1916; shared: CNE, "Annual Report," 1917, NYSA
3. "Unattended": G. Hunter Brown to Chas. I. Swift; Brown to Oakleigh Thorne, both March 31, 1903, ND&C; "shabbiest": clipping, *Winsted Citizen?* marked in ink May 6, 1892, McMA; inspector: NY State, Public Service Com'n, 2nd Dist., *An'l Rep.,* 1910, 2:325
4. NY State, Public Service Com'n, 2nd Dist., *An'l. Rep.,* 1911, 2:95;

Norfolk: clipping, marked Sept. 6, 1912, McMA; Ulster County: Anson, *Friends and Neighbors*, 1989, 29; "mechanics": *New Paltz Independent*, March 3, 31, 1911

5. Courteous: Stickles, "History of the Central New England Railway," 1970s?, 94a, STICK; "complaints": W. J. Martin to John W. Brock, Oct 8, 1897, BEAUJ; "new": CNE, *Summer Homes,*" 1899, 6
6. Late: Dutchess County Hist. Soc., *Yearbook*, 1969, 62; Botkin, *Treasury of Railroad Folklore*, 1953, 421, 506; NY State, Public Service Com'n, 2nd Dist., *An'l Reps.*, 1910-1912. Traffic: NY State, Bd. of Railr'd Com'rs, *An'l Rep.*, 1890, 2:177; 1900, 2:151; NY State, Public Service Com'n, 2nd Dist., *An'l Rep.*, 1917, 2:97; Nimke, *CNE*, 2:146.
7. *Poks. Evening Star*, Feb. 28, 1922; Kulp, *Railroads in the Lehigh River Valley*, 1962, 38; *Winsted Evening Citizen*, Jan. 17, 1927
8. NYNH&H, *Official List*, Nov. 1, 1926; *New Paltz Independent*, Nov. 19, 1925; *Poks. News-Eagle*, Nov. 2, 1926; NYNH&H and CNE, Time Table, June 13, 1926
9. Youth: Austin McEntee, interv., 1998; *Poks. Eagle-News*, Dec. 6, 1930; discontinued again: *Official Guide of the Railways*, Jan. 1931; July 1933, July 1934; Nimke, *CNE*, 2:125

CH. 13. THE ROOSEVELTS AND THE BRIDGE

1. For chapter: Jim Bishop, *FDR's Last Year*, 1974; Wm. D. Hassett, *Off the Record with FDR*, 1958; Henry La Cossitt, "He Takes the President on Tour," *Sat. Eve. Post*, June 16, 1951, 19ff.; Clara and Hardy Steeholm, *House at Hyde Park*, 1950; Geoffrey C. Ward, *Before the Trumpet: Young Franklin Roosevelt*, 1985; Ward, *A First Class Temperament, The Emergence of Franklin Roosevelt*, 1989; Bob Withers, *The President Travels by Train*, 1996
2. Subscribed: Steeholm, 52
3. On Jas. Roosevelt and the commission: *Poks. News-Telegraph*, March 12, 26, 1887; *Poks. Daily Eagle*, Feb. 22, 25, March 19, 1887. Activities of the commission: PEVS.
4. Union Bridge Co., "Poughkeepsie Bridge. . .Last Pin Driven," Aug. 30, 1888, flyer, marked "From Rosedale," FDRL; boasting: Grace Tully to Wm. F. Gaynor, Oct. 23, 1942, notes only, Railroad folder, President's Personal File, FDRL; "patroness": Poughkeepsie *News-Telegraph*, July 6, 1901
5. Poughkeepsie Bridge Co., *Bridging the Hudson*, 1875 (inscribed by FDR), FDRL
6. *Poks. Eagle-News*, Aug. 26, 1930
7. *Newburgh News*, June 27-28, 1938
8. Raymond Hegeman, interv., 1998; similarly, *New Paltz News*, Jan. 21, 1998
9. Bishop, 101-102, 135; Log of the President's Trip, July-August, 1944, FDRL; Hassett, Diary, 113, LC; La Cossitt, 151; Daniels, *Washington Quadrille*, 1968, 288-290
10. La Cossitt, 151
11. Hassett, Diary, 113, 115, LC; Bishop, 102; Withers, 176. Before giving his diary to LC, Hassett, evidently trying to protect the president's privacy, cut from it two pages, 111-112, about this trip. They are also missing in the copy at FDRL.

CH. 14. WEST POINT FOOTBALL SPECIALS

1. For chapter: US Corps of Cadets, athletic records, memos, orders, USMA; Tom Curtin, "Yale 14—Army 12, A Tradition Ends," NHRH&TA, *Shoreliner*, issue 3, 1991, 22:36-39; Peter McLachlan to CM, Dec. 18, 1997
2. US Corps of Cadets, memo, Oct. 29, 1923, USMA
3. *NY Times*, Oct. 29, 1922
4. *West Point Pointer*, Nov. 14, 1925, 8; Oct. 29, 1926, 20, 23; by bus: Judith Sibley, Archives, USMA, to CM, Apr. 29, 1998
5. "Interfere": G. W. Curtiss to Supt. R. M. Smith, et al., Oct. 7, 1932, box 2, file 6, PBEN; US Corps of Cadets, memo, Oct. 23, 1935, USMA
6. US Corps of Cadets, Order, Oct. 29, 1954, USMA

CH. 15. MILK TRAINS

1. For chapter: John J. Dillon, *Seven Decades of Milk: A History of New York's Dairy Industry*, 1941; Robert E. Mohowski, *The New York, Ontario & Western Railway and the Dairy Industry*, 1995
2. Ashman, *CNE*, 1972, 6
3. 1895 milk: W. J. Martin to John W. Brock, June 6, 1895, BEAUJ; ICC rates: Dillon, 51-53
4. Nathan Blodgett to Ed Colgan, Apr. 13, 1972; March 30, 1976, BLODG
5. Nimke, *CNE*, 2:66, 130-141; Blodgett to Ed Colgan, Nov. 11, 1975, BLODG; "work": Wm. Underhill to C. W. Fairchild, Sept. 1, 1904, ND&C
6. Millbrook milk: Underhill to D. H. McCoy, Aug. 4, 1904, ND&C; "Milk train": *Poks. Daily Eagle*, March 4, 1891; "largely":W. J. Martin to John W. Brock, Oct 11, 1901, BEAUJ
7. Dillon, 39, 327; Nimke, 2:131-141, 146, 178; 3: 154, 158; Mohowski, 31-33; Sam Christian, intervs., 1996-1997

CH. 16. BATTLE OVER THE SPRINGFIELD BRANCH

1. General for chapter: *The Press Against the Central New England Parallel* [May, 1901]; Robert B. Adams, "Battle for Springfield," NMRA *Bulletin*, Nov. 1978, 9-17; R. W. Nimke, *Central New England Railway*, 1995, 1: ch. 8 ; Gregg M. Turner et al., *Connecticut Railroads*, 1989, 148-150, 153
2. *NY Herald*, Apr. 7, 1901; Poughkeepsie Bridge Co., *Bridging the Hudson*, 1871, [4]
3. *NY Times*, Apr. 14, 1896; "benefit": PR&NE, Minutes, May 10, 1898, NYNH&H papers, vol. 228, HARV
4. *Waterbury American*, March 23, 1901; *Hartford Telegram*, Feb. 14, 1901
5. CNE, Minutes, Dec. 31, 1902, Feb. 25, 1903, Apr. 9, 1904, UCONN; *NY Herald*, Apr. 7, 1901
6. CNE, Time Tables, for employees, Oct. 7, 1906; June 25, 1917. *Official Guide of the Railways*, Dec. 1921, Jan. 1922

CH. 17. MAYBROOK GATEWAY

1. For chapter: Connecticut Valley Chapter, Natl. Railway Hist. Soc., *Maybrook Excursion*, May 4, 1947; Leslie Tyler, "Maybrook Line," *Transportation*, May, 1951, 5:1-10; NHRH&TA, *Shoreliner*, Spring, 1976, 7:3-21; Philip Simms, Albert Alexander, Samuel Christian, Antonio Marano, Joseph Spolverino, intervs.
2. Largest number: CT Valley Chapter, *Maybrook Excursion*, 1947, 6; double tracking: CNE, Minutes, 1909-1914, UCONN
3. Expediter: Nathan Blodgett to Ed Colgan, Apr. 16, 1973, BLODG; animals: Christian, intervs., 1996-1997
4. Immigrant: Marano, interv., 1998; Harris: *Newburgh News*, June 28, July 1, 1947; sitting: Christian, interv., 1999; Favaro: Frank Doolittle Jr., to CM, Jan. 31, 2000; Hess: *New Paltz Independent*, Aug. 19, 1948; "eat": Blodgett to Colgan, Jan. 20, 1977, BLODG
5. Albert Alexander, "Some Fascinating. . .Experiences," 1995, 3, MRM; Alexander, interv., 1996
6. *Newburgh Eve. News*, March 2, 1946
7. O&W-CNE Agreement, Sept. 9, 1921, NYNH&H Secretary's Files, UCONN; Blodgett to Colgan, Sept. 24, 1972, Oct. 5, Nov. 6, 1976, BLODG

8. Scoot: First Presbyterian Church, Highland, NY, *Time for Reflection*, 1992, 21; Blodgett to Colgan, Sept. 9, 1976, BLODG; "gym": Albert Alexander, talk, for O&W Railway Historical Soc., Middletown, NY, Nov. 1, 1997; "deep": *Middletown Record*, Sept. 1, 1970
9. "Virtues": NYNH&H's YMCA, New Haven, CT, *Railroad News*, March 1905, 111; Doolittle to CM, Jan. 31, 2000; *Walden Citizen Herald*, Dec. 11, 1980
10. Blodgett to Colgan, Sept. 14, 1977, BLODG; Christian, intervs., 1996-1998; Barbara A. Graf, interv., 1997; "saloon": *Walden Stewart Citizen*, Sept. 1, 1976; "grime": *Middletown Record*, Jan. 11, 1976; J. W. Swanberg, "Confessions of a Diesel Fireman," *Locom. & Rwy. Preservation*, Sept.-Oct. 1990, 55
11. Don Wallworth "A Five-Railroad Paradise," *Trains*, July 1996, 56-57 (quoted by permission of Kalmbach Publishing Co.)

CH. 18. FREIGHT TRAINS

1. For chapter: NY State, Bd. of Rail'd Com'rs, *An'l Reps.*, 1890-1906; NY State, Pub. Service Com'n, 2nd. Dist., *An'l Reps.*, 1907-1927; CT State, Rail'd Com'rs, *An'l Reps.*, 1888-1910; Central New England Rwy., An'l "Reports," 1910-1926 (manuscript), NYSA
2. 1905: Nimke, *CNE*, 2:10; 1921: *Poks. Eagle-News*, Jan. 27, 1921; 1943: CT Valley Chapter, *Maybrook Excursion*, 1947, 6; "heavy": *Engineering News*, March 8, 1917, 77:412; 1952: Abbey, "I Flew Across the Hudson on a Freight Train!" *Trains & Travel*, May, 1953, 26
3. W. J. Martin to Wm. Underhill, Aug. 17, 1898, Burr Collection, Box 4, HARV; Underhill to Howard Haight & Co., July 25, 1904, ND&C
4. Nathan Blodgett to Ed Colgan, Apr. 21, 1976, June 8, 1978, BLODG; clipping, *Winsted Citizen*, Nov. 28, 1900, McMA
5. *New Paltz Independent*, Aug. 14, 1896; June 24, 1892
6. Nimke, *CNE*, 2:144-145; Hudson Valley Bridge Association, *Report*, 1922, 13; Hurd, *Village of Clintondale*, 1959, 39
7. Deyo Diaries, DEYO; *New Paltz Independent*, Jan. 15, Oct. 8, 1925; clipping, marked May 20, 1937, PLATT. More on the Deyos' use of trains: Mabee, *Listen to the Whistle*, 1995
8. Nimke, *CNE*, 2:49-50; Blodgett to Colgan, Apr. 21, 1976, BLODG
9. W. J. Martin to John W. Brock, Feb. 4, 1902, Beaujon papers, UCONN; Blodgett to Colgan, Oct. 20, 1974, Apr. 21, ("throw") Nov. 6, 1976, March 23, 1977, BLODG
10. *Poks. Eagle*, 50th Anniversary Edition, 1911, no month or day, MRM; Arthur McComb, interv., 1996; to CM, July 1, 2000
11. Sam Christian, Austin McEntee, intervs., 1996-1999; Wm. D. Knauss, former Knauss Brothers Purchasing Agent, to CM, July 15, 24, 2000
12. *Poks. Eagle-News*, Jan. 27, 1921; Corwine, *History of the Poughkeepsie Bridge*, 1925, 2-3; *Poks. New Yorker*, July 22, 1951
13. NY State, Bd. of Rail'd Com'rs, *An'l Rep.*, 1891, 2:174; CNE, "Report," 1918, 508; 1926, 510-511; John Mylod, interv., 1997; to CM, March 12, 2001; Farrington, *Railroads at War*, 1944, 98-104
14. Nimke, *CNE*, 2:6, 93-94; Blodgett to Colgan, Feb. 19, 1972, June 8, 1978, BLODG; F. P. Sweeney, consist for train XC98, with engine 3368, Aug. 12, 1931, SWEE
15. Bixby, "Experiences on the L-1," NHRH&TA, *Shoreliner*, Spring, 1976, 7:22-23; John P. Sweeney to CM, Feb. 23, May 22, 1996

CH. 19. CENTRAL STATES DISPATCH

1. From Wallace W. Abbey, "Central States Dispatch," and "I Flew Across the Hudson on a Freight Train!" *Trains & Travel*, March 1952, 12:26-32; May 1953, 13:26-27 (quoted by permission of Kalmbach Publishing Co.). The two articles have here been combined and condensed.
2. W. J. Martin to John W. Brock, June 6, 1895, BEAUJ; W. Gifford Moore to CM, Nov. 3, Dec. 22, 1997

CH. 20. CIRCUS TRAINS

1. For chapter: Wm. Schnitzer, intervs., 1998; to CM, Apr. 7, 1998; clippings, MRM; Tom Parkinson et al., *The Circus Moves by Rail*, 1978; Tom Ogden, *Two Hundred Years of the American Circus*, 1993
2. Nathan Blodgett to Ed Colgan, Nov. 27, 1976, BLODG; *Poks. Daily Eagle*, *Poks. News-Press*, June 28, 1890. When the Poughkeepsie Bridge had scarcely opened to traffic in early 1889, it was announced that a Barnum circus train would cross the bridge to perform a circus in Poughkeepsie. It has often been claimed, as in Foster et al., *Splendor Sailed the Sound*, 1989, 210, that this train did cross the bridge, but in fact the Bridge Route not being ready to handle it, it did not (*Poks. Daily Eagle*, May 13-28, 1889).
3. *Poks. Daily Eagle*, July 1-3, 7, 1903
4. 1907 offer: Lord, *Country Depots in the Connecticut Hills*, 1996, 50; 1908 offer: CNE, "Circus Outfit" Tariff, June 24, 1908, SHUKR
5. Thomas A. Flynn, interv., 1998; Austin McEntee, interv., 1997
6. John Mylod, interv., 1997; to CM, March 12, 2000
7. Haight, *Pine Plains and the Railroads*, 1976, 46; C. L. Kimball to P. R. Seeley, Oct. 22, 1892, ND&C
8. Circus trains in Maybrook: Albert Alexander, Sam Christian, Robert J. Sandbothe, intervs., 1996-2000; B. J. Lawlor, telegram to various NYNH&H officials, May 22, 1968, MRM; *Walden Citizen*, Aug. 8, 1979

CH. 21. JUMPING OFF THE BRIDGE

1. "Oncet": Haviland, "Did Steve Brodie Really Jump off the Poughkeepsie Railroad Bridge?" *Hudson Valley*, Aug. 1973, 2:35
2. *National Police Gazette*, Oct. 13, 1888, 11; ("champion") Nov. 17, 1888, 7, 13; ("glare") Nov. 24, 1888, 14; "wildly": *Poks. Daily Eagle*, Nov. 10, 1888; "stuffed": William Doyle, letter, in *Poks. New Yorker*, Feb. 28, 1956
3. *NY Times*, Oct. 14, 1894; *Poks. Daily Eagle*, Oct. 15, 1894
4. "Rough," "here goes," "collapse," "alive," "pale": *NY Times*, Oct 28, 29, 1895; "painters," "passes," *Poks. Daily Eagle*, Oct. 28, 29, 1895

CH. 22. WALKERS ON THE BRIDGE

1. For chapter: Poughkeepsie Bridge Co., *Charter. . .and Amendments*, 1887
2. W. Van Benthuysen to John I. Platt, March 23, 1887, Poughkeepsie Bridge Co., Letters, v. 12, UCONN; *Poks. Daily Eagle*, March 27, 1889
3. Poks. Bridge Co., Minutes, March 5, 1890, UCONN; *New Paltz Times*, Sept. 17, 1890; Milton A. Fowler to C. E. Morgan et al., Feb. 9, 1891, BEAUJ; NY State, *Laws*, chap. 198, Apr. 16, 1891; *Poks. Daily Eagle*, Apr. 17-23, 1891
4. *Poks. Daily Eagle*, May 15, June 10, 1891; Nov. 29, 1892; *Poks. News-Telegraph*, Oct. 17, 1891
5. NY State, *Laws*, chap. 375, Apr. 13, 1893; *Poks. Daily Eagle*, Apr. 15, 1893; *Poks. News-Telegraph*, Apr. 22, 1893; *Kingston Daily Leader*, Sept. 7, 1893
6. *Poks. Daily Eagle*, Aug. 13, 1897
7. NY State, *Department Reports*, 1921, 24:240-250

CH. 23. A BRIDGE FOR AUTOMOBILES?

1. For chapter: Hudson Valley Bridge Association, *Report*, 1922; correspondence on the relation of the Poughkeepsie Bridge to the proposed new highway bridge, 1919-24, box 1, file 2; box 3, file 5, PBEN
2. *Poks. Courier*, Feb. 9, 1919; *Poks. Eagle-News*, Dec. 6-9, 1920; F. J. Pitcher to Edward Gagel, Jan. 24, 1921, box 2, file 4, PBEN
3. Lindenthal: *Poks. Eagle-News*, Jan. 27, 1921; Roosevelt: *Poks. Evening Star and Enterprise*, Dec. 14, 1922
4. *Poks. Eagle-News*, Oct. 9, 1925; Aug. 25, 26, 1930

CH. 24. HOBOES

1. *Poks. News-Telegraph*, Aug. 10, 1889
2. *NY Sun*, in *Poks. Daily Eagle*, March 4, 1891. According to CNE&W, Time Table, Sept 8, 1889, the Erie, CNE&W, and NY&NE together ran a daily through coach from Goshen, NY, via Campbell Hall, Poughkeepsie, Canaan, CT, and Hartford to Boston.
3. Nathan Blodgett to Ed Colgan, Nov. 26, 1977, BLODG
4. Arthur Lyons, Dorothy Phillips Gruner, intervs., 1996-1997
5. Antonio Marano, Ferris Davis, Albert Alexander, intervs., 1996-1998; Alexander, "Some Fascinating. . .Experiences," 1995, 8-9, MRM

CH. 25. POUGHKEEPSIE REGATTA

1. For chapter: Intercollegiate Rowing Assoc., *Programs*, 1910-1947; Thomas Phillips (West Shore Railr'd foreman), audiotaped interv., ca. 1982, GRUNR; Dorothy P. Gruner (Phillips' daughter), John K. Jacobs and La Verne Davis (spectators), Clark Bonesteel (West Shore Railr'd fireman), intervs., 1995-2000. The regatta was held in Poughkeepsie 1895-1949, with exceptions.
2. *Harper's Weekly*, June 22, 1895, 596; Lindenthal: *Poks. Eagle-News*, Jan. 27, 1921
3. Platt, Innis, O"Rourke: *Poks. Daily Eagle*, June 25, 1897; Roosevelt: *Poks. News-Telegraph*, July 6, 1901
4. Trains 1897: *Poks. Daily Eagle*, June 26, 1897; bets: Roger W. Mabie, interv., 1998
5. Astor, Vanderbilt, Morgan: *Poks. Daily Eagle*, June 26, 1896; June 25, 1897; Roosevelt: *Newburgh News*, June 27, 1938
6. *Poks. Daily Eagle*, June 22, 1895; ("steamed") June 26, (yell) June 27, 1896; fan: *Poks. News-Telegraph*, July 3, 1897
7. "Puff of smoke": Intercollegiate Rowing Assoc., *Program*, June 21, 1947, 9
8. Arthur Lyons, intervs., 1996-2000. By 1947, winners were announced by flags being lowered from both bridges, and by loud-speakers located on the Mid-Hudson Bridge, but bombs were no longer being dropped (*Poks. New Yorker*, June 21, 1947).
9. Peter H. Troy, chair, Intercollegiate Regatta Com., appeal, May 21, 1935, LLOYD; Wm. D. Knauss to CM, July 24, 2000

CH. 26. CONDUCTOR BLODGETT, STORY TELLER

1. "Inches": Nathan Blodgett, CNE recollections, audiotape, Aug. 16, 1970, BEAUJ; "stick": Blodgett to Ed Colgan, Nov. 6, 1976, BLODG
2. Blodgett to Colgan, Dec. 12, 1976, March 23, Oct. 6, 1977, BLODG
3. Blodgett to Colgan, Apr. 16, 1973 (including attachment), Nov. 11, 1975, Apr. 21, 1976, BLODG
4. Blodgett to Colgan, Sept. 11, 1973, Jan. 20, 1977, BLODG
5. Blodgett to Colgan, Sept. 5, 1975, BLODG. According to Leroy Beaujon, the "dinkey" engine had been built by Baldwin in 1881 for use on New York City's elevated railway system and had been sold in 1906 to the American Bridge Company, the contractor for the bridge strengthening.

CH. 27. STUDYING FOR EXAMS

1. For chapter: NYNH&H and CNE, "Examination for Operators, Signalmen, Levermen and Station Agents," printed, 59 pp., filled out by Winne Veeder Stover, no date [1914-1927], McEN; NYNH&H, "Examination for Conductors and Enginemen," printed, 38 pp., filled out by Albert J. Alexander, March 25, 1946, MRM
2. Nathan Blodgett to Ed Colgan, Aug. 28, 1974, Jan. 25, 1975, Apr. 21, 1976, BLODG
3. Alexander, intervs., 1997-1998

CH. 28. "THE CNE BOOMER": A FOLKSY TALE

1. "The CNE Boomer," original source and date unknown, marked "published before 1930," BEAUJ. Interpreted with help especially from Leroy Beaujon. He reports that locomotive 3213, mentioned in the story, was delivered to the CNE in 1918.

CH. 29. WATER BOY DI ROSA

1. For chapter: Tony Di Rosa, *El Pibe*, Buenos Aires, 1990; Di Rosa, interv., 1996. Di Rosa wrote CM, Nov. 12, 1996, giving permission to quote from his book. He died in 1997 in Florida, leaving a son in Hyde Park, NY.
2. According to a 1927 Poughkeepsie directory, Bartolo Barone not only lived and ran a grocery and delicatessen at 16 Cataract Place, but also next door ran a business in real estate and "private" banking. They were across the street from Mt. Carmel Church (what was called Cataract Square later came to be called Mt. Carmel Square after the church)
3. According to Holton, *Winslow Memorial*, 1888, 2:822-823, when the Poughkeepsie Bridge Co.'s President Winslow retired in the Poughkeepsie area, he settled on the Hyde Park road about one and a half miles north of Poughkeepsie on a beautiful spot overlooking the Hudson called "Woodcliff." It later became part of Woodcliff Pleasure Park.
4. Like other: Baily, *Immigrants in the Lands of Promise*, 1999, 208-209
5. A Poughkeepsie directory, 1939, lists "DeRosa [sic.], Anthony R." and wife Marie as having a house and coal dealership at 62-64 Gifford Ave.

CH. 30. BILL FELL, PAINTER ON THE BRIDGE

1. For chapter: William and Anita Fell, Ferris Davis, Tom Houston Jr., Paul Lown, intervs.
2. Methods used to paint the bridge varied. In 1919, workmen cleaned steel with a "wire brush," and applied the paint heated, which proved unsatisfactory (Nov. 13, 1920, box 3, file 7). By 1929 (June 12, box 2, file 5), where air pressure was available, men were using "air tools" for both cleaning and painting. In March, 1968, New Haven's specs for painting the bridge (box 2, file 7, PBEN) provided for scraping with metal brushes, scrapers, chisels, hammers, "or other effective means," and applying paint "by hand brushes" unless an engineer approved of "spraying."
3. The *Poks. New Yorker*, May 29, 1951, and A. R. Kottage, memo, Sept. 18, 1957 (box 1, file 5, PBEN) reported that while the bridge had traditionally been painted black, in this period it was being

painted with aluminum paint, making it silver-white. But the aluminum paint did not adhere well to the steel. In March 1968, New Haven specs called for the bridge to be painted black.

4. W. J. Backes to W. H. Moore, Feb. 13, 1917, file 7; F. J. Pitcher to E. E. Oviatt, Dec. 21, 1933, file 10 ; C. F. Comstock to Modjeski & Masters, Nov. 9, 1964, file 15, box 3, PBEN

CH. 31. LANTERN SIGNALING: THE CANSAS FAMILY

1. For chapter: Lola Cansas Casciaro, interv., 1996 (died 1997); Dorothy P. Gruner (friend of Lola) to CM, Oct. 9, 20, 21, 1997; Albert Alexander (co-worker of Cansas), interv., 1997. Rocky Cansas lived 1905-1963.

CH. 32. TOWERMAN BEAUJON

1. For chapter: Leroy Y. Beaujon, intervs. and letters to author, 1997-1999. Ski trains: clippings, *Winsted Citizen*, 1935-1940, McMA. While Beaujon now lives in Roseville, CA., he still has family in Canaan, CT.
2. Fan trip: NYNH&H, *Along the Line*, July 1947

CH. 33. MARY CARMODY, CREW CALLER

1. For chapter: Mary Carmody, intervs., 1997; Sam Christian, intervs., 1997-1998; clippings, MRM; Leuthner, *The Railroaders*, 1983, 76-79

CH. 34. DISDAIN FOR A PUSHER

1. For chapter: Pete McLachlan, "Diesel Pushers on Stormville Mountain," NHRH&TA, *Shoreliner*, 1995, Issue 3, 26:32-37

CH. 35. TOWERWOMAN COOPER

1. For chapter: women telegraphers, Tom Standage, *Victorian Internet*, 1998, 133-140; *Poks. New Yorker*, Aug. 19, 1956, quoted by permission of its successor, *Poks. Journal*
2. Mabee, *Listen to the Whistle*, 1995, 45; C. L. Kimball to Laura McCurdy, July 5, 7, 1892; Kimball to Minnie Burt, Apr. 15, 1893, ND&C; *Poks. News Telegraph*, Apr. 1, 1893
3. Nathan Blodgett to Ed Colgan, Oct. 20, 1974, BLODG; Nimke, *CNE*, 3:111; NYNH&H, *Official List of Officers, Station Agents*, Nov. 1, 1926, 65; May 1, 1941, 15; Abendshein, "J Tower & Peers," *Locom. & Rwy. Preservation*, Sept.-Oct. 1990, 26

CH 36. DRIVING DIESELS OVER THE BRIDGE

(No notes)

CH. 37. CONDUCTOR ALEXANDER, PROTECTOR

1. For chapter: Albert J. Alexander, intervs., 1996-2000; Alexander, "Some Fascinating, Hair Raising Experiences," 1995, MRM. Having retired from railroading in 1978, Alexander has recently been active in the Maybrook Railroad Museum.
2. In the 1940s and 1950s, journal boxes were packed with waste (rags) which were meant to be oiled frequently. Today journal boxes are more likely to be packed with roller bearings and need to be oiled only a couple of times a year. Also today along some tracks, detectors that are sensitive to heat are installed; they report hot boxes on passing trains.

CH. 38. TELEGRAPHY AWRY: A HEAD-ON COLLISION

1. On telegraphy: Carleton Mabee, *American Leonardo*, 2000; Tom Standage, *Victorian Internet*, 1998. On the collision: *Hartford Times*, Dec. 22, 1916; *Poks. Eagle-News*, Dec. 23, 1916; *Hartford Daily Courant*, Dec. 23, 24, 1916; ; Interstate Commerce Com'n, report on CNE collision of Dec. 22, 1916, NA
2. Poks. Bridge Co., Minutes, Apr. 23, 1889, UCONN
3. Highland: Wadlin, *Times and Tales of Town of Lloyd*, 1974, 190; two companies: CNE, Minutes, May 31, June 27, 1900, May 29, 1901, UCONN. By 1920, the CNE was using telephones to supplement the telegraph in dispatching trains. Telephones were available not only in stations but also in boxes along the tracks and on the bridge. The boxes were supplied with train order forms for conductors to fill out when they received train orders by phone, to read back to the operator to check for accuracy, to sign, and to deliver, all in accordance with the rules as if the orders had been received by telegraph (NYNH&H and CNE, Time Table, for employees, June 6, 1920, 140-141). In 1956, while the use of the telegraph was no longer likely, the NYNH&H, *Rules for the Government of the Operating Dept.*, Oct. 28, 1956, 203, still referred to operators and dispatchers as using either telegraph or telephone.
4. CT State, Railr'd Com'ners, An'l Rep., 1892, 120; NY State, Pub. Service Com'n, 2nd. Dist., An'l Rep., 1914, 2:115

CH. 39. CARRIE'S COW

1. The story, originally published by "Jack Jacobs" in *The Antiochian*, Dec. 1939, is republished here, slightly edited, by permission. For much of his life, Jacobs has lived adventurously at a distance, but he now lives again on his home farm.
2. While the tracks in question in the 1930s, when Jacobs wrote the story, were double and were essentially down grade all the way from Modena to the bridge, Jacobs has chosen, for his fictional purposes, to make the tracks single, as they were earlier, and to slope upward toward the bridge, as may once have been true in some locations.

CH. 40. SPECTACULAR SPILL

1. For chapter: *NY Herald Tribune*, Oct. 31, 1943; *Poks. New Yorker*, Oct. 31, Nov. 2, 1943; "spectacular": Highland *Mid-Hudson Post*, Nov. 4, 1943; "fun": *Highland News*, Nov. 4, 1943
2. Breuer, *Hitler's Undercover War*, 1989, 274-283

CH. 41. FOLLY ON THE BRIDGE

1. For chapter: Albert Alexander, intervs., 1996-1997
2. This curve was just off the New Paltz-Highland highway (Rt. 299), near the Central Hudson Gas and Electric Corp.'s storage yard.

CH. 42. BRIDGE SAFETY AND MAINTENANCE

1. For chapter: intervs. with bridge maintenance workers, Wm. Fell, Ferris Davis, John M. Reed, Tom Houston Jr.; section man Paul Lown who occasionally did bridge maintenance; train crewmen Sam Christian and Albert Alexander
2. "Except": CNE, Time Table, for employees, Sept. 24, 1916, 15; 1913: G. W. Clark to "All Concerned," Sept. 13, 1913, box 3, file 8, PBEN. When the bridge first came into use, speedometers not being available on locomotives, bridge speed limits were often expressed in minutes required to cross the bridge. Speedometers

were still not available on the New Haven's 1400 engines, when they came on the tracks in 1956, but by 1961 they were.
3. "Tail ends," "great": F. C. Keim et al. to Modjeski & Masters, Oct. 21, 1938, box 2, file 11; "light construction": F. J. Pitcher to E. E. Oviatt, May 15, 1934, box 3, file 8, PBEN
4. Fell, intervs., 1998; Pavlucik, *New Haven Railroad*, 1978, 58
5. "Few words": C. A. Fowler to E. G. Buckland, Nov. 18, 1916, box 2, file 2; lights: A. E. Cawood, memo, July 12, 1957, box 3, file 13, PBEN
6. "Rapid": W. H. Moore to Edward Gagel, June 1, 1918, box 3, file 7, PBEN; Davis, interv., 1999
7. C. M. Shearer to J. A. Droege, Aug. 5, 1930, box 2, file 6; P. B. Spencer to E. E. Oviatt, Feb. 4, 1932, box 2, file 6, PBEN; NYNH&H, trustees' mtg., Feb. 3, 1941, NYNH&H's Secretary's files, UCONN
8. Accident: clippings, letters, Sept.-Oct. 1950, box 3, file 13, PBEN. Davis believes that later in the 1950s, as river traffic declined, the New Haven stopped keeping the tug available.
9. *Poks. Daily Eagle*, Dec. 14, 1907; W. H. Moore to Edward Gagel, Jan. 14, 1909, box 2, file 2, PBEN
10. Modjeski used the masculine form of his mother's name. As a young man, he served as his mother's stage manager in California, and studied piano; he played the piano daily all his life. His firm of consulting engineers, founded 1893, went by slightly different names at different times, and kept its offices at such different sites as Harrisburg, Phila., and NY.
11. E. J. Pearson to A. E. Clark, Sept. 19, 1927, NYNH&H's Secretary's files, UCONN. In 1959, the gauntleting was eliminated, a single track being installed instead.
12. "Hand rope": M. A. Hayden, report, Feb. 22, 1947, box 1, file 6; Ralph Modjeski to E. J. Pearson, Feb. 1, 1924, box 1, file 2, PBEN
13. "First class": Modjeski to Edward Gagel, May 21, 1928, and B. J. Dietrich to Modjeski, June 12, 1929, box 2, file 5, PBEN; "renewing," "certain": NYNH&H, Bd. Mtg., Sept. 11, 1935, NYNH&H's Secretary's files, UCONN; "frozen": Modjeski, Inspection, 1941, box 4, file 3; "roller nests," "lubricated": Modjeski, Inspection, 1943, box 4, file 4, PBEN
14. F. C. Keim et al. to Modjeski & Masters, Sept. 24, 1946, box 2, file 15, PBEN
15. Christian, Houston, Davis, intervs., 1997-2000. Pin replacements were performed with the use of counterweights in 1957 and 1959 (box 3, files 13, 16, PBEN).
16. Davis, Houston, intervs., 1999-2000; underwater inspection reports, 1950, 1969, 1972, box 4, file 7, PBEN
17. Boy: Thomas A. Flynn, interv., 1998; Poughkeepsie-Highland Railroad Bridge Co., *Year of the Railroad Bridge*, 1994, 32. Goring: *Poks. Eagle-News*, July 9-11, 1919; J. L. Rippey to R. L. Pearson, July 8, 1919, box 3, file 7, PBEN; Oliver (the original Oliver's son), Ann, and Richard Goring (the original Oliver's grandson), intervs., 1998. Doolittle: Frank Doolittle Jr. to CM, Jan. 31, 2000
18. Davis, Reed, intervs., 1998-2000; state recommendations, 1951, box 3, file 13, PBEN
19. Boy: Flynn, interv., 1998; Henry Stanton, recalling his father was stationed on such a vessel, talk, for the Walkway, Poughkeepsie, Sept. 23, 1998; fence: S. R. Rusiackas et al. to Modjeski & Masters, Oct. 15, 1942, box 2, file 13, PBEN; dogs: NYNH&H, *Rider's Digest*, Jan. 1943; "thrilling": Hedin, *Great Machine*, 1996, 133; Fell, intervs., 1998
20. Supt., NYNH&H, "Poughkeepsie Bridge," May 17, 1954, WALK; Houston, intervs., 1998-2000
21. Albert Alexander, "Some Fascinating," 1995, 11-12, MRM; Alexander, interv., 1997

CH. 43. DECLINE

1. For chapter: Willard B. Rogers, W. George Cook, Albert Alexander, intervs.
2. J. W. Swanberg, "Confessions of a Diesel Fireman," *Locom. & Rwy. Preservation*, Sept.-Oct. 1990, 50-57; carped: Albert Alexander, talk, for the O&W Rwy. Historical Soc., Middletown, NY, Nov. 1, 1997
3. Grant, *Erie Lackawanna*, 1994, 161-162; Penn Central, Poughkeepsie region, Territorial Assignments, Deliveries, both 1971, COOK
4. Peter McLachlan to CM, Dec. 5, 1997. The NYNH&H issued rules for two-way radios in its *Rules for the Governance of the Operating Dept.*, Oct. 28, 1956, 195-198, but it was slow to install them on its trains. In 1957, the L&HR installed such radios, making communication possible on its trains end to end, train to train, and train to wayside. By the 1960s, such radios were common on American trains, if scarcely on the NYNH&H (*Warwick Valley Dispatch*, June 1, 1960; Vincent Reh, *Railroad Radio*, 1996, 6-11).
5. "Deliberate": Frank M. Merlino Jr., May 4, 1976; "poorly": Robert S. McKernan, Apr. 19, 1976, both to *Hartford Courant*, box 5, file 3, PBEN
6. "Pointless": O. F. Sorgenfrei to W. H. Jenkins, Feb. 21, 1969, box 5, file 1; Penn Central, Bridge Roster, Sept. 19, 1969, box 3, file 16; watchmen dismissed 1972: J. A. Wettstone to F. D. Day, May 10, 1974, box 4, file 8; Brotherhood: Thos. M. Rome to Wm. G. Galloway, Sept. 12, 1973, box 5, file 1, PBEN; *Poks. Journal*, July 3, 17, Nov. 5, 1973

CH. 44. THE BRIDGE BURNS

1. On the fire: Amtrak, Telex report of the fire, May 8, 1974, box 4, file 9, PBEN; *Poks. Journal*, May 9-11, 1974; Sam Christian, flagman on the last train, interv.; notes by an observer of the fire, Heyward Cohen (COHEN); Bob Mohowski, "Except That the Poughkeepsie Bridge Is Burning, " *Railfan*, Winter, 1974, 18-25; Poughkeepsie-Highland Railroad Bridge Co., *Year of the Railroad Bridge*, 1994
2. "Difficult:" *Poks. Journal*, May 9, 1974
3. *Hudson Valley Hornet*, Apr. 27, 1990, 3, 9; *New Haven Register*, May 12, 1981; Penn Central, insurance claim, May 20, 1974, box 4, file 9, PBEN

CH. 45. RESTORE OR DEMOLISH?

1. Penn Central, "Estimate to Repair," May 9, 1974, box 4, file 9, PBEN; "congestion": *Poks. Journal*, Sept. 7, 1974; Ella Grasso to Edward G. Jordan, Apr. 22, 1976, box 5, file 3, PBEN
2. "Gateway": Robert H. Eder, pres., P&W, to Donald C. Cole, Oct. 12, 1979, COHEN; *Poks. Journal*, May 2, 1980; May 11, 1981
3. "Against": *Poks. Journal*, Sept. 30, 1975; ("fall") March 6, 1979; Sen. Abe Ribicoff et al. to Edward G. Jordan, Apr. 9, 1976, box 5, file 3, PBEN; "classic": *Wall Street Journal*, Sept. 13, 1976; DeLeuw, Cather, & Co, *Inspection and Rating of the Poks. Railr'd Bridge*, Final Report, for US Dept. of Transp., v. 1, Feb. 1980; DeLeuw, Cather, & Co., *Report on Poks. Railr'd Bridge Approach Trackage*, for same, v. 3, March, 1980; "tragedy": *Poks. Journal*, Aug. 1, 1980
4. Arthur G. Adams, intervs., 1998; Adams, *The Hudson Through the Years*, 1983, 224-26; letters, 1981-83, ADAMS
5. "Desolate": Robert S. McKernan to *Hartford Courant*, Apr. 19, 1976, box 5, file 3, PBEN; *Walden Citizen Herald*, Nov. 6, 1980; "too bad": *New Paltz Herald*, March 31, 1983

6. Modjeski & Masters to E. J. Burns, Apr. 24, 1978, box 4, file 16; B. B. Wilson, Conrail, memo, Jan. 24, 1983, box 5, file 3, PBEN
7. One-dollar offer: Richard Wachenfeld to US Army Corps of Engrs., June 28, 1983, box 5, file 4; rent: 1981 items, box 5, file 3, PBEN; DeLeuw, Cather & Co., *River Span Evaluation of the Poks. Railr'd Bridge. . .for Central Hudson*, March, 1983; *Poks. Journal*, May 6, 1983; Sept. 7, 1984
8. B. J. Gordon to C. W. Owens, Sept. 27, 1983; J. M. Kepics to Jet Research, Sept. 30, 1983, box 5, file 4; Jet Research Center, "Engineering Plan," for Poks. Bridge [1983?], box 4, file 15, PBEN; *Poks. Journal*, Sept. 7, 1984. In 1977, Christopher J. Teasdale, a youthful volunteer for the Dutchess County Landmarks Association, prepared a preliminary application to enter the bridge on the National Register of Historic Places. It was edited 1978 by NY State's Division for Historical Preservation, Parks, and Recreation, and approved 1979 by the National Park Service, National Register of Historic Places (application, with approvals, WALK).

CH. 46. ALTERNATIVE PROPOSALS

1. For chapter: Arthur G. Adams, Leroy Beaujon, W. Geo. Cook, Edmond Loedy, John Mylod, Donald L. Pevsner, Bill Sepe, intervs. and correspondence, 1997-2001; Henry Stanton, talk, for the Walkway, Poughkeepsie, Sept. 23, 1998
2. Wm. F. Buckley Jr. to Arthur Erickson, Dec. 13; Richard H. Chamberlain to Pevsner, July 15, 1977; Chamberlain to Erickson, March 8; Lowell Thomas to Pevsner, May 31, 1978, PEVS. On Pevsner: Buckley, "Crusader With a Sense of Fun," *Reader's Digest*, May 1980, 169-174.
3. For one dollar: Pevsner to Don Phillips, Feb. 7, 1985, PEVS; "outrageous": *Poks. Journal*, Oct. 7, 1984; "kooky": Adams, Notes on the Save the Bridge Com. mtg., Jan. 18, 1985, ADAMS
4. *Poks. Journal*, Nov. 6, 16-17, Dec. 7, 1984; "albatross": *Newburgh Eve. News*, July 24, 1986; *Hudson Valley Hornet*, Apr. 27, 1990, 9
5. Adams, Notes on Save the Bridge Com. mtgs., Feb. 22, March 22, 1985, ADAMS; Pevsner to Gordon S. Miller, Dec. 28, 1984; to Chamberlain, Oct. 11, 1985, PEVS
6. Miller to W. C. Heming, July 20, 1993, WALK; *Crossings*, Summer, 1993; Miller to Loedy, March 25, 1993, LOEDY; *Poks. Journal*, Oct. 31, 1995
7. *New Paltz News*, March 14, July 25, 1984; *Poks. Journal*, Dec. 6, 1984, June 10, 1991; Apr. 19, 1992; March 26, 2000
8. *Poks. Journal*, Nov. 14, 1991; June 20, 21, 26, July 2, 1992
9. NY *Times*, Nov. 1, 1996

CH. 47. WALKWAY

1. For chapter, intervs. beside those for ch. 46: Thos. Aposporos, Heyward Cohen, Rich'd Coller, Ferris Davis, Tom Houston Jr., John K. Jacobs, Klaus Jonietz, Bernard Rudberg, Vivian Wadlin; *Crossings* [Walkway newsletter], 1992-
2. "Authorized": W. C. Heming, Coast Guard, to Jonas P. Lenktaitis, Sept. 27, 1991, LLOYD. The committee incorporated as a non-profit organization called Poughkeepsie-Highland Railr'd Bridge Co. (Certificate of Incorporation, July 1992, WALK). By Sept. 1994, it also called itself Walkway Over the Hudson.
3. Lichtenstein Engineering Associates, "Feasibility Study Report for the Preservation and Rehabilitation of the Poughkeepsie Railr'd Bridge," Dec. 2, 1992, JACBS; "Panoramas": Sepe to Sen. Daniel P. Moynihan, March 6, 1992, LLOYD; "quiet": *Poks. Journal*, May 24, 1992; meeting with Loedy: Sepe to Rich'd I. Cantor, Sept 28, 1992, WALK
4. "Criminal": Moreno, Counter Claim, Town of Lloyd vs Moreno et al., ca. March 10, 1996, WALK; county resolutions, Feb. 9, March 10, 1994, ADRI
5. Walkway's policy to raise funds only from private sources adopted as amendment to its Certificate of Incorporation, Sept. 28, 1995; became supporters: Fred W. Schaeffer to Town of Lloyd Zoning Bd. of Appeals, Nov. 2, 1995, both WALK
6. *Poks. Journal*, Oct. 21, 1995; July 13, 1997; July 16, 1999
7. "Respect": Vincent G. Bradley, decision, to Sean Murphy, March 24, 1997, WALK; Walkway, "Newsletter," Summer, 2000; *Highland Mid-Hudson Post*, Feb. 7, 2001
8. "Outstanding": Peter Harnik to Rich'd Cantor, June 2, 1993, WALK

SOURCES CITED

A. MANUSCRIPT, CLIPPING, TAPE, AND PHOTO COLLECTIONS CITED:

ADAMS— Arthur G. Adams, Mahwah, NJ: relating to his proposals for use of the bridge, clippings, letters, notes, photos

ADRI—Local History Collection, Adriance Memorial Library, Poughkeepsie, NY: clippings, letters, pamphlets, photos

ALEX—Albert Alexander, conductor, Maybrook, NY: photo

BEAUJ—Leroy Y. Beaujon, towerman, Roseville, CA: especially on the CNE, including letters, time tables, pamphlets, rosters, photos

BENJ—Charles Benjamin, Poughkeepsie, NY: especially on the New Paltz- Poughkeepsie train-trolley line: letters, time tables, photos

BLODG—Nathan Blodgett, conductor, Canaan, CT (dec.): his letters of recollection from the 1970s, held by Leroy Y. Beaujon, Roseville, CA

BONE—Clark Bonesteel, engineer, Port Ewen, NY (dec.): photo

CHRIS—Samuel Christian, brakeman, Maybrook, NY: clippings, photos

COHEN—Heyward Cohen, Amenia, NY: letters, clippings, postcards, photos, notes

COOK—W. George Cook, trainmaster, Valatie, NY: notes, correspondence

DAVIS—Ferris Davis, Hopewell Jct., NY: photos on bridge maintenance work

DEYO—Deyo family diaries kept especially by Agnes Deyo (Mrs. Andrew L. F. Deyo), 1890-1930: held by Carleton Mabee, Gardiner, NY

DiROS—Anthony Di Rosa, water boy, Ft. Lauderdale, FL (dec.): photo

FDRL—F. D. Roosevelt Library, Hyde Park, NY: correspondence, photos, travel log, pamphlets, flyers

FULLR—Allyn Fuller, Canaan, CT (dec.): photos, rosters, held by Leroy Y. Beaujon, Roseville, CA

GRUNR—Dorothy Phillips Gruner, Highland, NY: clippings, photos, audiotape

HARV—Baker Library, Harvard Business School, Boston, MA: NYNH&H papers, including Hudson Connecting Railr'd, Minutes, 1887-1889; PR&NE, Minutes, 1891-1898, CNE Journals, 1903-1906; Clifford Burr Collection, including PR&NE correspondence

HOUST— Thomas Houston Jr., Hyde Park, NY: photos, NYNH&H rule books

HVRTA—Hudson Valley Rail Trail Assoc., Highland, NY: map

JACBS—John K. Jacobs, Highland (Clintondale), NY: reports, photos

KNAU—Wm. D. Knauss, Ticonderoga, NY: photo

LC— Library of Congress, Washington, DC: Wm. D. Hassett Diary

LILL—Roger Liller, Leeds, NY: photos, time tables, pamphlets, rosters

LLOYD —Town Historian's collection, Lloyd Town Hall, Highland, NY: clippings, photos, letters

LOEDY—Edmond G. Loedy, Millbrook, NY: on his proposals for the bridge, clippings, letters, photos

McCO—Arthur B. McComb, Poughkeepsie, NY: photo

McEN—Austin McEntee, Poughkeepsie, NY: extensive, including articles, letters, pamphlets, clippings, time tables, photos

McMA—Joseph McMahon, Winsted, CT, & NY City (dec.): clippings on railroads in the Winsted region, 1890-1968, held by Leroy Beaujon, Roseville, CA

MRM—Maybrook Railroad Museum, Maybrook, NY: manuscripts, photos, clippings, rosters, rule books

NA—National Archives, College Park, MD: Interstate Commerce Commission, accident reports

ND&C—Beacon Historical Society, Beacon, NY: Newburgh, Dutchess, & Connecticut Railroad, bound volumes of correspondence, 1879-1904

NHRH&TA—New Haven Railroad Historical & Technical Assoc., publisher of *Shoreliner*: photo

NYSA—New York State Archives, Albany, NY: CNE, Annual "Reports" to NY State Public Service Comm'n (manuscript), 1906 to 1926

O'KEEF—Dennis O'Keefe, New Paltz, NY: photo

O'ROUR— Innis O'Rourke, III, Glen Cove, NY: clippings, photos, legal documents about his great grandfather, John F. O'Rourke

PBEN—Poughkeepsie Bridge Engineering Records, held by Walkway Over the Hudson, Poughkeepsie, NY. These manuscripts, in five boxes, dated about 1906-1980s, were preserved by the railroads which successively owned the bridge, in accordance with engineering custom, and were turned over by Conrail to the Walkway in 1998 when the Walkway acquired the bridge.

PEVN—Donald L. Pevsner, Ulster Park, NY, and Miami, FL: photo

PEVS—Donald L. Pevsner papers, including his correspondence on alternative uses of the bridge: held by the Walkway, Poughkeepsie, NY

PLATT—Town Historian's collection, Plattekill Town Hall, Plattekill, NY: clippings, photos, maps

ROEBL—John A. Roebling Family Collection, Rensselaer Polytechnic Institute, Troy, NY: Brooklyn Bridge clippings, trustee minutes, reports

SHAU—Jim Shaughnessy, Troy, NY: photos

SHUKR—Kenneth A. Shuker, Cornwall, NY: photos, postcards, correspondence

SIMMS—Philip S. Simms, Montgomery, NY: maps, photos, time tables

SMITH—Transportation Collections, Smithsonian Institution, Washington, DC: photo

STICK—Dudley J. Stickles, Torrington, CT, and Norwell, MA (dec.): manuscripts, pamphlets, photos, held by Leroy Y. Beaujon, Roseville, CA

Sources Cited

SWANB—J. W. Swanberg, railroad fireman, photographer, Branford, CT: photos

SWEE—John P. Sweeney, son of conductor F. P. Sweeney, Port Jervis, NY: his father's papers

UCONN—Dodd Research Center, University of Connecticut, Storrs, CT: extensive railroad collections, including Poughkeepsie Bridge Co., Journal, 1873-90, its Letters, 1886-1887, its Board Minutes, 1871-1892; Poughkeepsie Bridge Railroad Co, Board Minutes, 1888-1907; CNE, Board Minutes, 1899-1927; NYNH&H Railroad's Secretary's file, 1921-1944; Leroy Y. Beaujon papers (not to be confused with BEAUJ above); Charles B. Gunn papers, photos

USMA—Archives and Library, US Military Academy, West Point, NY: including atheltic records, memos, orders, diagram

WALK—Walkway Over the Hudson, Poughkeepsie, NY: especially on the campaign from 1991 to remake the bridge into a walkway, correspondence, pamphlets, photos

WEBER—Carol Harris-Weber, Wayne, NJ, of the Donaldson family: photo

WILLM—Archives, Williams College, Williamstown, MA: records of Charles S. Maurice, class of 1861

YALE—Archives, Yale University Library, New Haven, CT: photo

B. BOOKS, PAMPHLETS, AND ARTICLES CITED:

Abbey, Wallace W., "Central States Dispatch," and "I Flew Across the Hudson on a Freight Train!" *Trains & Travel*, March 1952, 12: 26-32; May 1953, 13:26-27

Abendschein, Frederic H., "J Tower & Peers," *Locomotive & Railway Preservation*, Sept.-Oct. 1990, 21-28+

Adams, Arthur G., *The Hudson Through the Years*, 1983. Comprehensive.

Adams, Robert B., "Battle for Springfield," National Model Railr'd Asso. (NMRA), *Bulletin*, Nov. 1978, 9-17. A patient, experienced railroad scholar, Adams unfortunately did not live to complete his detailed history of the CNE.

—— "Decline and Fall of the CNE," Lakeville, CT, *Journal*, Oct. 16, Nov. 6, 26, 1980

—— "Poughkeepsie's Unusual 'Rapid Transit,'" National Model Railr'd Asso., *Bulletin*, Apr., 1975, 8-15

The American Railway: Its Construction, Development, Management, and Appliances, NY, 1889

Anson, Shirley V., *Friends and Neighbors*, Clintondale, NY: Clintondale Friends Meeting, 1989

Ashman, Robert, and Charles Milmine, *Central New England Railroad*, Salisbury, CT.: Salisbury Association, 1972. No index. Ashman presents his memories, Milmine a brief history.

Baily, Samuel L., *Immigrants in the Lands of Promise: Italians in Buenos Aires and New York City*, Ithaca, NY: Cornell University, 1999

Biographical Dictionary of American Civil Engineers, NY: Amer. Soc. of Civil Engineers (ASCE), v. 1, 1972; v. 2, 1991

Bishop, Jim, *FDR's Last Year*, NY, 1974

Bixby, Arthur, Sr., "Experiences on the L-1," New Haven Railroad Historical & Technical Asso., *Shoreliner*, Spring, 1976, 7:22-23

Boston & Maine Railroad Historical Society, *The Central Mass.* [sic.], Reading, MA, 1975

Botkin, B. A., and Alvin F. Harlow, *Treasury of Railroad Folklore*, NY, 1953. Rich, satisfying.

Breuer, William, *Hitler's Undercover War: The Nazi Espionage Invasion of the U.S.A.*, NY: St. Martin's, 1989

Buckley, Wm. F., Jr., "Crusader With a Sense of Fun," *Reader's Digest*, May 1980, 169-174

Central New England Railway (CNE), *Poughkeepsie Bridge Route*, 1903

—— *Summer Homes among the Mountains*, 1899-1901

Clarke, Thomas Curtis, "Architects and Architecture," *Christian Examiner*, Sept. 1850, 49: 278-286

—— "Building of a Railway," *Scribner's Magazine*, June 1888, 3:643-670

—— "Engineering," in Alfred R. Wallace, *Progress of the Century*, NY, 1901, 421- 452

—— "European and American Bridge-Building Practice," *Engineering Magazine*, Apr. 1901, 21:43-58

—— "Hudson River Bridge at Poughkeepsie," *Scientific American Supplement*, May 19, 1888, 25:10311-10312

—— "Railway-Engineering in the US," *Atlantic*, Nov. 1858, 2:641-656

—— "Vincent Bourne and the Modern Latinists," *Christian Examiner*, Sept. 1848, 244-245, 257

Committee of the Boston Board of Trade, *Boston and the West*, Feb. 1875

Connecticut Valley Chapter, National Railway Historical Society, *Maybrook Excursion*, May 4, 1947

Corwine, Wm. R., *History of the Poughkeepsie Bridge and its Railroad Connections*, for Poughkeepsie Chamber of Commerce, Feb. 2, 1925, 41 pp.

Crossings (title varies slightly), 1992—, newsletter of Poughkeepsie-Highland Railroad Bridge Co., also known as Walkway Over the Hudson

Curtin, Tom, "Yale 14-Army 12, A Tradition Ends," New Haven Railroad Historical & Technical Association, *Shoreliner*, issue 3, 1991, 22: 36-39

Daniels, Jonathan, *Washington Quadrille*, Garden City, NY, 1968

Deleuw, Cather, & Co., *Inspection and Rating of the Poughkeepsie Railroad Bridge, Phase II, Final Report*, for US Dept. of Transp., 2 vs., Feb. 1980

—— *River Span Evaluation of the Poughkeepsie Railroad Bridge. . .for Central Hudson*, March, 1983

—— *Report on Poughkeepsie Railroard Bridge Approach Trackage*, for US Dept. of Transp., v. 3, March 1980

Dillon, John J., *Seven Decades of Milk: A History of New York's Dairy Industry*, NY, 1941

Di Rosa, Tony, *El Pibe*, Buenos Aires, no publisher, 1990

Farrington, S. Kip, Jr., *Railroads at War*, NY, 1944

Federal Writers' Project, Works Progress Administration, *Dutchess County*, Phila., 1937

Federal Writers' Project, Works Progress Administration, *Italians of New York*, NY, 1938

First Presbyterian Church, Highland, NY, *Time for Reflection*, no publisher, 1992

Foster, George H., and Peter C. Weiglin, *Splendor Sailed the Sound: The New Haven Railroad and the Fall River Line*, San Mateo, CA: Potentials, 1989

Gardner, Ed, *Central New England Railway: A Pictorial Review*, Mountaintop, PA, 1981. No index.

Gould, J. P., *A Review of a Project for Bridging the Hudson River at Poughkeepsie*, Poughkeepsie, 1876 (in a pamphlet file, Newburgh Free Library)

Grant, H. Roger, *Erie Lackawanna: Death of an American Railroad*, Stanford, CA: Stanford University, 1994

Haight, Lyndon A., *Pine Plains and the Railroads*, Pine Plains, NY, 1976

Hassett, William D., *Off the Record with F. D. R.*, New Brunswick, NJ, 1958

Haviland, Jim, "Did Steve Brodie Really Jump off the Poughkeepsie Railroad Bridge?" *Hudson Valley*, Aug., 1973, 2:35-36

Hedin, Robert, *The Great Machine, Poems and Songs of the American Railroad*, Iowa City: Univ., of Iowa, 1996. Delightful.

Holton, David P. and Frances K., *Winslow Memorial: Family Records*, 2 vs., NY, 1877, 1888

Howells, William Dean, *The Sleeping-Car and Other Farces*, Boston, 1904

Hudson Valley Bridge Association, *Report. . .on Proposed Free Highway Bridge Across the Hudson River*, 1922 (at FDRL)

Hurd, Jerome and Elizabeth, *Village of Clintondale from Its Beginning*, 1959

Intercollegiate Rowing Association, *Programs* (titles vary), 1910-1947

Jacobs, Jack [John K.], "Carrie's Cow," Antioch College, *The Antiochian*, Dec. 1939, 11-14

Keystone Bridge Co., *Descriptive Catalogue of Wrought Iron Bridges*, Phila., 1874

Kulp, Randolph L., *Railroads in the Lehigh River Valley*, Allentown, PA, 1962

La Cosssitt, Henry, "He Takes the President on Tour," *Saturday Evening Post*, June 16, 1951, 19-21+

Laying the Corner-stone of the Great Bridge to Span the Hudson at Poughkeepsie, Dec. 17, 1873 (in Harvey Eastman scrapbook, Vassar College)

Leuthner, Stuart, *The Railroaders*, NY, 1983. On railroad workers. A quality book.

Lord, Robert F., *Country Depots in the Connecticut Hills*, New Hartford, CT: Goulet, 1996

Mabee, Carleton, *American Leonardo: A Life of Samuel F. B. Morse* (Revised), Fleischmanns, NY: Purple Mountain Press, 2000

—— *Listen to the Whistle, An Anecdotal History of the Wallkill Valley Railroad*, Fleischmanns, NY: Purple Mountain Press, 1995

Macdonald, Charles, "Address," Amer. Soc. of Civil Engineers, *Transactions*, Dec. 1908, 61:544-552

—— "The Six-Hundred Ton Testing Machine at the Works of the Union Bridge Company at Athens, Pa.," Amer. Soc. of Civil Engineers, *Transactions*, Jan., 1887, 16:1-29

McDermott, William P., *Dutchess County Railroads*, Clinton Corners, NY: Clinton Historical Society, 1996

McLachlan, Pete, "Diesel Pushers on Stormville Mountain," New Haven Railroad Historical & Technical Assoc., *Shoreliner*, 1995, Issue 3, 26:32-37

Moffett, Glendon L., *Down to the River by Trolley: The History of the New Paltz-Highland Trolley Line*, Fleischmanns, NY: Purple Mountain Press, 1993

—— *To Poughkeepsie and Back: The Story of the Poughkeepsie-Highland Ferry*, Fleischmanns, NY: Purple Mountain, 1994

Mohowski, Robert ("Bob") E., "Except That the Poughkeepsie Bridge is Burning, Everything is Fine," *Railfan*, Winter, 1974, 18-25

—— *The New York, Ontario & Western Railway and the Dairy Industry*, Laurys Station, PA: Garrigues, 1995

Mott, Edward H., *Between the Ocean and the Lakes: The Story of Erie*, NY, 1899

New York, New Haven, & Hartford Railroad, *Official List of Officers, Station Agents*, Nov. 1, 1926; May 1, 1941

New York, New Haven, & Hartford Railroad, *Rules for the Government of the Operating Department*, Oct. 28, 1956

New York, New Haven, & Hartford Railroad and Central New England Railway *Rules for the Government of the Operating Department*, May 17, 1914

Nimke, R. W., *Central New England Railway*, 3 v., Westmoreland, NH: The author, 1995-1996. Rich in photos and technical detail. Poor in clarity and sources. No index.

Nutt, John J., *Newburgh*, 1891

Ogden, Tom, *Two Hundred Years of the American Circus*, NY: Facts on File, 1993

O'Rourke, John F., "Construction of the Poughkeepsie Bridge," Amer. Soc. of Civil Engineers, *Transactions*, June 1888, 18:199-215

Parkinson, Tom, and Charles P. Fox, *The Circus Moves by Rail*, Boulder, CO, 1978

Pavlucik, Andrew J., *New Haven Railroad: A Fond Look Back*, New Haven, 1978

Pennsylvania Railroad, *History: Passenger. . .Service*, Chicago, 1974

Philadelphia, Reading, & New England Railroad (PR&NE), *Summer Homes Among the Mountains*, 1895-1898

Platt, Edmund, *The Eagle's History of Poughkeepsie*, Poughkeepsie, 1905

Poughkeepsie Bridge Co., *Bridging the Hudson at Poughkeepsie*, Poughkeepsie, 1871, 1875

—— *Charter. . .and Amendments*, Poughkeepsie, 1887

—— *The Poughkeepsie Bridge: A New Route*, Feb. 1887

Poughkeepsie-Highland Railroad Bridge Co. (Walkway over the Hudson), *Year of the Railroad Bridge*, Poughkeepsie, 1994

Poughkeepsie, The Bridge, Poughkeepsie Eagle Souvenir Number, Oct. 9, 1889 ("Issued on the occasion of the opening. . .of the Poughkeepsie Bridge")

The Press Against the Central New England Parallel [pamphlet], no publisher, no place [May 1901]. Selections from over 50 CT and MA newspapers, 1896 to 1901, opposing the CNE's proposed Springfield Branch, 42 pp. [at CT State Library, Hartford]

Reh, Vincent, *Railroad Radio: Hearing and Understanding Railroad Radio Communications*, Grand Isle, VT: Byron Hill, 1996

Rensselaer Polytechnic Institute (RPI), *Biographical Record of the Officers and Graduates*, Troy, NY, 1887

Rinaldi, John K., and Lawrence A. Perretta, *History of the Italian Center . . .Poughkeepsie*, Poughkeepsie, 1981

Robinson, Winfield W., "Locomotives of the New Haven Railroad," Railway & Locomotive Historical Society, *Bulletin*, Oct. 1939, 7-27

Skinner, Frank W., "Great Achievements in Modern Bridge Building," *McClure's*, Jan. 1901, 16: 249-258

Standage, Tom, *Victorian Internet, The Remarkable Story of the Telegraph and the Nineteenth Century's On-Line Pioneers*, NY: Walker, 1998

Steeholm, Clara and Hardy, *House at Hyde Park*, NY, 1950

Swanberg, J. W., "Confessions of a Diesel Fireman on the Old New Haven," *Locomotive & Railway Preservation*, Sept.-Oct. 1990, 49-59

Tingley, Richard H., "Bridging the Mighty Hudson," *Railroad Man's Magazine*, Aug. 1918, 36:726-730

Turner, Gregg M. and Melanchthon W. Jacobus, *Connecticut Railroads*, Hartford: Connecticut Historical Society, 1989. Careful, comprehensive.

Tyler, Leslie, "Maybrook Line," *Transportation*, May 1951, 5:1-10

Wadlin, Beatrice H., *Times and Tales of Town of Lloyd*, [Highland, NY], 1974

Waddell, John A. L., *Bridge Engineering*, 2 vs., NY, 1916

Wall, Joseph F., *Andrew Carnegie*, Pittsburgh: Univ. of Pittsburgh, 1989

Wallace, Alfred R., *Progress of the Century*, NY, 1901

Wallworth, Don, "A Five-Railroad Paradise," *Trains*, July 1996, 56-57

Ward, Geoffrey C., *Before the Trumpet: Young Franklin Roosevelt*, NY: Harper, 1985

—— *A First Class Temperament, The Emergence of Franklin Roosevelt*, NY: Harper, 1989

Ward, James A., *J. Edgar Thomson: Master of the Pennsylvania*, Westport, CT, 1980

Weibust, Patricia S., *Italians. . .in Connecticut*, Storrs, CT, 1976

Withers, Bob, *The President Travels by Train*, Lynchburg, Va.: TLC Publishing, 1996

Zuccarello, Louis C., "The Catholic Community in Poughkeepsie," Dutchess County Historical Society, *Year Book*, 1987, 112-115

C. INTERVIEWS CITED. All took place from 1995 to 2001.

ADAMS, ARTHUR G., Mahwah, NJ
ALEXANDER, ALBERT J., Maybrook, NY
APOSPOROS, THOMAS, Poughkeepsie, NY

Sources Cited

BEAUJON, LEROY Y., Roseville, CA
BENJAMIN, CHARLES, Poughkeepsie, NY
BONESTEEL, CLARK, Port Ewen, NY (dec.)
CANINO, ANTHONY, Highland, NY
CARMODY, MARY, Walden, NY (dec.)
CASCIARO, LOLA CANSAS, Highland, NY, (dec.)
CHRISTIAN, SAMUEL and DORIS, Maybrook, NY
COHEN, HEYWARD, Amenia, NY
COLLER, RICHARD, Staatsburg, NY
COOK, W. GEORGE, Valatie, NY
DAVIS, FERRIS, Hopewell, Jct., NY
DAVIS, LA VERNE, Poughkeepsie, NY
DI ROSA, ANTHONY, Fort Lauderdale, FL (dec.)
FELL, WILLIAM L. and ANITA, Hyde Park, NY
FLYNN, THOMAS A., LaGrangeville, NY
GORING, OLIVER AND ANN, Hyde Park, NY
GORING, RICHARD, Kingston, NY
GRAF, BARBARA ANN, Gardiner, NY
GRUNER, DOROTHY PHILLIPS, Highland, NY
HEGEMAN, RAYMOND, Highland, NY
HOUSTON, THOMAS F., JR, Hyde Park, NY

JACOBS, JOHN K., Highland, NY
JONIETZ, KLAUS, Highland, NY
LOEDY, EDMOND G., Millbrook, NY
LOWN, PAUL, Hyde Park, NY
LYONS, ARTHUR J., Highland, NY
MABIE, ROGER W., Port Ewen, NY
MARANO, ANTONIO, Montgomery, NY
McENTEE, AUSTIN, Poughkeepsie, NY
McCOMB, ARTHUR B., Poughkeepsie, NY
MYLOD, JOHN, Poughkeepsie, NY
PEVSNER, DONALD L., Ulster Park, NY, Miama, FL
REED, JOHN M., Poughkeepsie, NY
ROGERS, WILLARD B., Germantown, NY (dec.)
RUDBERG, BERNARD, Wappingers Falls, NY
SANDBOTHE, ROBERT J., Walden, NY
SCHNITZER, WILLIAM. L., New Paltz, NY
SEPE, WILLIAM, Poughkeepsie, NY
SIMMS, PHILIP S., Montgomery, NY
SPOLVERINO, JOSEPH, Maybrook, NY
WADLIN, VIVIAN, Highland, NY

INDEX

Abbey, Wallace W., 140
Accidents, Safety, 244-245; chs. 26, 27, 30, 34, 36-42; 1974 bridge fire, 246-249
Adams, Arthur G., 254, 262-263, 266
Adriance, John P., 10, 12-13, 54-55
Alexander, Albert J., 119-120, 123, 168-169; his exams, 181-183; as brakeman and conductor, 211-215, 227-228, 239
Allen, Charles H., 25
Allen, Horatio, 12-13, 16, 36
American Bridge Co., 25, 27, 29, 38, 180, 232
American Society of Civil Engineers, 35, 45
Ancram, NY, 108, 129, 205
Ancram Lead Mines, NY, 109
Anson, Robert Sr., 270
Aposporos, Thomas, 253, 259
Appleton, Julius H., 50, 113
Arlington, NY, 212
Astor, John Jacob, 172
Astor, Vincent, 175

Baker, Roy, 239
Baldwin, Mike, 271
Baltimore & Ohio (B&O), 71, 73, 76, 98-99, 140-141, 144
Barnard, Jack, 271
Barnum, P. T., and his circuses, 147, 150-152
Barnum, William H., 50
Barone, Bartolo, 188-190
Bartholomew, George M., 25
Beacon, NY, 87, 103-104, 132, 243 (see also Ferries, Fishkill Landing, and Matteawan)
Bear Mountain Bridge, 104, 163
Beaujon, Leroy Y., 197-200; compiler of H&CW locomotive roster, 276-277
Beebe, Jack, 181
Bennett, Lewis, 219
Bethel, NY, 152
Billings, NY, 90
Black, David, 272
Blacks, 59, 131, 192
Blodgett, Nathan, 109, 121, 123, 132-133, 147, 166-167, 169, 181; his stories, 176-180
Bloomfield, Ct., 129
Booth, Oliver H., 13, 25

Boston, Ma., committee from, investigates need for a bridge at Poughkeepsie, 20-25; Boston-Harrisburg express, 71-74, Boston-Washington express, 71-76, 79; destination for freight, 130, 138, 140, 143, 145-146, 206, 247
Boston & Albany Railroad (B&A), 21, 112-114
Boston & Maine Railroad (B&M), 21, 65, 71-73, 75, 77, 112-113
Boston & South Mountain Railroad, 20
Boston Corners, NY, 78, 87, 176
Boston, Hartford, & Erie, 11
Botsford, Ct., 215
Brescia, Frank and Joe, 226
Brewster, NY, 69, 185, 212
Bridgemen, (watchmen, maintenance men), chs. 30, 42; photos, 1950, 239; 1973, 244; listing of (table), 1969, 244 (see also Workers)
Bridgeport, Ct., 138, 140, 152, 171
Bridges, Hudson River, table of, 163
Brinkerhoff, NY, 90
Brock, Arthur, 65, 72, 113
Brock, John W., 50, 65, 113-114

Brodie, Steve, 153-155, 157, 264
Brooklyn Bridge, 17, 27, 38-39, 55, 153, 156, 164
Buckley, William F., Jr., 261
Burger, Bill, 193
Burgess, Edward F., 123
Burns, Ed, 121
Burnside, NY, 132

Callahan, Patrick, 156-157
Campbell Hall, NY, 59-60, 62, 64-65, 76, 87, 111, 117, 124, 131, 136, 139; Campbell Hall-Hartford express, 78-79
Canaan, Ct., 69, 78, 87, 132-133, 137, 148, 176, 178, 181, 197-200
Cansas, Rocky, and family, 194-196
Canton, Ct., 218
Carmody, Mary, 201-202, 205
Carnegie, Andrew, 15-16, 36, 45
Carpenter, B. Platt, 19
Carter, Jimmy, Pres., 253
Castleton Bridge (Thruway Extension), 163
Central Hudson Gas and Electric Corp., its electric cables on bridge, 193, 237, 250; brought in lumber by the Bridge Route, 243; considered buying bridge but declined, 250, 259, 262, 268; removed its cables from bridge, 259, 271; hindered bridge Walkway, 271, 273-275
Central New England & Western (CNE&W), 89, 126, 160, 183, 216, 276-277; incorporated 1899 combining several railroads into the Bridge Route, succeeded in 1892 by the PR&NE (table), 64; maps, 56, 60; opening the route, 61-63; passenger trains, 71-74
Central New England Railway (CNE), 126, 161-164, 166, 176-179, 181, 184-185, 192, 195, 197-198, 231, 243; acquired the Bridge Route, 1899, 76; controlled by the New Haven from 1904, 79, (tables) 135, 146; its passenger trains, 81-84, 86-91; time tables, 1903, 87; 1916, 111; its vital statistics, 1917 (table), 90; its road mileage (table), 92; its wages for road crews (table), 93; its milk service, 108-111; its Springfield Branch, 112-115; its directors, 1900 (table), 113; its Maybrook Yard, 117-119, 121-122, 125; its freight, 127-138, (table) 135; its station agents (table), 137; its circus trains, 147-149; its telegraph system (tables) 216-219; its shops at Hartford, 277; its chronology (table), 135
Central Railroad of New Jersey, 64-65, 72, 117, 120, 130, 138, 141-142, 243
Central States Dispatch, 137, 140-146
Chapman, Charles, 239
Charder, Clair, 193
Christian, Samuel, 118, 123-125, 134, 234, 253
Circus trains, 147-152
Clark, John I., 25
Clark, G. W., 163, 177, 179
Clarke, Thomas Curtis, 32, 35-41, 47, 54-55, 275
Cleveland, Grover, Pres., 74
Clinton Corners, NY, 109, 129, 178
Clintondale, NY, 62, 78, 88, 90-91, 130, 132, 199-200, 206, 210, 220, 243
Coal, 189-190; as a factor in campaigning to build the bridge, 9-11, 19-20, 21-25, 33; constructing connecting lines to Pennsylvania for, 60, 62-65; its traffic, 115, 120, 124, 127-128, 134-136, 138; its tonnage, table, 135
Coller, Richard, 270, 275
Collingwood, James, 13
Collinsville, Ct., 112
Cone, Henry D., 57-58
Connecticut Western Railroad (CW), 11, 17, 22, 37, 276
Conrail (Consolidated Railroad Corp.), became owner of bridge, 251-252, 269; declined to sell bridge-related lines to potential competitor, 254; abandoned bridge-related lines, 255-258; tried to sell bridge and related rights of way and did, 250, 259, 262-264, 269; ceased to exist, 266
Coolidge, W. G., 27, 29, 36, 39
Cooper, Elizabeth and Edward, 205-207
Copake, NY, 108, 129
Cornell, Thomas, 50
Covert, Frank, 193
Crane, L. Stanley, 262-263
Crowell, R. Lee, 269-270, 275
CSX (Chessie, Seaboard) Corp., 266
Cuineen, J. W., 90, 177

Danbury, Ct., 58, 64, 69, 87, 111, 129, 148, 185, 198, 200, 211, 213, 217, 225, 245, 253
Daniels, Jonathan, 99
Davis, Ferris, 168, 192-193, 230, 234, 236-237, 239, 244-245
Dean, William, hotelier, 154, 156-157, 194
Decker, Jon, 270
Delano, Warren, and family, 96

Delaware & Hudson Canal, 12, 21, 25, 27, 51
Delaware & Hudson Railroad, 97
DeLeuw, Cather, & Co., 259, 268
Dennis, A. L., 15-16, 18-19
Deyo, Andrew L. F., and family, 85-86, 130-132
Dickinson, Pomeroy P., 10-13, 25, 29, 36, 38, 50, 53, 62, 154, 158
Dingee, William, 219
Di Rosa, Anthony (Tony) and Marie, 186-190
Divers, checking the bridge, 29, 43, 231, 235-236
Donaldson, Daniel, 130
Doolittle, Frank, Sr. and Jr., 123, 124, 181, 237
Downs, telegrapher, 218-219
Doxie, Charles, 154
Dumaine, Frederick C., 134
Dutchess County Railroad, 58, 64, 92, 135
Dyseven, Joseph, 239, 244

Eads, James B., 13, 23, 36
East Canaan, Ct., 178, 198
East Walden, NY, 69
Eastman, George, 10
Eastman, Harvey G., 10-13, 15, 18-19, 21-25; died, 31
Economou, Jack, 267, 270
Erie Lackawanna Railway, 242-243, 245, 247, 254, 266
Erie Railway (also Railroad) (New York, Lake Erie, and Western) (NYLE&W), 12, 15, 22-23, 27, 86, 90, 125, 128; potential connection for bridge, 9, 10, 25, 59; maps, 60, 77; joint owner of L&HR, 117, 120; its bridge traffic, 61, 64, 118, 131, 134, 136, 138-139, 143; merged into Erie Lackawanna, 242
Examinations, railroad, 181-183, 213

Favaro, Phil, 119
Federal Express, 76, 79, 164
Feeding Hills, Ma., 115
Fell, Bill and Anita, 191-193, 238
Ferries, Hudson River, 9, 162; Kingston-Rhinecliff, 21; Poughkeepsie-Highland, 17, 44, 46, 83, 127, 159, 161, 194-196; Newburgh-Beacon (Fishkill Landing), 21-23, 33, 128-129, 172; NY City, 21, 76, 86
Field, George S., 34, 36, 45
Fire, closing the bridge (1974), 246-249
Fishkill Landing, NY, 22, 128-130 (see Beacon, Ferries)
Fisk, Jim, 23

Index

Fleming, Sandford, Sir, 27
Flynn, Thomas A., 148
Foley, Dennis, 92, 198
Forest Glen, NY, 205
Forth Bridge, 39, 48
Fowler, Milton A., 159
Fowler, Thomas P., 61
Frank, Theodore, 267
Freight trains, Springfield Branch, ch. 16; Maybrook yard, ch. 17; freight at large, ch. 18; Central States Dispatch, ch. 19; circus trains, ch. 20 (see also Coal, Fruit, Iron, Meat, Milk)
Fruit service, 118, 129-132, 138

Gallaudet, Peter W., 34
Gardiner, NY, 22-23, 58, 85-86, 130-131
Gaylord, George R., 13, 25
George Washington Bridge, 163
Giambu, track gang foreman, 187-188
Gibbs, William W., 33-34, 48, 50, 54, 61-62, 65, 124
Goethals, George W., 164, 165
Gorden, Tom, 193
Goring, Oliver, 236-237
Goshen, NY, 86, 131
Gould, J. P., 25
Grand Trunk Railway, 76, 136
Grasso, Ella, Gov., 251
Grattan, J. Kenny, 244
Green, Frederick S., 165
Gridley, Judy, 272
Griffin, Bert, 123
Griffin, Ct., 129
Grisinger, "Big John," 119
Grogan, Louis V., 210
Guimaro, Sal, 236

Hansut, Robert, 270
Harlem Line, 78
Harris, Jack, 119
Harrisburg, Pa., Harrisburg-Boston express, 71-74
Hartford, Ct, 19, 23-24, 77, 86, 166, 171, 197-198, 217, 225; Hartford-Campbell Hall express, 78, 87; freight business, 109, 111, 129, 138, 148-149, 177, 239; uneasy with Springfield Branch, 112-115; CNE&W and CNE shops at, 179, 277; dispatcher at, 178, 183
Hartford & Connecticut Western Railroad (H&CW), 48, 53, 57, 61, 63-64, 112, 114-115; its officers, 1887-1888 (table), 50; its locomotives, 276-277
Hartford, Providence, & Fishkill Railroad, 22, 176
Hasbrouck, Abram, 13

Hasbrouck, Joseph E., 138
Hassett, William D., 99-101
Hayes, Edmund, 34, 36
Heady, Pete J., 193, 244
Healy, Fred, 239
Hegeman, Raymond, 98
Hell Gate Bridge, 79, 98, 164, 225
Hess, George B., 119
Hibernia, NY, 108, 183
Hicks, Norman, 180
Highland, NY, 45, 52, 54, 63, 66, 69, 78, 89,133, 138, 143, 147, 180, 185, 189, 192, 208, 210; trolley service over the bridge, 80-83; relation to F. D. Roosevelt, 98-101; site for jumping off the bridge, 154, 156-157; hoboes in, 166-168; regatta in, 171-172, 174; Dean's Hotel in, 154, 156-157, 194; derailment near, 225-226; after the 1974 fire, abandoning bridge-related tracks in, 256; proposals for use of the bridge, 259; 269 ; issue of walking on the bridge, 159-161, 269-271, 273, 275 (see also Ferries, Mid-Hudson Bridge)
Hoboes, 166-169
Holly, Alexander H., 50
Hopewell Branch, 56, 77, 87
Hopewell Junction, NY, 56, 58, 110, 128-130, 138, 185, 203-204, 207, 209, 213-214, 243, 252-253, 255, 258, 264
Hospital Branch (to Poughkeepsie State Hospital and Hudson Line), 88, 134, 252
Houston, Tom F., Sr. and Jr., 234, 239, 244-245
Howells, William Dean, 74
Hoyt, Mary, 205
Hudson Connecting Railroad, 59, 64
Hudson Line, 10, 53, 55, 62, 66, 83, 95, 97-98, 101, 109-110, 128, 130, 148, 156, 166, 171, 187, 246-247, 252 ; shrinking, taken over by Penn Central, 241, 243
Hudson River Railroad, 10, 17 (see also Hudson Line)
Husted, James William, 18, 48, 50
Hyde Park, NY, 95-101

Indians, American, 192
Innis, Aaron, 13, 41, 97
Innis, George, 10, 12-13, 25, 41, 54
Innis, Katherine B., 41, 61
Iron and lime service, 133, 135

Jackson Corners, NY, 90
Jacobs, John K., 220-224
Jenkins, Rufus, 85

Jones, Casey, 184-185
Jonietz, Klaus, 269-270
Jumping off the bridge, 153-157, 264-265

Kane, Matt, 176
Kelly, William J., 199-200
Kepler, Johnny, 167
Keystone Bridge Co., 14, 16-17
Kindlen, John, and Mrs., 50
Kingston, NY, 21-22, 25, 42, 225
Kingston-Rhinecliff Bridge, 163
Kjeldsen, Rick, 269
Knauss Brothers, meatpackers, 134

Lackawanna Railroad, 117
LaGrange, NY, 109, 111
Lake, Theodore, 180
Lakeville, Ct., 108, 148
Lehigh & Hudson River (L&HR), 60, 63-64 99, 101, 117, 121, 125, 138, 152, 245, 251, 253; its passenger service, 72-73, 77, 79, 90-91; its chronology (table), 120; part of Central States Dispatch, 140-144; bankrupt, 242-243
Lehigh & New England (L&NE), 62, 90-91, 117; its chronology (table), 124; 143, stopped running, 241-242
Lehigh Valley Railroad, 64, 65, 76, 117, 120, 130, 243
Leion, Florence M., 205
Lenktaitis, Jonas, 267-268
Lepre, Tom, 267
Lichtenstein Consulting Engineers, 268
Lindenthal, Gustav, 37, 134, 162, 164, 171
Linville, Jacob H., 16-18, 36
Lloyd, NY, station, 62, 86, 90; town of, 264, 269, 273-275
Locomotives, H&CW roster, 276-277; before 1900, ten wheelers were common on Bridge Route, 76, 90, 126, 231; after the bridge was strengthened in 1917-1918, the heavier Schenectady-built Santa Fe's became common and remained so until diesels replaced them in 1940s, 203, 231; in 1947, Santa Fe's were still in Maybrook's hump yard service, 200; using diesels meant less need for pushers, 203-204; what rail fans looked for in Mayrook, 1940s, 125; in 1952 on the Central States Dispatch, the B&O and Western Maryland still used steam, but the Reading, L&HR, and New Haven used diesel, 142-145; in the 1960s, the New Haven used Alco FA's, 241
Loedy, Edmond and Ann, 260-261, 263-264, 266, 268, 271

Long, Dewey, 99
Loop, Chet, 227-228
Lossing, Benson J., 19
Lucas, E. M., 253
Lyons, Arthur, 167, 174-175

MacCracken, Henry N., 175
Macdonald, Charles, 32, 35-36, 38-39, 41, 50, 54, 66
Mackessey, John, 178
Magee, Michael, 264-265
Malley, Art, 212
Manchester Bridge, NY, 206, 243, 252
Manhattan Bridge Co., 34, 46, 50, 54
Maps, proposed connections for bridge, 1875, 20; eastern rail connections for bridge, 1889, 56; western rail connections for bridge,1889, 60; Bridge Route map, PR&NE, 1895, 77; route of West Point-New Haven football specials,104; Poughkeepsie regatta race course, 172; Conrail's sale of abandoned rail lines, 1984, 258; proposed rail trail connections for Walkway, 1998, 273
Marano, Antonio, 167-169
Marcy, Frank, 178-179
Margraf, W. J., 132
Martin, Montague, 155-156
Martin, W. J., 110, 128
Matteawan, NY, 128, 138, 205 (see Beacon)
Maurice, Charles S., 34-35
Maybrook, NY, 62, 69, 88, 91, 176, 188, 197-200, 205- 207, 210, 247; its yard, 116-125; yard chronology (table), 125; freight in, 133, 136, 138, 142-143, 203, 245; circuses in, 148-149, 151-152; hoboes in, 167-168; boomer in, 184; woman crew caller in, 201-202; Alexander as brakeman or conductor in, 181, 211-213, 215, 217, 220, 225, 227; its yard inefficient, declining, taken over by Penn Central, 242- 243, 249; question of what to do with yard, 252, 253; part of yard sold as truck terminal, 255; yard's future, 266
Maybrook Line (term), 103
McComb, Arthur B., 134
McCormick, Jim, 121
McCurdy, Laura, 205
McEntee, Austin, 148, 210, 246
McLachlan, Peter G., 203-204, 209-210
McLeod, A. Archibald, 64-65
McMahon, Mike, 178-180
McMann, Jack, 193
McNamee, Graham, 173

Meat service, 118, 127-128, 134, 138, 220-224; table, 135
Mellen, C. S., 115
Merritt, Austin, 49
Mid-Hudson Bridge (F. D. Roosevelt Mid-Hudson Bridge), 91, 97, 101, 174, 196, 199, 263, 271; its creation, 162-165
Milk service, 108-111
Millbrook, NY, 88, 109, 110, 205
Miller, Gordon Schreiber, 262-264, 269, 271
Millerton, NY, 84, 128-129, 152, 179
Modena, NY, 22, 62, 85, 86, 88, 272; freight business, 78, 108, 111, 130-131, 136, 138, 243; abandoning bridge-related lines in, 256-257; wreck at, 277
Modjeski, Ralph, and his engineering firm, 163-165, 231-233, 235-236, 238, 244, 258
Montague, Charles C., 114
Moore's Mills, NY, 90
Moran, Daniel E., 164-165
Moreno, Vito, 264, 268-269, 271, 274
Morgan, J. P., 65, 114, 172
Morse, Samuel F. B., 10; his telegraph instruments, 216; his code, 217
Murray, Ray, 232
Myers, John, 123
Mylod, Frank V. (father) and John (son), 148-150, 268

Neide, Theodore, 159
Nelson, Homer A. 13, 25
New Hamburg, NY, 187
New Hartford, Ct., 219
New Haven, Ct., West Point football excursions to, 102-107; as freight destination, 138, 245; 145, 176, 183, 198-200, 204, 206, 227-228, 242
New Haven Railroad (NY, New Haven, and Hartford Railroad, NYNH&H), its early competition with the Bridge Route, 65, 72; its chronology, including its absorbing of the Bridge Route (table), 146; its passenger service, 76, 79, 83, 87, 89-92; its West Point football excursions, 102-107; its war against the Springfield Branch, 113-115; its Maybrook yard, 117-125, 201-202; its freight service, 127, 129, 133-136, 138, 141-146; its station agents, 1926 and 1941 (tables), 137, 139; its circus trains, 150, 152; its issue of remodelling the bridge for autos, 162-164; its exams,181-183; its track and bridge maintenance, 188-189, 191-193, 231-233, 239; its railmen at work, 197-200, 203-206, 208-209, 211-215, 225, 227-228; its decline and merger into Penn Central, 240-243, 248
New Jersey Midland, 22, 58
New Paltz, NY, 58;trolley service over the bridge from, 69, 81-83, 86, proposed light rail line to bridge from, 262-263; proposed rail trail connections, 273-274
New York & Massachusetts Railroad, 53, 57-58
New York & New England Railroad, 22, 58, 64-65, 75, 77, 128-129, 146, 166, 176
New York Central Railroad, 10-11, 15, 21, 53, 76, 85, 87- 88, 90-91, 98, 102-105, 117, 128, 134, 136, 139, 145, 188-189, 205, 266; its Hudson Line connected with the Bridge Route at Rhinecliff, Poughkeepsie (via the Hospital Branch), and Beacon (Dutchess Jct. and Fishkill Landing), 109-110, 130, 252 ; its decline and merger into Penn Central, 241-243 (see also lines it controlled, including Hudson, Harlem, Wallkill Valley, West Shore)
New York, Lake Erie, & Western, see Erie
New York metropolitan region, its hostility to the bridge, 24; its railroad tunnels, 71, 98, 105, 163; as a market for milk, 108-111; as a market for fruit, 129-131; its congestion and expense, 21, 33, 105, 115, 140; recruiting railroad workers in, 66, 68, 187; its need for better rail freight access, 266 (see also Ferries)
New York, Ontario, & Western Railway, see Ontario & Western
Newburgh, NY, 21-23, 33, 96, 128-129, 225 (see also Ferries)
Newburgh, Dutchess & Connecticut Railroad (ND&C), 58, 69, 78, 88, 90, 92-93. 110, 128, 132, 152, 205; acquired by CNE, 1907, 109, 129, 135
Newburgh-Beacon Bridge (Hamilton Fish Bridge), 163
Norfolk, Ct., 61, 78, 84, 86, 88, 111, 178, 198
Norfolk Southern Railway, 266

Odell, Homer, 239
Odlum, Robert, 153
Olmsted, Frederick Law, 268
Ontario & Western Railway (NY, Ontario, & Western) (O&W), 15, 59-61, 77, 87, 90, 105, 107, 117, 121, 132, 143, 168, 171; stopped, 241-242

Index

Opdyke, S. B., 72
O'Rourke, John F., 36, 41, 44-45, 48-50, 61, 66-67, 79, 154-155, 162-164, 171, 275

Paine, Arthur B., 36, 46, 50
Passenger Trains, long distance, 71-79, local, 80-92
Patch, Sam, 153
Pattison, Lucille, 263
Pearson, E. J., 163
Peekskill, NY, 42; promoted a proposed rival bridge over the Hudson, 12, 21-22, (Storm King bridge) 33, 48
Pelton, George P., 13, 25
Penn Central, absorbs the New Haven, 1968, thus taking ownership of the bridge, (table) 146; its policies, 202, 242-245; its freight (table), 1969, 243; its bridge maintenance men (table), 1969, 244; its bridge fire (1974), 247-249; bankrupt, it merges into Conrail, 1976, giving the bridge a new owner, 251-252, (table), 269
Pennsylvania, Poughkeepsie, & Boston (PP&B), 62, 65, 73; maps, 60, 77; its chronology (table), 124 (see South Mountain)
Pennsylvania Railroad, 22-23, 64, 72, 76, 79, 95, 98-99, 101, 117, 120, 141, 143, 205, 266; pushed to build the bridge, but failed, 15-16, 18-19; merged into Penn Central, 1968, 242
Pevsner, Donald L., 261-263, 266
Philadelphia & Reading, 72-73, 91; its shops at Reading, Pa., 277
Philadelphia, Reading, & New England (PR&NE), 126, 160, 277; incorporated 1892, 64; its directors, 1893 (table), 65; its passenger trains, 71, 74, 76, (map) 77, 80-81; its stations, agents, and passes, 87, 89, 91, 93; its struggle over the Springfield Branch, 113-114; succeeded by CNE, it ceased to exist, 1899 (table), 135
Phillips, Dorothy (Mrs. Gruner), 167-168
Phillips, Thomas J., 167-168, 173-174
Phillipsburg, NJ, 253
Pine Meadow, Ct., 219
Pine Plains, NY, 78, 88, 92, 109, 128, 138, 152
Platt, John I., 10-11, 13, 54, 62, 158-159
Plattekill, NY, 264
Pleasant Valley, NY, 109
Pomeroy, R. M., 25
Port Jervis, NY, 245, 247, 254
Potter, Engineer, 166, 169
Poughkeepsie, NY, 78, 123, 127, 130, 134, 144-145, 154, 178, 180-181, 191, 209, 234, 238-239; its early agitation for a bridge, 10-13; conerstone laying in, 1873, 17- 19; Boston committee supports a bridge at, 21-22; placing piers in the river at, 27-31; bridge opens in, 53-55; immigrant laborers in, 66-69, 187-190; its station, 61, 89, 209; trolley line over the bridge to, 81-83; shopping in, 85, 88; economic impact of bridge route on, 134-135; circus trains in, 147-152; site for jumping off the bridge,155-156; issue of walking on the bridge, 158-161; creating the Mid-Hudson Bridge, 162-165; hoboes in,166; regatta in, 170-175; control tower in, 205-207, 217, 228; freight traffic in, 132-134, 138, 243, 252; bridge fire in (1974), 246-249; after the fire, at first continued use of connecting lines in, 252; clean-up of fire debris in, 254-255; abandoning lines in, 256-257; and post-fire proposals for the bridge, 259; and proposed light rail system for the bridge, 263; rail trail reaching out from, 264; campaign for walkway on the bridge in, 267-268, 270-272, 275 (see also Ferries, Hospital Branch)
Poughkeepsie & Connecticut, 57, 64
Poughkeepsie & Eastern (P&E), 9-11, 13, 17, 22, 25, 53, 57, 90, 93, 109, 128-129, 135
Poughkeepsie Bridge Co., 8, 11-12, 15-16, 19-20, 23-25, 29 ,33-34, 38, 46, 48, 50, 53-54, 61, 64; its incorporators (table), 13; its officials, 1875, 25; its directors, 1887(table), 34; and issue of walking on the bridge, 158-161
Poughkeepsie Bridge Route, origin of term, 61 (see also Maybrook Line)
Poughkeepsie Railroad Bridge, its creation, chs. 1-9; laying its cornerstone, 1873, ch. 2; its leading design and construction engineers (table), 36; evolution of its design (table), 38; its cost (table), 38; its official opening, 1889, ch. 7; steps in opening it (table), 62; its vital statistics (table), 55; its use by passenger trains, chs. 10-14; its use by freight trains, chs. 15-20; its other uses, chs. 21-25; proposals to adapt it to highway use,162-165, 258-259; railroaders at work on it, chs. 26-37; risks, accidents, safety on it, chs. 38-42; its decline, ch. 43; decline of related railroads (table), 242; its closing by fire, 1974, ch. 44; the question of how to use it since then, ch. 45-47; registered as a Historic Place, 259; its changing ownership from 1968 (table), 269; (see also under the names of the railroads which successively owned it, CNE&W, PR&NE, CNE, New Haven, Penn Central, Conrail)
Pratt's Mills, NY, 81, 160, 225
Providence & Worcester (P&W), 251, 254

Queensboro Bridge, 37

Rail Trails, abandoned bridge-related rail lines becoming, 264, 273-275
Ramsdell, Homer, 23, 29-31
Rapelje, Lawrence C., 13
Reading Railroad, 64-65, 74-75, 117, 120, 141-142, 144, 243; its shops at Reading, Pa., 277
Reagan, Ronald, Pres., 253
Red Hook, 129, 191
Reed, John M., 237
Reel, William J., 91
Regatta, Intercollegiate Rowing, Poughkeepsie, 97, 170-175
Relyea (also Relyeas) (New Hurley), NY, 88, 136
Rensselaer Polytechnic Institute (RPI), 12, 35, 39
Rhinebeck & Connecticut Railroad, 21, 25
Rhinecliff Branch, 87, 109, 132, 277 (see Ferries)
Ribicoff, Abraham, 252
Ringwood, Thomas, 247
Rip Van Winkle Bridge, 163
Robbins, Frank A., 148
Robinson, Ernest, 226
Roebling, John A. and Washington A., 17, 39, 41, 66
Rogers, Willard B., 243
Roosevelt, Eleanor, 94, 97-101, 165
Roosevelt, Franklin D., Pres., 18, 67, 69, 94-101, 145, 164-165, 172-173, 175, 191-192, 263
Roosevelt, James (father of FDR) and family, 18, 95-97, 157
Roosevelt, John A., 96-97, 172
Roosevelt, Theodore, Pres., 97, 122
Rouse, James, 262
Rozelle, Fred, Capt., 238
Russell, William E., Gov., 75
Rust, H. A., 29
Rutherford, Lucy, 99-101

St. Louis, Mo., Eads Bridge at, 13, 23, 27, 39

Salisbury, Ct., 88, 132, 177-178
Salt Point, NY, 78, 91, 108, 177-178
Sanchez, Ingrid, 272
Schaeffer, Fred W., 270
Scott, Sam, 211
Selkirk-Castleton Bridge (Alfred H. Smith Bridge), and its accompanying Selkirk rail yard, 163, 241-243, 248-249, 251-252
Sepe, Bill, 267-268, 270
Seymour, Joseph, 158-159
Sheahan, J. F. (Rev.), 161
Shekomeko, NY, 91, 109, 152
Sherwood, Jas. K. O., 113
Silvernails, NY, 132, 183
Simsbury, Ct., 78, 87, 218-219
Sisson, Everett, 92, 218-219
Slatington, Pa., 62-63, 65
Smith, Edward, 126
Smith, James, and the Smith Brothers cough drops, 18, 61, 134
Snow, Fred W., 91
South Mountain Railroad, 22, 62, 124 (see PP&B)
Spencer, P. B., 231
Sperry, N. D., 25
Springfield, Ma., 19, 112-115, 138, 239
Springfield Branch, 87, 92, 112-115, 146
Stanfordville, NY, 59, 61, 178
Stanton, John Clarke, 33, 48, 58, 62-63, 65
Stations and station agents, 87-88, 91, 205, in 1926 (table), 137, in 1941 (table), 139 (see also time tables, 73, 75, 87, 111)
Stissing, NY, 132
Storm King bridge (proposed), see Peekskill
Stormville, NY, 203, 243
Stover, Winne Veeder, 181-182
Swanberg, J. W., 240-242
Sweeney, F. P., 138
Sweeney, John L., 252
Sweeney, John P., 139
Sweetman, W. H., Capt., 31, 230

Tappan Zee Bridge, 37, 163
Tariffville, Ct., 37, 53, 88, 112, 114, 277
Telegraphers, ch. 38; 181, 183, 200, 205, 216-219; telegraph call letters and telegraph lines (tables), 218, 219 (see also Western Union)
Tennyson, Edson L., 263
Thomas, Lowell, 261
Thomson, J. Edgar, 15-16, 18-19
Tingley, Richard H., 62-63
Tower operators, chs. 32, 35 (see also Telegraphers)
Towners, NY, 139
Trainmen (engineers, conductors, brakemen, flagmen), preparing for exams, 181-183; at work, chs. 26-37; facing risks and accidents, chs. 38-42 (see also Workers)
Trepton, William, 212, 253
Trolley cars, New Paltz-Poughkeepsie, pass over the bridge, 80-83, 86-87
Trowbridge, Joe, 179-180
Trump, Donald, 262
Turner, N. R., 72
Tyson, Herbert, 193

Underhill, William, 129
Union Bridge Co., 34-35, 41, 45

Valentino, Joseph, 270
Van Benthuysen, Watson, 34, 158
Vanderbilt, Cornelius, and family, see New York Central
Vanderbilt, Frederick W., 172
Vassar, Matthew, Sr. and Jr., 10, 12-13
Vassar College, 10, 11, 50, 263
Venier, Jeff, 264-265
Verbank, NY, 109, 111

Waddell, J. A. L., 37
Wadlin, Vivian, 270
Walden, NY, 202 (see also East Walden)
Walking on the bridge, 11, 16, 50-51, 54-55, 158-161, 192, 206-207, 237, 267-275
Walkway Over the Hudson, 267-275; its directors, 1994-1995 (table), 270

Wallingford, Ct., 138
Wallkill Valley Railroad, 10, 22-23, 25, 57-58, 60, 86, 91, 117, 130, 136, 205; turned into a rail trail, 273-274
Washington, DC, Washington-Boston express, 71-76, 79
Waterbury, Ct., 138, 147, 184
West Norfolk, Ct. 178; wreck near, 277
West Pawling, NY, 150
West Point Military Academy, football excursions, 102-107
West Shore Railroad, 41-42, 45, 66, 105, 145, 167, 194-196; connection to Bridge Route proposed but not built, 58, 64, 105; and the F. D. Roosevelts, 97-101, 192; and jumpers from the bridge, 154, 156-157; and Regatta, 171-173, 175
West Winsted, Ct., 178-179
Western Maryland Railway, 141-142, 144
Western Union Telegraph Co., 61, 88, 216-217
Whitehouse, John O., 25
Willimantic, Ct., 138, 166
Wilson, George, 25
Wilson, John S., 61, 72
Wilson, Malcolm, Gov., 251
Winslow, John Flack, 12-13, 23-25
Winsted, Ct., 68, 78, 87-88, 177, 179, 218
Women, not bridge builders, track layers, or railroad directors, 50, 205; as yard workers, 201-202; as station agents, telegraphers, 205; as tower operators, 205-207
Woodin, Fred, 177-178, 218-219
Workers, railroad, building the bridge, 42-44, 47-51; building connecting rail lines for the bridge, 57, 59, 62-63; blacks, 59; immigrants, 66-69, 187-190 (see also Bridgemen, Divers, Indians, Telegraphers, Tower operators, Trainmen, Women)

PURPLE MOUNTAIN PRESS, LTD., established 1973, is a publishing company committed to producing the best original books of regional and maritime interest as well as bringing back into print significant older works. For a free catalog, write Purple Mountain Press, Ltd., P.O. Box 309, Fleischmanns, New York 12430-0309, or call 845-254-4062, or fax 845-254-4476, or email purple@catskill.net. Visit the website at http://www.catskill.net/purple